T0310007

Autonomic Intelligence Evolved Cooperative Networking

Autonomic Intelligence Evolved Cooperative Networking

Wiley Series on Cooperative Communications

Michał Wódczak

This edition first published 2018

Registered Office(s)
John Wiley & Sons, Inc., 111 River Street, Hoboken, NJ 07030, USA
John Wiley & Sons Ltd, The Atrium, Southern Gate, Chichester, West Sussex, PO19 8SQ, UK

Editorial Office
The Atrium, Southern Gate, Chichester, West Sussex, PO19 8SQ, UK

For details of our global editorial offices, customer services, and more information about Wiley products visit us at www.wiley.com.

Wiley also publishes its books in a variety of electronic formats and by print-on-demand. Some content that appears in standard print versions of this book may not be available in other formats.

Library of Congress Cataloging-in-Publication Data:

Names: Wódczak, Michał, author.
Title: Autonomic intelligence evolved cooperative networking / by Michał Wódczak.
Description: Hoboken, NJ : John Wiley & Sons, 2018. | Series: Wiley series on cooperative communications | Includes bibliographical references and index.
Identifiers: LCCN 2017050777 (print) | LCCN 2018000675 (ebook) | ISBN 9781119215981 (pdf) | ISBN 9781119215998 (epub) | ISBN 9781118325414 (cloth)
Subjects: LCSH: Wireless communication systems. | Autonomic computing.
Classification: LCC TK5103.2 (ebook) | LCC TK5103.2 .W583 2018 (print) | DDC 004.6–dc23
LC record available at https://lccn.loc.gov/2017050777

Cover Design: Wiley
Cover Image: © RBFried/iStockphoto

Set in 10/12pt WarnockPro by SPi Global, Chennai, India
Printed in Singapore by C.O.S. Printers Pte Ltd

10 9 8 7 6 5 4 3 2 1

Contents

About the Author

Michał Wódczak holds a PhD in Telecommunications from Poznań University of Technology, obtained under the umbrella of the European Union Sixth Framework Programme, as well as an Executive MBA from Aalto University School of Business, distinguished by the Triple Crown of AACSB, AMBA, and EQUIS accreditations. Currently, he is with Samsung Electronics, while prior to that he was also with Telcordia Technologies, formerly known as Bellcore or Bell Communications Research, and with Ericsson, following the Telcordia merger. He served as an Editorial Board Member of IEEE ComSoc Technology News, as well as ran standardisation activities as Vice Chairman and Rapporteur of ETSI ISG AFI, the Industry Specification Group on Autonomic network engineering for the self-managing Future Internet, established under the auspices of European Telecommunications Standards Institute. He is also a Senior Member of the IEEE Communications Society and, apart from this book, he has published two single-authored scientific books with Springer, co-authored and co-edited two industrial standardisation specifications with ETSI, and, overall, authored or co-authored more than 60 peer-reviewed journal, magazine, and conference papers, as well as book chapters. He has also contributed to over 40 scientific reports within European Union FP6 and FP7 projects IST-2003-507581 WINNER I, IST-4-027756 WINNER II, IST-2004-507325 NEWCOM, INFSO-ICT-215549 EFIPSANS, and SEC-242411 E-SPONDER. In addition, he holds Postgraduate Diplomas in Managerial Studies and Psychology of Management, both from Poznań School of Banking, as well as a BA in English Philology and a Postgraduate Diploma in Translation and Interpreting, both from Adam Mickiewicz University. In this capacity, he also acted as an Executive Board Member of the Association of Polish Translators and Interpreters.

Preface

In this book the concept of Autonomic Intelligence Evolved Cooperative Networking is proposed, building on top of both the previous books by the author, where the technological developments in the form of autonomic cooperative networking and autonomic computing enabled cooperative networked design were outlined. In fact, while the former emphasised the aspects of the Open Systems Interconnection Reference Model and the latter elevated the perspective of the Generic Autonomic Network Architecture, not only does the idea of Autonomic Intelligence Evolved Cooperative Networking provide a substantially expanded, but also the most comprehensive and consolidated account in this respect. In other words, a fully-fledged Autonomic Cooperative Networking Architectural Model is presented encompassing the relevant workings both of the layers of the Open Systems Interconnection Reference Model and the levels of the Generic Autonomic Network Architecture. Given the lack of direct correspondence between these two dimensions, as such classified to form the Vertical Technological Pillars and Horizontal Architectural Extensions, the mechanisms of the Autonomic Cooperative Node, Autonomic Cooperative Behaviour, and Autonomic Cooperative Networking Protocol are deployed along with all the pertinent architectural extensions thereto. What is more, during the entire endeavour, the notion of autonomic computing becomes naturally elevated and transposed into autonomic intelligence, as explained later in the book, on the one hand, to better reflect the value it brings to cooperative networking in general, and, on the other hand, to prepare the ground for possible further conceptual advancements.

Acknowledgements

The ultimate design outlined in this book was inspired by the prior involvement of the author in the European Union Sixth Framework Programme IST-2003-507581 Integrated Project: Wireless World Initiative New Radio I (WINNER I), the European Union Sixth Framework Programme IST-4-027756 Integrated Project: Wireless World Initiative New Radio II (WINNER II), the European Union Sixth Framework Programme IST-2004-507325 Network of Excellence in Wireless COMmunications (NEWCOM), the European Union Seventh Framework Programme INFSO-ICT-215549 Integrated Project: Exposing the Features in IP version Six protocols that can be exploited/extended for the purposes of designing/building Autonomic Networks and Services (EFIPSANS), and the European Union Seventh Framework Programme SEC-242411 Integrated Project: A Holistic Approach Towards the Development of the First Responder of the Future (E-SPONDER), as well as the Industry Specification Group (ISG) on Autonomic network engineering for the self-managing Future Internet (AFI) established under the auspices of the European Telecommunications Standards Institute (ETSI). In the light of the above, for the sake of transparency, the utmost attention was paid to clearly highlight the author's unique contribution to the state-of-the-art advancement in the major theme of this book, as well as to ensure its proper separation from any externally referenced background and context-setting information. At the same time, it is crucial to note that all the evaluation results presented by the author were obtained with the use of a dedicated simulation environment designed exclusively for the needs of the preparation of this book, while all the views presented in this book are of the author's.

Acronyms

3GPP	3rd Generation Partnership Project
AAC	Address Auto-Configuration
AAC-OLSR	Address Auto-Configuration OLSR
AB	Autonomic Behaviour
AC	Autonomic Computing
ACB	Autonomic Cooperative Behaviour
ACL	Autonomic Control Loop
ACN	Autonomic Cooperative Node
ACNAM	Autonomic Cooperative Networking Architectural Model
ACNP	Autonomic Cooperative Networking Protocol
ACRR	Autonomic Cooperative Re-Routing
ACS	Autonomic Cooperative Set
ACSAM	Autonomic Cooperative System Architectural Model
ACT	Autonomic Cooperative Transmission
ADME	Autonomic Decision-Making Element
AE	Autonomic Element
AF	Amplify-and-Forward
AFI	Autonomic network engineering for the self-managing Future Internet
AI	Artificial Intelligence
AIECN	Autonomic Intelligence Evolved Cooperative Networking
ALD	Angular Diversity
AM	Autonomic Manager
AN	Autonomic Networking
ANO	Autonomic Node
ANCS	Autonomic Networked Computing System
ANS	Autonomic Nervous System
AO	Autonomic Overlay
AP	Access Point
AR	Autonomic Routine
ARP	Address Resolution Protocol
AS	Autonomic System
ASS	Autonomous System
ATS	Agent System
AUF	Autonomic Function
AWGN	Additive White Gaussian Noise

B2B	Business-to-Business
B2C	Business-to-Customer
BBF	Broadband Forum
BER	Bit Error Rate
BS	Base Station
CA	Coding Advantage
CAdDF	Complex Adaptive Decode-and-Forward
CAS	Computer-Assisted Simulation
CB	Coherence Bandwidth
CBR	Constant Bit Rate
CCG	Channel Capacity Gain
CCI	Co-Channel Interference
CD	Coherence Distance
CDF	Cumulative Distribution Function
CDR	Code Rate
CFR	Chief First Responder
CG	Coding Gain
CHB	Channel Bandwidth
CHG	Channel Gain
CI	Characteristic Information
CLI	Command Line Interface
CM	Channel Matrix
CMDE	Cooperation Management Decision Element
CNR	Conventional Relaying
COD	Complex Orthogonal Design
CODE	Cooperation Orchestration Decision Element
COR	Cooperative Relaying
COT	Cooperative Transmission
CPR	Computing Process
CPU	Central Processing Unit
CPX	Cyclic Prefix
CRDE	Cooperative Re-Routing Decision Element
CRO	Cooperative Routing
CRR	Cooperative Re-Routing
CSI	Channel State Information
CT	Coherence Time
CTP	Control Plane
CTDE	Cooperative Transmission Decision Element
D2D	Device-to-Device
DAA	Duplicate Address Avoidance
DAD	Duplicate Address Detection
DCP	Decision Plane
DMP	Dissemination Plane
DSP	Discovery Plane
DE	Decision Element
DF	Decode-and-Forward
DG	Diversity Gain

DHCP	Dynamic Host Configuration Protocol
DME	Decision-Making Element
DMN	Decision-Making Entity
DN	Destination Node
DO	Diversity Order
DTP	Data Plane
DR	Decode-and-Reencode
DSTBC	Distributed Space-Time Block Coding
DTD	Delayed Transmission Diversity
DVB-T	Digital Terrestrial Video Broadcasting
ECMP	Equal Cost Multipath Protocol
ECN	Emergency Communications Network
EDSTBE	Equivalent Distributed Space-Time Block Encoder
EFIPSANS	Exposing the Features in IP version Six protocols that can be exploited/extended for the purposes of designing/building Autonomic Networks and Services
EGC	Equal Gain Combining
ELC	Extended Link Code
ELM	Extended Link Mask
EMS	Evolved Messaging Structure
EOC	Emergency Operations Centre
EREACT	Extended Routing information Enhanced Algorithm for Cooperative Transmission
E-SPONDER	A Holistic Approach Towards the Development of the First Responder of the Future
ES	Economic Science
ETSI	European Telecommunications Standards Institute
EU	European Union
EVMIMO	Equivalent Virtual Multiple-Input Multiple-Output
EVMISO	Equivalent Virtual Multiple-Input Single-Output
FB	Functional Block
FI	Future Internet
FLD	Frequential Diversity
FNL	Function Level
FMDE	Fault Management Decision Element
FMPR	Flooding Multi-Point Relay
FP	Framework Programme
FP6	Sixth Framework Programme
FP7	Seventh Framework Programme
FR	First Responder
FRN	Fixed Relay Node
FRR	Fast Re-Routing
GANA	Generic Autonomic Network Architecture
GCOD	Generalised Complex Orthogonal Design
GR	Generic Receiver
GS	Group Specification
GT	Generic Transmitter

GVAA	Generalised Virtual Antenna Array
HACL	Hierarchical Autonomic Control Loop
HAE	Horizontal Architectural Extension
HANS	Human Autonomic Nervous System
HRP	Horizontal Reference Point
IANA	Internet Assigned Number Authority
ICI	Inter-Channel Interference
IETF	Internet Engineering Task Force
IoT	Internet of Things
IP	Internet Protocol
IPv6	Internet Protocol version 6
ISG	Industry Specification Group
ITS	Intelligent Transport System
ITU-T	International Telecommunication Union – Telecommunications
KNP	Knowledge Plane
L3DF	Layer-3 Decode-and-Forward
LNK	Link Layer
LoA	Level of Abstraction
LOS	Line-of-Sight
LSB	Least Significant Bit
LSRP	Link-State Routing Protocol
LSTC	Layered Space-Time Coding
LTE	Long Term Evolution
LV	Link Verification
M2M	Machine-to-Machine
MAC	Medium Access Control
MAD	Multiple Address Declaration
MANET	Mobile Ad hoc Network
MAS	Multi-Agent System
MCS	Modulation and Coding Scheme
ME	Managed Element
MEA	Multi-Element Array
MEN	Managed Entity
MEOC	Mobile Emergency Operations Centre
MIMO	Multiple-Input Multiple-Output
MISO	Multiple-Input Single-Output
MLD	Maximum Likelihood Detection
MLSE	Maximum Likelihood Sequence Estimator
MMIMO	Massive Multiple-Input Multiple-Output
MMSE	Minimum Mean Square Error
MN	Mobile Node
MNP	Management Plane
MPR	Multi-Point Relay
MRC	Maximal Ratio Combining
MRN	Mobile Relay Node
MRRC	Maximal Ratio Receive Combining
MSB	Most Significant Bit

NC	Network Coding
ND	Neighbour Discovery
NDL	Node Level
NE	Network Element
NET	Network Layer
NEWCOM	Network of Excellence in Wireless COMmunications
NFV	Network Function Virtualisation
NGMN	Next Generation Mobile Networks
NGN	Next Generation Network
NLOS	Non-Line-of-Sight
NOA-OLSR	No Overhead Auto-Configuration OLSR
NTL	Network Level
OBU	On-Board Unit
OFDM	Orthogonal Frequency-Division Multiplexing
OFDMA	Orthogonal Frequency-Division Multiple Access
OLSR	Optimised Link State Routing
OSI	Open Systems Interconnection
OSPF	Open Shortest Path First
PDAD-OLSR	Passive Duplicate Address Detection OLSR
PDF	Probability Density Function
PHY	Physical Layer
PI	Process Interaction
PLD	Polar Diversity
PSK	Phase-Shift Keying
PSN	Public Safety Network
PTL	Protocol Level
QO	Quasi-Orthogonal
QoS	Quality of Service
QPSK	Quadrature Phase-Shift Keying
RA	Reference Architecture
RAP	Radio Access Point
RBCD	Repetition-Based Cooperative Diversity
REACT	Routing information Enhanced Algorithm for Cooperative Transmission
REC	Relay-Enhanced Cell
RF	Radio Frequency
RFP	Reference Point
RM	Reference Model
RME	Routing Mechanism
RMPR	Routing Multi-Point Relay
RN	Relay Node
RND	Reception Diversity
RPA	Reference Point Architecture
RSDE	Resilience and Survivability Decision Element
RTB	Routing Table
SA	Software Agent
SAdDF	Simple Adaptive Decode-and-Forward

SAS	Single-Agent System
SBA	Service Based Architecture
SC	Selection Combining
SCD	Scanning Diversity
SDE	Sub-Decision Element
SDN	Software-Defined Networking
SDO	Standards Development Organisation
SDR	Software-Defined Radio
SFAAC	Stateful Address Auto-Configuration
SIMO	Single-Input Multiple-Output
SINR	Signal-to-Interference-plus-Noise Ratio
SISO	Single-Input Single-Output
SLAAC	Stateless Address Auto-Configuration
SLD	Spatial Diversity
SN	Source Node
SNR	Signal-to-Noise Ratio
SOA	Service-Oriented Architecture
SON	Self-Organising Network
SPR	Single-Path Relaying
STBC	Space-Time Block Coding
STBD	Space-Time Block Decoder
STBE	Space-Time Block Encoder
STC	Space-Time Coding
STCCD	Space-Time-Coded Cooperative Diversity
STP	Spatio-Temporal Processing
STTC	Space-Time Trellis Coding
SVD	Singular-Value Decomposition
SWC	Switched Combining
TC	Topology Control
TCM	Trellis-Coded Modulation
TCO	Total Cost of Ownership
TCP	Transmission Control Protocol
TCP/IP	Transmission Control Protocol/Internet Protocol
TDD	Time Division Duplex
TLD	Temporal Diversity
TMF	Telemanagement Forum
TND	Transmission Diversity
TTL	Time To Live
TS	Technical Specification
UDP	User Datagram Protocol
UF	Utility Function
UT	User Terminal
VAA	Virtual Antenna Array
VANET	Vehicular Ad hoc NETwork
VCS	Virtual Cooperative Set
VMIMO	Virtual Multiple-Input Multiple-Output
VRP	Vertical Reference Point

VTP	Vertical Technological Pillar
WI	Work Item
WINNER I	Wireless World Initiative New Radio I
WINNER II	Wireless World Initiative New Radio II
WRR	Weighted Round-Robin

Notation

$ACS(x, n^{(2)})$	Set of Autonomic Cooperative Nodes providing the capability of cooperative transmission between the source node x and the destination node $n^{(2)}$.
$C(n^{(2)})$	Set of channel coefficients between the members of $ACS(x, n^{(2)})$ and $n^{(2)}$ itself.
E_Y^X	Equivalent distributed space-time block encoder, where X_Y denotes a specific space-time block coding scheme, e.g. G_2, G_3, G_4, H_3, or H_4.
GT_X	Generic transmitter X, where $1 \leq X \leq N$.
GR_Y	Generic receiver Y, where $1 \leq Y \leq M$.
$h_{i,j}$	Channel coefficient between the ith transmitting antenna and jth receiving antenna, where $i = 1, \dots, N$ and $j = 1, \dots, M$.
$\|h\|^2$	Channel gain.
$H_{N \times M}$	Channel matrix for N transmitting antennae and M receiving antennae.
I_N	$N \times N$ identity matrix.
$L(d)$	Path loss defining a given radio propagation model.
$L_{j,i}^{t-1}$	Buffer load in the previous cycle $t - 1$, where $j = 1, \dots, 3$ denotes a slot, while $i = 0, \dots, 4$ corresponds either to the base station, indicated by the $i = 0$, or a fixed relay node, assigned a value of 1 through 4.
$MPR(x)$	Set of multi-point relays of a given source node x.
$MPR^i(x)$	Redundant, i.e. secondary, ternary, and so on, set of multi-point relays of a given source node x, where $i \geq 1$; $MPR^1(x)$ is equivalent to $MPR(x)$.
m_{Rel}	Number of transmitting antennae deployed at a relay node.
m_{Rx}	Number of transmitting antennae deployed at a destination node.
m_{Tx}	Number of transmitting antennae deployed at a source node.
$N(x)$	Set of one-hop neighbour nodes of a given source node x.
n	One-hop neighbour node belonging to the set $N(x)$.
$N^{(2)}(x)$	Set of two-hop neighbour nodes of a given source node x.
$n^{(2)}$	Two-hop neighbour belonging to the set $N^{(2)}(x)$.
P_{T}	Transmitted signal power.
P_a^b	Power level of the signal transmitted by a given network node a as observed by its neighbour node b.
σ^2	Noise power.
X^{H}	Hermitian transpose of a matrix X.

1

Introduction

The opening, context-setting, chapter introduces the background behind the concept of autonomic computing, and accounts for its convergence with modern networked systems. A conceptual analysis is then carried out in order to draft a fully-fledged framework depicting the scientific advancement in this respect. In essence, the general vision and the state of the art in the field of autonomic computing are approached from the viewpoint of the related mechanisms inherent in the functioning of the human autonomic nervous system. Given its importance, this consists in the analysis of the key dimensions of self-configuration, self-optimisation, self-healing, and self-protection, altogether known to constitute the notion of self-management, and is extended to cover the pertinent architectural assumptions and variations complemented with insight into the overlapping nature of autonomic computing and agent systems. Then, the ultimate question of convergence between autonomic computing and autonomic networking is addressed, and, thus, the ground for the discussion of the role of self-awareness is settled, with the eventual goal of introducing the target Autonomic Cooperative Networking Architectural Model. In order to make this possible, first the investigation of the most recent incarnation of the Generic Autonomic Network Architecture is characterised with special attention paid to the explanation of the role of decision elements and hierarchical autonomic control loops, along with their respective levels of abstraction, presented in an incremental order, starting from the lowest protocol level, through the function level and node level, up to the top network level.

Once the related ground has been settled, the scope of the Autonomic Cooperative Networking Architectural Model is examined in more detail through the introduction of the Vertical Technological Pillars and the Horizontal Architectural Extensions. In fact, the layers of the Open Systems Interconnection Reference Model are made perpendicular to the levels of the Generic Autonomic Network Architecture in order to identify the key architectural challenges to be addressed by the ultimate Autonomic Cooperative Networking Architectural Model. For this reason, an incremental conceptual outline is presented involving the key architectural components in the form of the Autonomic Cooperative Node, Autonomic Cooperative Behaviour, and Autonomic Cooperative Networking Protocol, as well as the major decision elements of relevance. In particular, first, the protocol level cooperative transmission decision element is presented with its responsibility for virtual multiple input multiple output channel based and distributed space-time block coding enabled cooperative relaying. Next, the function level cooperative re-routing decision element is deployed, with its role of being a trigger for transmission resiliency driven cooperative re-routing. Moving forward, the

Autonomic Intelligence Evolved Cooperative Networking, First Edition. Michał Wódczak.

node level cooperation management decision element is introduced in order to facilitate the integration between cooperative relaying and routing mechanisms. Last, but not least, the network level cooperation orchestration decision element is presented as being accountable for comprehensive oversight of the overall system. All in all, a high-level blueprint of the Autonomic Cooperative Networking Architectural Model is drafted to be further advanced in the chapters to follow.

The third chapter follows on with specific architectural considerations. In particular, the presentation is started with the foundations of the protocol level spatio-temporal processing, where the initial emphasis is laid on developments related to the multiple-input multiple-output channel to provide a good understanding of its workings. Then, the pertinent diversity-rooted origins of spatio-temporal processing are discussed, so that it becomes possible to clearly justify its role and the necessity for its later deployment. Moreover, the question of radio channel virtualisation is visited, where the singular-value decomposition theorem is explained in order to introduce the notion of an equivalent virtual multiple-input multiple-output radio channel to be deployable among Autonomic Cooperative Nodes. The related radio channel capacity is incorporated into the bigger picture of the opening analysis to account for its linear scaling with the number of so-called generic transmitters or generic receivers. Finally, a specific model for radio channel coefficient calculation, to be referenced throughout this book, is described, and the difference between coding gain and diversity gain is addressed for the sake of clarity. Given such a context, the focus moves towards space-time coding techniques, to account for their superiority over the above-mentioned diversity techniques and to pave the way for their later use in networked configurations, where the concept of distributed space-time block coding is expected to prevail.

In particular, the most baseline approach to space-time coding is presented with special attention paid to space-time block coding, where the question of its being perceived more as a modulation rather than a coding technique is visited. Then, the derivation process of the decoding metrics for a selected set of space-time block coding matrices is outlined with the aim, among others, of clarifying certain inconsistencies the author came across in the referenced source materials. Based on this, an extension towards space-time trellis coding is also presented, where additional coding gain becomes clearly visible. Eventually, after all the aforementioned technological aspects have been analysed, their relation to the protocol level control logic is discussed in the light of the prospective architectural integration aspects. To this end, the notion of an Autonomic Cooperative Node is introduced as one of the major building blocks of the proposed concept. Not only is the relation between autonomics and cooperation discussed further, but the internal structure of the Autonomic Cooperative Node is scrutinised. Next, the cooperative transmission decision element is brought into the global picture as belonging to the protocol level, while being mostly responsible for the interaction with the routines of the physical layer. Given such a context, not only is the role and notion of the concept of a protocol addressed, but a pertinent adaptive logic is presented, where the relevant code matrices are switched on the basis of the radio channel parameters. Finally, all the architectural integration aspects of relevance are outlined and the way is prepared for further extensions.

In the fourth chapter, the topics of both conventional and cooperative relaying are addressed from the classificatory perspective; the two approaches are characterised, and the forwarding strategy and protocol nature of the latter are further investigated. Following this, the focus is redirected towards the question of supportive and collaborative protocols, introduced as subcategories of a generic cooperative protocol. Such an approach means that the former shall be considered as a preparatory phase for the latter, making the interaction between the two highly correlated. Going further, the concept of virtual antenna arrays is outlined on the basis of its most versatile multi-tier incarnation, where, assuming a generalised cooperative transmission scheme, its special operation mode of distributed space-time block coding is discussed as being clearly intended to play a crucial role for all the further developments to be discussed in this book. Given such a context, attention is directed towards a fixed deployment concept, where both the conventional and cooperative relaying techniques could become equally applicable, yet the plot is advanced on the assumption that the subject of subsequent analyses will be the mobile deployment concept. In particular, the grid-based Manhattan scenario is initially outlined to underline that as much as the pattern formed by the buildings could become critically important for the suppression of interference among the fixed relay nodes, it would make it literally impossible to exercise any cooperative relaying based on virtual antenna arrays.

In essence, the evaluation effort is carried out to highlight that, despite limitations related to cooperative relaying, certain link layer and network layer performance optimisations would still be possible. To this end a specific adaptation strategy is proposed with regard to the framing structure and the buffer memory, so that, using the process interaction simulation method, it becomes possible to observe improved packet throughput at the network layer. Similarly, a cooperation-enabled relay-enhanced cell indoor scenario is analysed, where the major emphasis is put on the link layer aspects, keeping in mind its applicability to any later mobile deployment concept considerations. Eventually, the focus is shifted towards the function level overlay logic, where, first of all, the roots of Autonomic Cooperative Behaviour are outlined to account for its role and complexity, including its enablers – the equivalent distributed space-time block encoder in particular. Then, the rationale behind the cooperative re-routing decision element is presented, including its transition from the node level to the function level and the logic behind cooperative re-routing involving the role of the fault management decision element and the place of the resilience and survivability decision element. Last, but not least, the architectural integration aspects are discussed to account for the general dependencies between the routines of all three layers of interest, as well as to provide a more detailed insight into the architectural relations driven by the pairing of the link layer and the function level, complemented by the introduction of a specifically extended version of the Autonomic Cooperative Node.

In the fifth chapter, first of all the workings of the experimentation-related version of the Optimised Link State Routing protocol are discussed, with special emphasis on its functional and structural characteristics related to the field of applicability and the assumed messaging structure. Apart from the proactivity-driven relevance to mobile ad hoc network scenarios, special attention is paid to the multi-point relay station selection heuristics with the incorporation of certain small alignments. Additionally, the information storage repositories are analysed in order to provide the required context for further developments, and specifically to introduce new elements in the form of both the VAA selector set and its related VAA selector tuples, intended to become the enablers of the target concept of enhancing cooperative transmission with routing information. What follows directly are the developments originating from the routing information enhanced algorithm for cooperative transmission, conceived by the author as a method for applying the additional information collected by the Optimised Link State Routing protocol inherent in the network layer, and its modified version in particular, for the sake of both enabling and orchestrating cooperative transmission at the link layer. To this end, the justification for the introduction of the routing information enhanced algorithm for cooperative transmission is provided with particular emphasis on the relevant algorithmic description, where, additionally, certain elements and nomenclature of the Optimised Link State Routing protocol are assumed, predominantly because of a fairly direct usage of the outcome of the multi-point relay station selection heuristics.

Given such a context, the elevated concept of the extended routing information enhanced algorithm for cooperative transmission is outlined along with the evolved messaging structure in order to lay the groundwork for the target Autonomic Cooperative Networking Protocol. In this respect, both the very vital topics of address auto-configuration and duplicate address detection are considered, before the focus shifts more towards the umbrella formed by the function level overlay logic. In this way the workings of the Autonomic Cooperative Networking Protocol are outlined, covering the role of the extended routing information enhanced algorithm for cooperative transmission in its conception, as well as justifying the place of the evolved messaging structure in the process of Autonomic Cooperative Node preselection, along with the layout and the reasoning for the related design of the routing table. The extended algorithmic description defining the logic of the cooperation management decision element is then examined in reference to what was previously outlined for the original routing information enhanced algorithm for cooperative transmission. Based on this it becomes possible not only to evaluate the advantages thereof by means of simulation analysis, but also address the overhead aspects of the evolved messaging structure. Finally, the entire analysis is elevated even further to conclude with aspects of the architectural integration, covering the roots of the Autonomic Cooperative Networking Protocol, the conceptual transitions, and the related dependencies among its architectural entities.

In the final chapter the standardisation-orientated design is introduced, assuming a research and investment driven perspective, in order to explain the origins of the Autonomic Cooperative Networking Architectural Model by touching on issues related to the standardisation of the Open Systems Interconnection Reference Model, as well as emphasising the role of prestandardisation related to the Generic Autonomic Network Architecture. What naturally follows is a description of the staged instantiation of the Generic Autonomic Network Architecture Reference Model, depicting the progression of various levels of abstraction in an incremental manner. This introductory part is concluded with certain cross-specification-related considerations intended to incorporate select concepts from software-defined networking, machine-to-machine communications, and intelligent transport systems into the bigger context of the Autonomic Cooperative Networking Architectural Model. Then another, highly practical, deployment scenario in the form of an emergency communications network is considered, which becomes especially interesting because of its being driven by a combination of specifically tailored requirements, where safety appears to take priority over the latest technological advancements. In particular, it is emphasised that the system operation becomes bound to exercise the hierarchy between chief first responders and their respective first responders, as implied by human established relations. In this respect, the relevant network topologies are discussed along with the related configurations of chief first responders.

The way is thus prepared for further incorporation of autonomic routines, since, after the cooperative mode of operation has been introduced and the proactive and reactive resiliency process has been outlined, the integration of the emergency communications network into the ultimate Autonomic Cooperative Networking Architectural Model may be discussed. Following the complementary justification for the cooperative enhancement in question, supported with performance evaluation analysis, the related network level overlay logic is introduced to the overall picture to encompass any still outstanding or not comprehensively addressed workings of the Autonomic Cooperative Networking Architectural Model. In this way the mutual relation between the Autonomic Cooperative Networking Protocol and the Autonomic Cooperative Behaviour is presented from the perspective of the priority between the two, on the grounds of their being inherent in the respective dimensions of the Open Systems Interconnection Reference Model and the Generic Autonomic Network Architecture. Based on this, the notion of the cooperation orchestration decision element is introduced in a way emphasising more tangibly when the Autonomic Cooperative Behaviour may be prioritised over the Autonomic Cooperative Networking Protocol. In particular, the relay-enhanced cell scenario is revisited under certain additional assumptions allowing for a more accurate evaluation of the second hop. Finally, the architectural integration aspects are raised to address the mutual operation of all the discussed decision elements to introduce additional synergy to the fairly exhaustive depiction of the Autonomic Cooperative Networking Architectural Model.

2

Autonomically Driven Cooperative Design

2.1 Introduction

This opening chapter begins with an analysis of the rationale behind the introduction and the role of the visionary concept of autonomic computing for the needs of accounting for its rapid translation into and convergence with the domain of state-of-the-art networked systems. Although it is of an introductory or context-setting nature, this part of the book provides an extensive commentary on and insight into the entire design to allow a comprehensive view of the current status of and future advancements in the realm of autonomics. For this reason, not only are the major aspects of the original approach detailed, but the relevant conceptual and architectural changes are indicated to allow for the introduction of the concept of Autonomic Intelligence Evolved Cooperative Networking. In particular, once the general vision has been introduced, the main emphasis shifts towards an explanation of the workings of the classic approach, advocating the adoption of the mechanisms governing the human autonomic nervous system into the architecture responsible for the management of complex networked systems. As such, it involves discussion of the aspects of self-configuration, self-optimisation, self-healing, and self-protection, together constituting the notion of self-management, while the pertinent architectural assumptions and variations follow, developing into a discussion regarding the complementary nature of autonomics and agent systems. In particular, the ultimate issue of convergence is emphasised together with the role of self-awareness with regard to the design principles driving the respective architectural considerations.

In fact, based on such an introductory analysis, originally intended to provide a comprehensive and consistent explanation of the current role and place of autonomics, thereby primarily covering autonomic computing, yet also being highly pertinent to autonomic networking per se, the focus then shifts to address the investigation and presentation of the latter, with special emphasis on state-of-the-art architectural advancements of relevance, to prepare for the introduction of the idea of the ultimate Autonomic Cooperative Networking Architectural Model. To this end, the Generic Autonomic Network Architecture, playing a referential role, is introduced and scrutinised as the baseline solution for the further incremental conceptual development to be outlined in the remainder of this chapter. In this respect, first, the origins of the Generic Autonomic Network Architecture are detailed, with special attention being paid to account not only for its composition based on functional planes, but also for the decision element driven orchestration processes. What follows is a presentation of the key notion of hierarchical autonomic control loops, being tightly correlated with their corresponding levels of abstraction, as well as an explanation of their role from the bottom up, i.e. from the protocol level, through the function level and node level, up to the network level. The entire discussion is carried out keeping in mind the notion of legacy autonomic networking, yet assuming a rather more modern perspective outlined in the Group Specifications which the author of this book coauthored and coedited under the auspices of the European Telecommunications Standards Institute.

Moving towards the end of this chapter, the concept of the Autonomic Cooperative Networking Architectural Model is drafted in more detail, with a special emphasis on the enabling Vertical Technological Pillars and the relevant Horizontal Architectural Extensions. To this end certain assumptions are made, according to which the orientation of the layers of interest belonging to the Open Systems Interconnection Reference Model, including the physical layer, link layer, and network layer, is changed so that they become perpendicular, if not orthogonal, to the levels inherent in the Generic Autonomic Network Architecture, most of all for design transparency reasons. Such a positioning allows to identify the key architectural challenges to be addressed throughout the entire book. What is more, an incremental conceptual outline follows, where the reader is acquainted with the major notions, including the Autonomic Cooperative Node, Autonomic Cooperative Behaviour, and Autonomic Cooperative Networking Protocol, as well as all the decision elements of relevance are introduced. In particular, first comes the cooperative transmission decision element of the protocol level intended to orchestrate distributed space-time block coding. Next is the cooperative re-routing decision element of the function level, being the main enabler of increased reliability and resiliency of transmission in general. The cooperation management decision element of the node level follows, to be responsible for the integration of cooperative relaying with the related routing mechanisms. Last, but not least, comes the cooperation orchestration decision element of the network level, responsible for comprehensive system oversight.

2.2 Biologically Inspired Autonomics

2.2.1 Rationale and Vision

As much as it may be perceived a sign of the times, the current profound interest in the visionary concept of autonomic computing (AC) appears to be primarily attributable to the related advancement in networked system design (Wódczak, 2014). Even though there is no denying that the natural drive for automation had been followed well before the conception of autonomic computing per se, it also transpires that the very foundations laid with the emergence of the same opened a completely new era in the development of its networking-related counterpart (Wódczak, 2012). Most obviously, this phenomenon is rooted in the rapid progression of unprecedentedly complex distributed systems featuring a multitude of interconnected appliances, each overseen by sophisticated software and, thus, altogether posing a critical demand in terms of being properly configured and maintained with, virtually, no delay. In order to provide an exhaustive explanation in this respect, it is necessary to investigate the non-technical, biological origins of autonomics, and to analyse its business-orientated and technologically varying enablers. In this way one may not only be able to discern the technological shift, but also obtain a much better insight into the relevant convergence-driven and hardware-based similarities between computing and networking in general. Yet, no sooner may the entire picture become more complete than the prevailing role of software has been disguised in the sense that it is not only responsible for the overseeing of its underlaying hardware, but especially accountable for the orchestration of its own routines (Wódczak, 2014).

Looking from the intrinsic perspective, as indicated by Kephart and Chess (2003), the origins of the idea behind legacy autonomic computing are undoubtedly profoundly rooted in inspiration drawn from the functioning of the human autonomic nervous system (HANS),[1] naturally intended to orchestrate the behaviour of the internal organs through the monitoring of numerous related parameters, such as heart rate or body temperature, for the sake of releasing the brain itself from controlling them in a fully aware manner. This appears to be the key point as to the explanation of the actual rationale behind autonomics, especially given the fact that, even though there is an abundance of such parameters that are considered to be crucial for the uninterrupted functioning of a human organism, the said process of orchestration is carried out in a somewhat detached, if not distributed, fashion, as explained by Paulson (2002). For this reason, the structure of such an autonomic system (AS) is said to resemble a recursively arranged hierarchy, where each of the numerous entities that display the capability of self-management at a given level of abstraction (LoA) should encompass the similarly numerous and conceptually alike, yet functionally varying, components one level below, and so forth, as advocated by Kephart and Chess (2003). Attempting to apply a more descriptive terminology, one could equally well approximate the above-mentioned conceptual construction by instantiating it with a rather generic pattern, where the structure itself would originate from the lowest molecular level, just to move forward through human markets and societies, in order to approach the top level of world's global socio-economy (Kephart and Chess, 2003).

Yet, assuming a more extrinsic viewpoint, it seems that the business model has undergone so substantial a change, over little more than a decade, that it became the major enabling factor in the said respect. In fact, originally, the objective of autonomics[2] was to address large commercial computer systems, clearly bringing the connotation of and indicating an institutional or corporate context, where the key interested parties could be identified as network or system administrators, without any specific participation of the end users. However, as it is to be demonstrated throughout the remainder of this book, the most recently observable progress in the field of autonomics tends to result from a shift of the previously so popular in-house computing, based on business applications, more towards the use of ubiquitously mobile and interconnected consumer devices, thereby replacing the former business-to-business (B2B) model with one that is more business-to-customer (B2C) orientated. At the same time, the major technological stress is still being placed on network providers forming, in many cases, a kind of a mediation layer between the customers, chiefly understood as the end users, and the service providers, who are frequently the network operators themselves. This

---

1 One might note that an equivalent notion is sometimes expressed under the plain name of autonomic nervous system (ANS).

2 As of now, the term 'autonomic computing' will be mostly referred to as 'autonomics' in order to purposely decouple it from computing per se, and to anticipate the introduction of the target context of its networking-related flavour.

way a business-driven market demand has arisen with the intention of advancing the technological and architectural entities of relevance so that a synergetic and convergent approach could materialise, where the realms of both autonomic computing and autonomic networking (AN) become welded very tightly together, as indicated by Wódczak (2014).

In the light of the above-mentioned non-technical origins and technologically variable enablers of autonomics, one may conclude that the reason for its emergence might be highly correlated with the question of the continually increasing complexity of computer systems (Wódczak, 2014). In fact, such an observation brings the immediate connotation of the related demand for timely operation and proper maintenance, which, following Paulson (2002), may be directly translated into the application of certain self-monitoring and self-recovery techniques in order to ensure the ability of the system to function in a fully sustainable manner, without reliance on human-driven intervention, at least over the majority of the time of observation. A reverse analysis of the subsequent stages of the development of autonomics might seem to trace the origins of its concept to the famous manifesto referred to by Kephart and Chess (2003). In particular, scrutinising the related works towards this direction, one might eventually come across a fairly repeatedly used notion of the said self-healing applied not with regard to computer systems, as perceived from the macroscopic perspective, but rather embedded into the microscopic picture of a central processing unit (CPU). Interestingly, as depicted in Figure 2.1, this leads to the conclusion that, initially, the term 'autonomics' might have been semantically capacious enough to encompass even concepts such as redundant microprocessor modules (Paulson, 2002). While this understanding could still hold today, it appears that currently the macroscopic context prevails more conspicuously.

Nonetheless, the presented notions should by no means be understood as contradictory, but, quite the opposite, rather perceived in an evolutionary way since the primary and mirror units visible in Figure 2.1 already form a kind of a connected setup very much resembling the rationale behind a networked system, especially because of the fact that the intermediary parity check and recovery units appear to be responsible for establishing a relevant communication and data exchange protocol between the interconnected elements. What is more, following Paulson (2002), it may similarly be claimed that

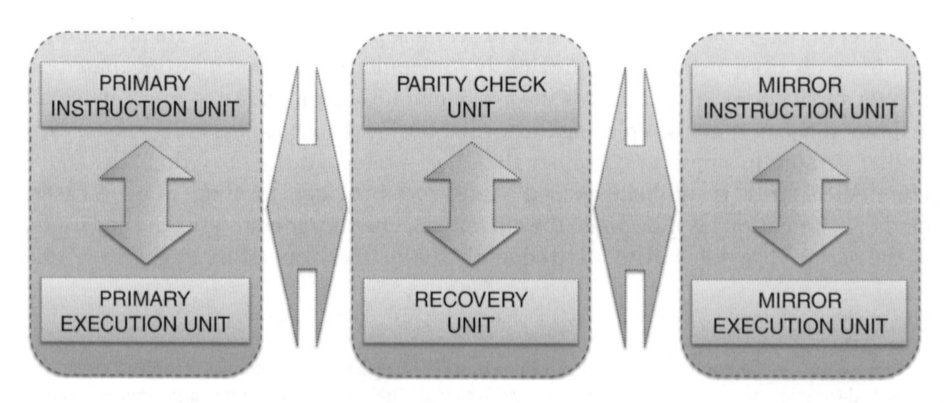

Figure 2.1 Roots of autonomics. Adapted from Paulson (2002).

such hardware introduces intelligence,[3] expected to be accompanied by relevantly smart applications, which brings up the very practical question of assuring the most appropriate composition of an autonomic networked system design in terms of both the deployed software and hardware. This, in turn, extends to the very rationale behind and the vision ahead of autonomics, as in the case of designing a fully-fledged autonomic networked setup one would need to accommodate the necessity for embedding not only a self-managing operating system and software, but also redundant memory and processors featuring self-healing firmware (Paulson, 2002). Most importantly, however, one should note that the value of software becomes critically important even in the case of networked systems, as otherwise autonomics could be mistakenly perceived as being realised with hardware redundancy only, while, in general, it would be primarily expected to manifest itself through accurate software self-management, where network layer protocols could be encompassed accordingly, just to follow Wódczak (2014).

As much as this perception might be a conceptual enabler of a proper positioning of the role of such software for the sake of autonomic system balancing, especially in the field of distributed computing, it is crucial to keep in mind that, in fact, nowadays software has already become so ubiquitous that it could be equally well named a regular commodity of a highly complex nature. This way, the autonomic flavour of such software appears to be twofold: on the one hand, its role is to maintain the proper operation of the underlying networked system; on the other hand, it needs to be able to sustain its own operation, which might make both aspects equally challenging, mostly for complexity reasons. This holds true especially for critical systems where any potential failure may bring catastrophic consequences, primarily, but not always, in financial terms. As pointed out by Hinchey and Sterritt (2006), the mitigation of such effects may cause an organisation to come under pressure to spend as much as 33 to 50 per cent of the total cost of ownership (TCO) of the computing and communications infrastructure for the sake of avoiding any unintended or unexpected system malfunction due to software failure. It therefore became highly desirable to find a holistic approach at the overall system level that would be of value in such cases and, as a result, the idea of self-managing software was proposed to handle self-direction, self-governance, and self-adaptation (Hinchey and Sterritt, 2006). Clearly, such an approach is tightly bound with self-managing hardware, where the above-mentioned principles of complexity hold both at the microscopic and macroscopic levels (Wódczak, 2014).

Thus, the concept of autonomics should be chiefly understood as an arrangement of networked devices, altogether taking the form of a widely distributed setup, where the notion of self-management interweaves the aforementioned software- and hardware-related aspects. Following Brazier et al. (2009), and assuming a more systematic perspective, one may enumerate the following three major enablers of autonomic system design. In the first place, the enabling role of the complexity of an autonomic networked computing system (ANCS) is considered, with the emphasis on topics such as highly demanding and advanced routines for database management, characterised by the nontriviality of orchestrating numerous parameters of high significance. Next, the commercial roll-out and widespread deployment of service-oriented architecture

3 Even though the reference to 'intelligence' follows the mood of the cited publication, it could be somewhat misleading, as its meaning would offer a different connotation compared to the semantic field of the term 'autonomics', if it was not for the notion of autonomic cooperative intelligence, as reflected in the title of this book.

(SOA) increases the complexity of the setup even further, adding yet another dimension to the above-mentioned issue, since the distribution of the pertinent system entities usually suggests they no longer fall under the umbrella of a single organisation. Eventually, there is the notion of the heterogeneity of both the hardware and software components, additionally expanding into and reshaping the services provided on top of them, which together with the enablers mentioned previously appear to transpose today's networked computing systems much more into a melting pot comprising a broad variety of concepts and approaches, thereby more and more profoundly calling for the incorporation of self-sustainable operation in order to facilitate the instantiation of the ultimate concept of autonomic networking (Wódczak, 2014).

All in all, recalling the biological origins of autonomics along with its business-motivated orientation yet technologically varying enablers, and taking into account the convergence-driven and hardware-based similarities between the computing and networking aspects, not to even mention the prevailing role of software in the sense of being responsible for overseeing its underlying hardware and especially for the orchestration of its own routines, a variety of not always fully justified understandings of autonomics have been established. In fact, the author has determined that, in practice, whenever a comprehensive definition of this term is attempted, sooner or later a mixture of connotations is brought up which only makes it even more obfuscated. Before this issue is to be scrutinised in detail in the section to follow, one should note that, in principle, not only does the idea of autonomics assume a virtually entire exclusion of any human-related involvement, even though it may still refer to the role of an administrator for the task of distributed computing system operation and maintenance, but it also requires that such a system not be perceived as cognitive in the sense that it cannot reason by itself. Following Kephart and Chess (2003), such a unique combination of functionalities pretends to be achievable exactly through an inherent ability of this kind of a system to self-manage with a special emphasis on continuous monitoring of internal and external conditions, as well as applying relevantly imposed policies and taking pertinent actions.

2.2.2 Nomenclatural Perspectives

In the light of the above context, as the extent of convergence between the worlds of computing and networking appears to have gone well beyond any conceivable boundaries, by no means could the nomenclatural perspectives on autonomics remain tightly restricted or focused, especially since it has started to become more difficult to tell the related core technologies clearly apart, not to mention the influence of the already introduced biological connections. In fact, the boundaries between computer systems and their related interconnecting telecommunications networks seem to have gradually disappeared, and hybrid solutions started to materialise, as indicated by Phan (2009), making attempts to draw a distinction between the two realms somewhat artificial, if adequate at all. Irrespective of the fact that computers and networks have come to be perceived as a unified entity, the notion of autonomics has started becoming more and more vague, especially due to the seemingly perpetual confusion over what should be meant by autonomy and automation (Wódczak, 2014). In order to shed more light on this aspect, as well as to account for the proper meaning of and differences between the related semantic fields, an attempt is made below to introduce autonomics through an explanation of the role of awareness. Then, a wider context of multi-agent systems (MASs) and decision-making elements (DMEs)[4] is highlighted, just to conclude with what should be really meant by the notions of being autonomic, autonomous, or cognitive.

Looking at the technological progress, one can discern not a steady but an exponential increase in the adoption of the more and more ubiquitous software-based components and entities into the world of networking. To emphasise the significance of this phenomenon, following Mainsah (2002), one could suggest that, should the aircraft industry experience growth of the same order, 'transatlantic flights would take no more than a few minutes and cost no more than a few dollars'. Attempting to pinpoint exactly what the main incentive for such development could be, it transpires that one of the key drivers for the related convergent advancement results directly from the need of an autonomic networked design to be self-contained in the sense of being capable of behaving in a self-aware manner. In an abstract sense one may claim that, as such, self-awareness[5] could go hand-in-hand with autonomics, because of the assumption that the system should exploit all the acquired knowledge to 'understand' its own status change and the implications. However, following the classification outlined by Vassev and Hinchey (2010), as illustrated in Figure 2.2, one may clearly notice that self-awareness is not a standalone feature, but is complemented by the concept of context-awareness, emphasising rather the external aspects (Chiti et al., 2016). From such a perspective, context-awareness may be understood more as the ability of a system to interact with its environment through communication and negotiation for the purposes of being able to predict any forthcoming situational changes in advance.

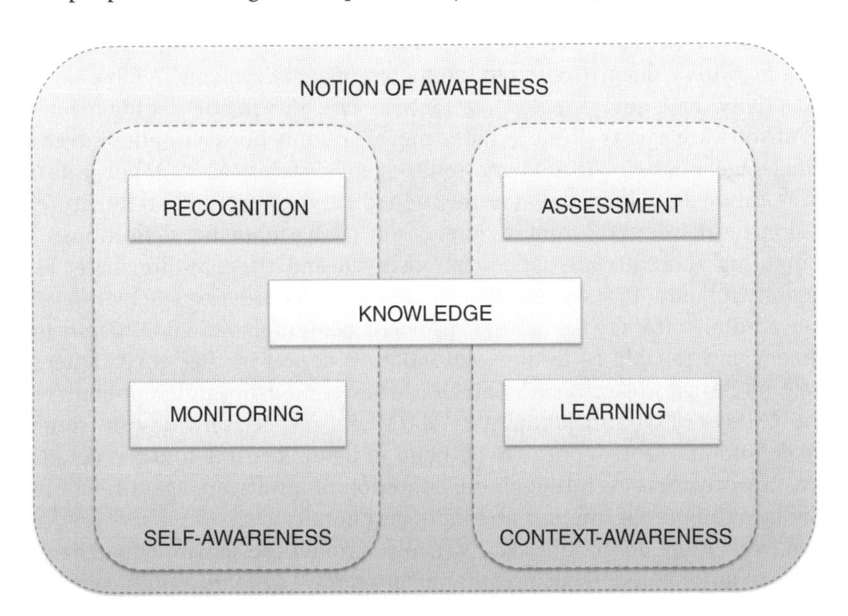

Figure 2.2 Position of awareness. Adapted from Vassev and Hinchey (2010).

4 A decision-making element should be perceived as being equivalent to the notion of a decision element (DE), at least as far as the Generic Autonomic Network Architecture (GANA) is concerned.
5 For additional context, the reader is also referred to Smirnov et al. (2009).

Most importantly, however, not only do both the notions of self-awareness and context-awareness revolve around the aforementioned focal component of autonomics in the form of knowledge, but each of them is, respectively, further composed of two functional entities referred to as recognition and monitoring, as well as learning and assessment (Figure 2.2). According to Vassev and Hinchey (2010), the task of recognition must be defined to correspond to the exploitation of knowledge for the purposes of tracking changes, while the function of monitoring provides all the data necessary for building such knowledge by means of collecting, aggregating, and processing the acquired information. The complementary role of assessment, then, relates to testing hypotheses and identifying situational schemata, while at the same time being responsible for triggering the process of learning, consisting in the formulation of similar new schemas and keeping track of the system evolution history (Vassev and Hinchey, 2010). Given such a context, and the supporting literature, one may conclude that there seems to be much more to self-awareness than the initially assumed direct correspondence with autonomics. What is more, the meaning of autonomics should not be confused with cognition and reasoning, because as long as self-awareness clearly involves the processes of learning, it may also imply a bit more than simple monitoring for purely mechanical context recognition. All this adds substantially to the emergence of the previously mentioned spectrum of different perceptions of autonomics, resulting in the lack of a commonly understood and agreed definition.

In essence, regardless of the increasing number of scientific and industrial works on autonomics, ranging from more conceptual ones to their purely standardisation-orientated counterparts, it continually appears that the positioning and nomenclature of this concept are still evolving, while carrying a certain element of vagueness. In particular, as argued by Kephart and Chess (2003), the very concept of autonomic computing is claimed to be deeply rooted in the theory of agent systems (ATSs), since autonomy, proactivity, and goal-directed interactivity are the major distinguishing characteristics of software agents (SAs). What is more, not only does autonomic computing span single-agent systems (SASs) and multi-agent systems (MASs), but it also encompasses the rationale behind the aforementioned service-oriented architecture. Such a conclusion results directly from a comparison of the following definitions of autonomic computing and software agents, as observed and cited by Brazier et al. (2009), which contrast 'computing systems that can manage themselves given high-level objectives from administrators' with 'an encapsulated computer system, situated in some environment and capable of flexible, autonomous action in that environment in order to meet its design objectives'. As much as they come from different sources of Kephart and Chess (2003) and Jennings et al. (1998), respectively, despite some contrast, these definitions clearly appear to go hand in hand. Reading into the details one may notice, however, that even though the definition of a software agent refers to the notion of being autonomous, there is no hint of autonomics.

Going further, while these definitions clearly revolve around the semantic field originally attributable to automation, the expression of autonomy[6] appears to induce more the lack of human control, while that of autonomics clearly draws from the employment

6 One should also bear in mind that in an environment comprising many networks, the concept of an autonomous system (ASS) denotes a network 'operating independently of all the others', as defined by Tanenbaum and Wetherall (2011), where the network could be also perceived as a network domain for a better understanding of the concepts to be further analysed in this book.

of the principles behind the functioning of the human autonomic nervous system. This immediately implies that, even though there might exist a similar temptation to put both the notions of being autonomic and agent-driven under the same umbrella, a certain dose of prudence is necessary to avoid any unintended introduction of an accidental bias related to the proper understanding of the term of autonomics. What is more, seeking direct correspondence between the two concepts, one could rely on the existence of individual self-managing decision-making elements, making it possible to perceive the autonomic computing system in question from the perspective of a multi-agent system, as such a system would be composed of multiple subsystems or services, while an agent system would comprise multiple agents (Brazier et al., 2009). This is, in fact, where the related service-oriented architecture context appears on the horizon and enters the global picture of the sought-after synergy between autonomic computing and multi-agent systems. This way, decision-making elements could be claimed to fairly well resemble software agents in the sense that, normally, the latter would similarly act on the basis of both monitoring the data from sensors and applying the imposed policies (Brazier et al., 2009). Still, despite all the tight links between these approaches, one needs to be careful to avoid any warping of the sense of autonomics by attributing to it what should not really belong there (Wódczak, 2014).

In fact, even though the role of semantic fields could be regarded as an ignorable detail in this respect, a more thorough analysis unfortunately makes this no longer appear to be so (Wódczak, 2014). In particular, even from a purely linguistic perspective, the word 'autonomous' seems to bring two major connotations of relevance to the argument. As such, the pertinent definitions derived from the market-leading dictionaries tend to classify the notion of being autonomous as 'having the ability to work and make decisions by' oneself and 'without any help from anyone else', in other words independently, according to Gadsby (2003), or making one's decisions on one's own 'rather than being influenced by somebody else', to follow Sinclair (1997).[7] Following the above course of thinking, the quality of being autonomous may apparently be attributed to a system of a standalone type in the sense that it is self-sufficient and, thus, may operate on its own, without any need for external intervention or orchestration, i.e. as a fully separate entity. Most obviously, this perception does not necessarily need to imply a full semantic overlap with autonomics, as depicted in Figure 2.3. Quite the contrary: one may also consider a more cognitive flavour, where the system is autonomous in the sense that it exposes an advanced capability for reasoning. Such a quality, in turn, could manifest itself through the ability to make decisions driven by some hidden logic, just as if the system were steered by a brain-like device. Not surprisingly, this connotation is also not particularly useful, as autonomics detaches the features of the human autonomic nervous system from any learned decisions the brain could be responsible for (Wódczak, 2014).[8]

7 The connotations outlined in this section stem from relevant industrial and academic discussions the author participated in during his standardisation-related activity within the Industry Specification Group (ISG) on Autonomic network engineering for the self-managing Future Internet (AFI) under the auspices of the European Telecommunications Standards Institute (ETSI).

8 In general, however, despite the passage of time and increased awareness, one may come across recent publications, for example Alippi et al. (2016), where an explicit autonomous design appears to be rightly referred to by the authors, even though the reader may have an equally well justified, though vague, impression of the existence of certain reminiscence of autonomics.

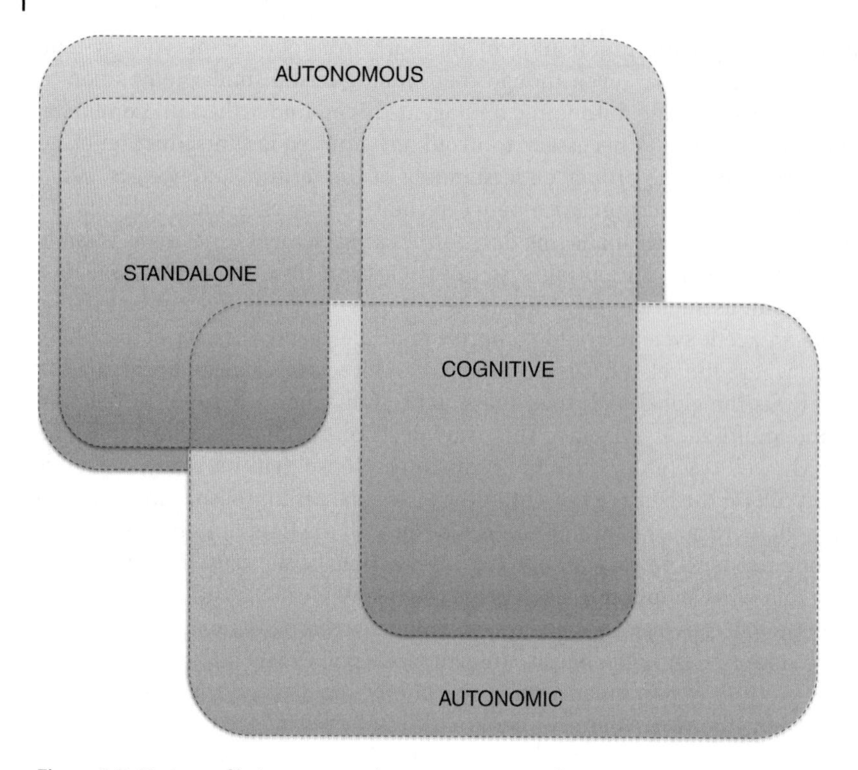

Figure 2.3 Notions of being autonomic, autonomous, and cognitive.

Given the above context, and progressing with the argument, the notion of being autonomic seems somewhat intangible in the sense that it may be understood in a number of sometimes clearly disjoint ways, thereby making the overall process of establishing a widely acceptable definition even more difficult, if possible at all. In fact, one may account for such a situation by analysing the potential reasons leading to the related terminological confusion, chiefly resulting from unjustified attempts at perceiving some other flavours of orchestration as if they were supposed to incarnate the autonomic one, while not really qualifying to be considered so. Thus, one may immediately stumble across misunderstandings related to being autonomous that may be classified as belonging to a disjoint semantic field, and consequently understood in a different way, having little or nothing to do with autonomics. This is because the perception of 'being autonomous' most obviously leans towards semantically overlapping with, or being partially equivalent to, being independent in the sense of functioning in a standalone or detached manner. Yet, while such a characteristic feature could be justified as an enabler of autonomics, it does not necessarily need to be so. What is more, the very fact of being autonomous could go beyond the notion of awareness, more or less directly translating itself into the capability of acting in a cognitive manner in the sense that a system possessing such an ability would be entitled to reason in a human-like manner, solely relying on brain-resembling logic to take decisions.

2.2.3 Towards Self-Management

This variety of impressions of being autonomous does not even come close to the proper perception of autonomics, even though cognition could well work together with autonomics, so long as a clear distinction is made between the two. And yet, expanding the scope of analysis, one may inevitably come across the notion of an automated system where the difference from autonomics appears to be even more conspicuous, as what should be normally understood in this respect clearly boils down to the more and more passé approach of script-language-based programming. In other words, the confusion in this respect may apparently arise from the fact that the grammatical form of being automated may be understood either as if it were reflexive, where some internal drive could be expected to exist, or as if it were triggered externally, where the overall operation would be prescribed well in advance. It so happens that, while the former appears to be a reasonable justification for making the semantic field of the word 'automated' even closer to that of' autonomic', yet still not overlapping, the latter explanation is, in fact, what the experts in the field of network management would actually mean when using it. For the sake of shedding more light in this respect, this section opens with a wider context, going beyond what one could name as the domain of networking or computing. Next, given the observation made by Kephart and Chess (2003) that the 'essence of autonomic computing systems is self-management', the major pillars of self-configuration, self-optimisation, self-healing, and self-protection are scrutinised. Finally, the placement of autonomic features on proper levels of assignment is discussed for organisational purposes.

In essence, looking from a wider perspective, maybe even a bit unexpectedly at first sight, yet most naturally in general, similar notions of a relevant form and type appear to have been deployed similarly way beyond networking, for example in the remote exploration of outer space, where one may come across a definite distinction between the notion of autonomics and autonomy, perfectly addressing the purpose of this explanation.[9] In fact, as claimed by Truszkowski et al. (2006), 'while autonomy supports cost-effective accomplishment of mission goals, autonomicity supports survivability of remote mission assets, especially when tending by humans is not feasible'. One should note, however, that as much as it is useful to account for the notion of self-management, the above citation, somewhat unintentionally, brings up yet another issue, thoroughly discussed in standardisation and pertaining to the word 'autonomicity', which does not appear to exist in the major dictionaries of the English language. This term was similarly coined within the specifications released by the aforementioned ISG AFI under the auspices of ETSI. Unfortunately, as observed by the author, even though the notion of 'autonomicity' is gaining more and more recognition among the experts, who are continually trying to find the most accurate expression to reflect the hardly tangible meaning, this word appears to be not very well received by native speakers of the English language and should be used rather carefully, with a proper commentary fully justifying the act of doing so, as indicated by Wódczak (2014).

---

9 Unfortunately, before attempting to prove such 'definiteness' the reader is referred to Behringer et al. (2015), where no sooner have so-called 'autonomic functions' been defined 'on a system level' than the entire setup is referred to as 'system-level autonomy'. This only shows that the already highlighted profound vagueness in nomenclature can be chiefly attributed to the most natural linguistic fluctuations.

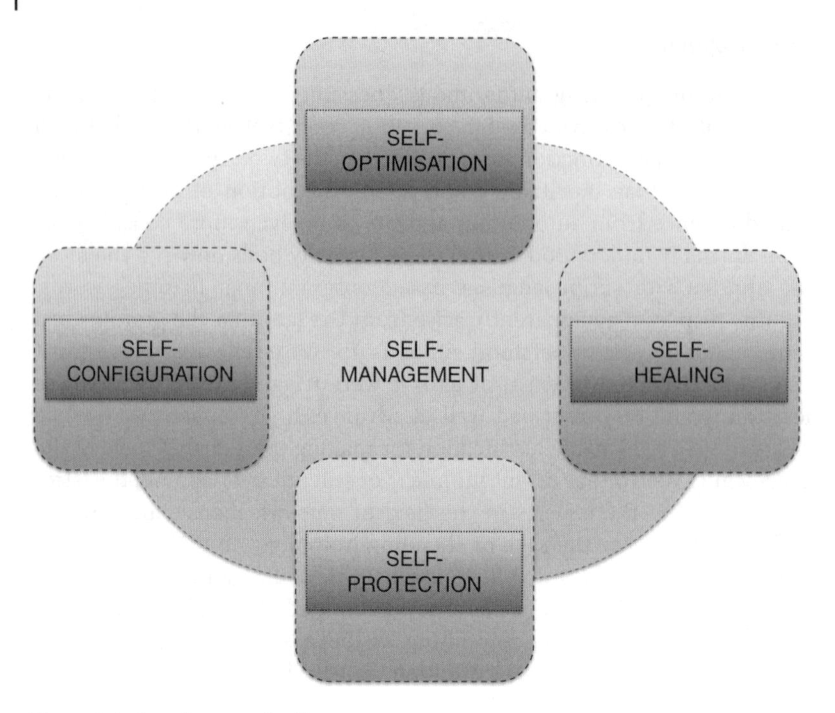

Figure 2.4 Constituents of self-management.

Since self-management turns out to follow the pattern of a circular process, as outlined in Figure 2.4, it appears fairly obvious to commence the analysis of its constituents with the component of self-configuration related to the initial orchestration of an entire selection of software and hardware entities, potentially devised and provided by various vendors, with the inclusion of, yet not implying any limitation to, elements such as databases, routers, or servers (Kephart and Chess, 2003). Bearing in mind the aforementioned complexity of currently deployed networked computing systems, the most appropriate configuration of such entities may become critical from the time consumption perspective, especially in the light of any initially unforeseeable, and not immediately solvable, interoperability issues (Wódczak, 2014). This aspect becomes substantial for distributed configurations, where an overwhelmingly high number of network nodes makes any manual effort towards their holistic orchestration highly impractical, particularly cost-wise. Consequently, it becomes of prime importance for an autonomic computing system to identify its own capabilities in order to apply a selection of relevant predefined high-level policies to perform proper self-configuration. Once the above task has been accomplished, it might be necessary to either further align the existing setup or react to any fluctuations in software or hardware configuration dynamically with the aid of the next component – self-optimisation – generally perceived more as being responsible for tuning specific parameters through the adjustment of software and hardware settings.

In fact, when the case of self-optimisation is concerned, the parameters inherent in software should not be mistakenly confused with those residing in hardware, as the process of tuning either might be perceived as being entirely rooted in software

(Wódczak, 2014). As indicated by Kephart and Chess (2003), since it is not uncommon knowledge that every single system release drastically increases the number of such parameters, it transpires that, as for self-configuration, the task of manual optimisation becomes virtually impossible to be accomplished without the incorporation of some relevantly designed Autonomic Routines. This is where self-optimisation comes into the overall picture of self-management, as it is not only expected to be triggered when a new software or hardware component has been installed, but as a reoccurring operation evoked in a manner resembling the functioning of the human autonomic nervous system in order to allow modifications to be performed on the fly according to additional guidelines stemming from the overlay formed by a specifically designed management plane (MNP) (Wódczak, 2014). Once an autonomic computing system has executed and successfully accomplished both the tasks of self-configuration and self-optimisation, it might require further monitoring to accommodate another key feature of self-management, known as self-healing, an even more dynamic process of analysing the current system status in relation to previously acquired data, so that it is possible to react to issues in virtually no time and guarantee uninterrupted operation.[10]

The reason for such continuous monitoring is related to the fact that the higher the complexity of deployed networked computing systems becomes, the more difficult it appears to identify, trace, and determine the root causes and symptoms of a given incident or a group thereof, especially when additional and unexpected mutual interrelations arise, to follow Wódczak (2014). In fact, as indicated by Kephart and Chess (2003), any nonautonomic approaches attempting an exhaustive analysis of the said root causes may consume several weeks of demanding effort and still result in a rather vague or intangible diagnosis, not to mention the case when the root cause may have suddenly ceased to exist. As such, the described approach to self-healing belongs to the class of predominantly resilience-related solutions, where the system itself is capable of observing any relevant incidents well in advance, which means way before they have caused or contributed to any issue, so that it is possible to react properly beforehand to potentially avoid any otherwise imminent implications, as advocated by Wódczak (2011b). Thus, the requirement of self-healing may become perceived as substantially correlated with self-protection, at the same time exposing certain similarities with the management of system faults. In fact, due to the mutual relations between resilience and fault management, the approach based on inference from symptoms may undoubtedly be even more related to the ultimate component of self-protection.

Given the above context, and keeping in mind the circular nature of the process of self-management, it is possible to shift the point of view a little to look at Figure 2.4 not from the above, but from the side. Then, all the four constituents of self-management could be imagined to reside on the same level and, once separated more significantly, they would naturally form a much more linear configuration, as depicted in Figure 2.5. Such a conceptually transposed representation, where the viewpoint is clearly located somewhere between self-configuration and self-protection, so exactly at the point where the process of self-management reiterates, taking the form of a presumably horizontal arrangement, makes it possible to further subdivide each of the components of self-configuration, self-optimisation, self-healing, and self-protection into two

---

10 As might already have been noted, all these tasks appear to form a rather continuous process, as the configuration usually changes dynamically and adaptively over time.

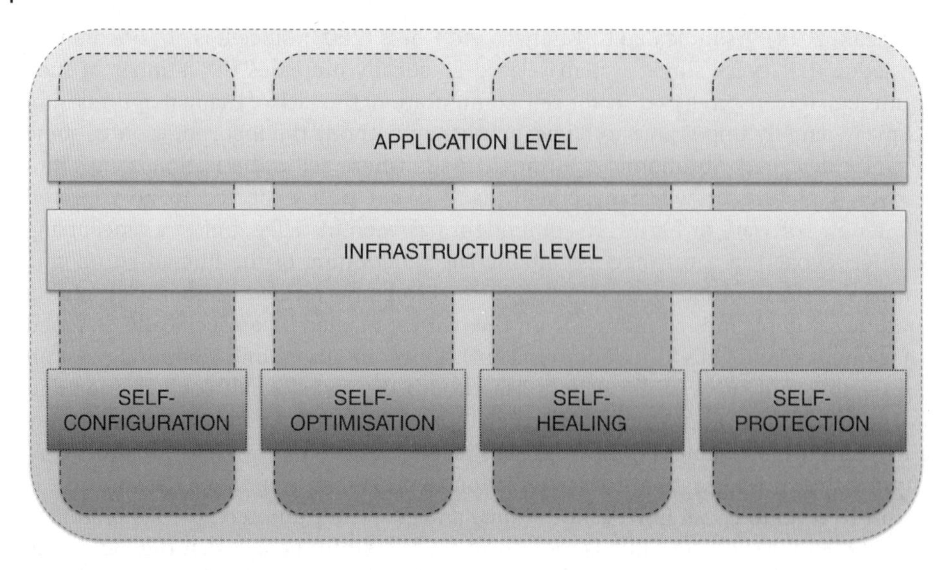

Figure 2.5 Placement levels of autonomic features.

sublevels. In fact, as classified by Brazier et al. (2009), autonomic functions (AUFs) naturally need to be performed at both the application and infrastructure levels, thereby creating a logical separation within the previously described tight, or even overlapping, relation between the realms of computing and networking, where certain aspects advocating for the deployment of automation encompass the baseline concepts behind autonomic system design as a whole. The main driver for such a separation is to maintain sufficient distinction between software and hardware, which, clearly, may pose certain additional challenges, as, especially nowadays, more and more hardware devices appear to carry the 'software-defined' label (Wódczak, 2014).

All in all, the system would need to autonomically detect and resolve any unexpected malicious behaviour or cascading failures (Kephart and Chess, 2003), so that the data collected during the monitoring phase could be properly filtered in order to extract any potential root causes and, thus, reason about as yet unknown, but sufficiently probable, potential issues. In fact, while self-protection should provide the measures necessary for failure avoidance, consisting chiefly in problem mitigation, self-healing appears to be more of a remedial nature (Wódczak, 2011b). This would mean, in turn, that both mechanisms should interact and base their functioning on the aforementioned tasks of self-configuration and self-optimisation. Obviously, such functionality may only be attained gradually, and it is assumed that initially all of the components would need to be considered and deployed separately. However, with the passage of time, as self-configuration, self-optimisation, self-healing, and self-protection become more and more welded together, this process could require the evolution of the self-management component in a more generic way, i.e. as a consistent concept (Kephart and Chess, 2003). Thus, autonomics would first solely provide help in collecting all the data an administrator could use to support the decision-making process. Then, the role of autonomic control processes would be expected to be elevated to the level of suggesting certain actions to the human, and finally such processes would function in an entirely standalone and detached fashion, basing their intelligent decisions on the actions of the other lower-level control routines (Kephart and Chess, 2003).

2.3 Emergent Autonomic Networking

2.3.1 Generic Autonomic Network Architecture

In the light of the inclination towards the ultimate convergence[11] of autonomic computing and autonomic networking, a relevant concept in the form of the Generic Autonomic Network Architecture (GANA)[12] was originally proposed by Chaparadza (2008), at that stage still not encompassing any comprehensive cooperative design, and then further advanced through standardisation, as described by Wódczak et al. (2011).[13] Even though the initial objective of the Generic Autonomic Network Architecture originally revolved around autonomics while remaining distinct from the notion of being autonomous, not to mention the aspect of cognition, its actual instantiation should be never perceived as entirely exclusive of the possibility of exercising any cognitive or autonomous operation. Quite the contrary: what has materialised as a result of a predominantly evolutionary approach could be referred to as a structured categorisation of all the pertinent and related notions for achieving a multifaceted system design. In order to prepare the ground for the Autonomic Cooperative Networking Architectural Model (ACNAM), it is necessary to introduce the baseline design principles adopted by the underlying concept of the Generic Autonomic Network Architecture. To this end, first of all, a more generic perspective of a planar approach to architectural building blocks will be introduced, to be followed by a still high-level blueprint, concluding with the rationale behind the intrinsic signalling tools and methods, before more advanced topics may follow.

At this stage of the analysis, the perspective of autonomics may seem to already have shifted towards autonomic networking, especially since the Generic Autonomic Network Architecture builds upon the 4D Architecture, introduced by Greenberg et al. (2005), where the following four major planar components are distinguished: the decision plane (DCP), the dissemination plane (DMP), the discovery plane (DSP), and the data plane (DTP).[14] As such, the Generic Autonomic Network Architecture fully adopts the above-mentioned planes, and also retains their respective naming structure, which further highlights its evolutionary flavour, as outlined in the ETSI-GS-AFI-002 (2013) Group Specification (GS). Moreover, there is also the knowledge plane (KNP), which is maybe not openly enumerated in the ETSI-GS-AFI-002 (2013) GS, but is

11 For further context, the reader is additionally referred to Wódczak (2011a).

12 Most recently, the acronym GANA, as previously defined, for example, by Chaparadza et al. (2009), has also been expanded as the 'Generic Autonomic Networking Architecture', just to follow Meriem et al. (2016). For the sake of ensuring full consistency with the prevailing majority of relevant publications to date, the original version will be maintained throughout this book.

13 At the present time, the Generic Autonomic Network Architecture embodies certain cooperation-orientated aspects, mostly introduced with the publications originating from the author of this book, which is written to contain the most fully-fledged description of the complementary Autonomic Cooperative Networking Architectural Model (ACNAM), being an evolved incarnation of the previously drafted Autonomic Cooperative System Architectural Model (ACSAM), as outlined by Wódczak (2014).

14 For further background information in this respect, not related to the Generic Autonomic Network Architecture per se, the reader is also referred to the survey by Movahedi et al. (2012).

inherent in its major drawings, chiefly in adding the said flavour of cognition[15] as defined by Clark et al. (2007). Given such a context, and looking in particular at the DTP, one may notice that its major responsibility is to make all the decisions related to the orchestration of the so-called autonomic nodes (ANOs), including the aspects of 'reachability, load balancing, access control, security, and interface configuration'; this is so because nowadays the DCP has replaced the already legacy management plane by operating in 'real time' and assuming a 'network-wide' perspective (ETSI-GS-AFI-002, 2013). In fact, such a view will be directly and fully attributable not only to the yet to be introduced notion of the levels of abstraction and their pertinent hierarchical autonomic control loops, but also the remaining three planes.[16]

Similarly, the dissemination plane encompasses additional protocols and mechanisms, as prescribed in the ETSI-GS-AFI-002 (2013) GS, to ensure sufficiently reliable communication capabilities, allowing for control information such as monitoring data to be safely and efficiently conveyed to the hierarchical autonomic control loops, so that well-informed decisions can be taken by the decision-making elements residing within the DCP. Equally important, yet differently positioned, is the DSP, which makes similar use of specific protocols and mechanisms to identify the physical and logical constituents of a given network or service. This feature is especially valuable when the question of capability discovery is taken into account, as despite the fact that network devices may use self-advertisement to disseminate their characteristics, there might still exist a need for mechanisms capable of collecting such data and exploiting the acquired information for the sake of autonomic networked system orchestration. Last, but not least, comes the DTP, which involves protocols and mechanisms intended to enhance the traditional network layer (NET) protocols such as the iconic Transmission Control Protocol (TCP) or User Datagram Protocol (UDP), using the control and orchestration provided by the DCP (ETSI-GS-AFI-002, 2013). In fact, all the discussed entities are immersed in the Generic Autonomic Network Architecture along with their related decision-making elements[17] and hierarchical autonomic control loops.

Moving forward, as described in the ETSI-GS-AFI-002 (2013) GS, one of the key aspects of the Generic Autonomic Network Architecture is related to the existence of the inherent functional blocks (FBs) and their respective reference points (RFPs), not only for the sake of internal orchestration, but particularly to enable open interaction with architectures defined by other standards development organisations (SDOs), such as the 3rd Generation Partnership Project (3GPP), for example, where the highly pertinent concept of self-organising networks (SONs) is being advanced (Meriem et al., 2016). Given the nature of the Generic Autonomic Network Architecture, as depicted in Figure 2.6, a staged roll-out of its major constituents was prescribed to be orchestrated in a top-down manner, originating from the KNP where network level (NTL) decision elements (DEs) would normally reside, down to the network elements (NEs), where all the

15 As already indicated, the notion of cognition may be perceived as disputable in the realm of legacy autonomics, yet the current trends fully justify such a shift.

16 To show the complexity of the potential routines to be subject to autonomic orchestration, one may think of the task of network renumbering, as explained by Beck et al. (2010).

17 While the notion of a decision-making element (DME) appears to be more descriptive, at this stage of the analysis the naming pattern may be reduced to the more common and equivalent form of a decision element (DE).

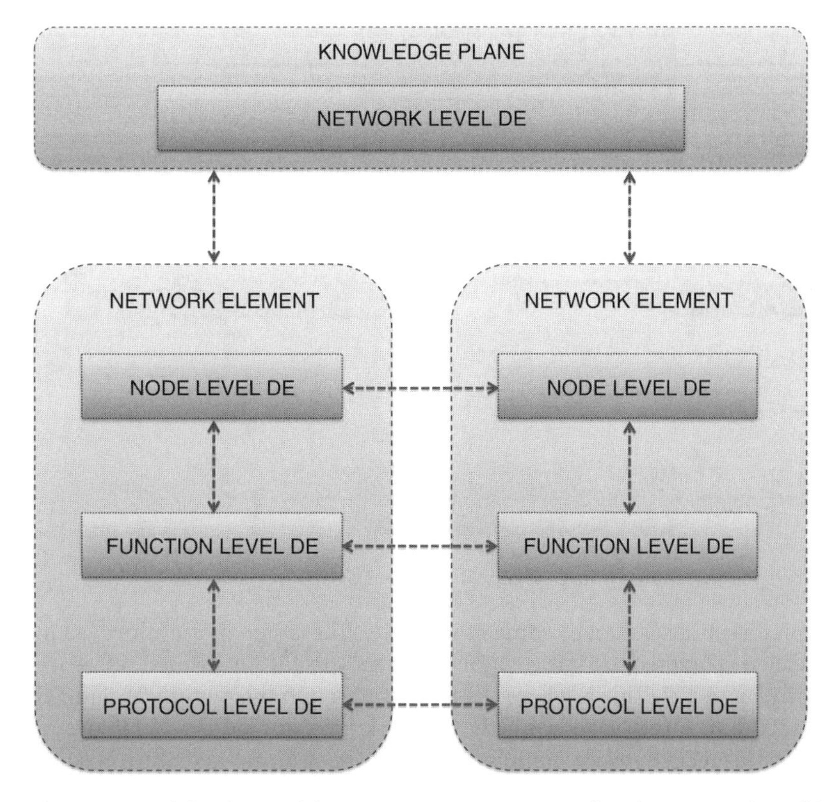

Figure 2.6 High-level view of the Generic Autonomic Network Architecture. Adapted from ETSI-GS-AFI-002 (2013).

remaining node level (NDL), function level (FNL), and protocol level (PTL) decision elements would be located.[18] Not only does such a composition facilitate the overall instantiation process, but it also allows for proper coordination of autonomic functions (AUFs) for the sake of stability and scalability provisioning, which requires that the hierarchical autonomic control loops be designed to guarantee non-coupling and non-conflicting behaviour, as outlined in the ETSI-GS-AFI-002 (2013) GS. Still, given the fact that the hierarchical autonomic control loops may condition any interaction between such autonomic functions, it is necessary to note the role of the relations between or among DEs, which extends to involving their respective managed elements, too.

In fact, the DEs, as further detailed in the section to follow, play an instrumental role in the instantiation of the related autonomic behaviour (AB), since they attempt to identify and determine a state of equilibrium, while being continually triggered by various signals calling for interaction, in the form of either commands from upper-level DEs or events from managed elements, as well as policies or data monitoring[19] (ETSI-GS-AFI-002, 2013). Consequently, in certain circumstances, the freedom of a DE to make a decision on the basis of information it already possesses may equally well be

18 The four levels, i.e. the network level, node level, function level, and protocol level, are normally referred to as levels of abstraction (LoAs). As the descriptive analysis develops, more and more emphasis will be put on relevant aspects of all these entities towards the end of the section.

19 For additional context in this respect, the reader may also refer to Liakopoulos et al. (2008).

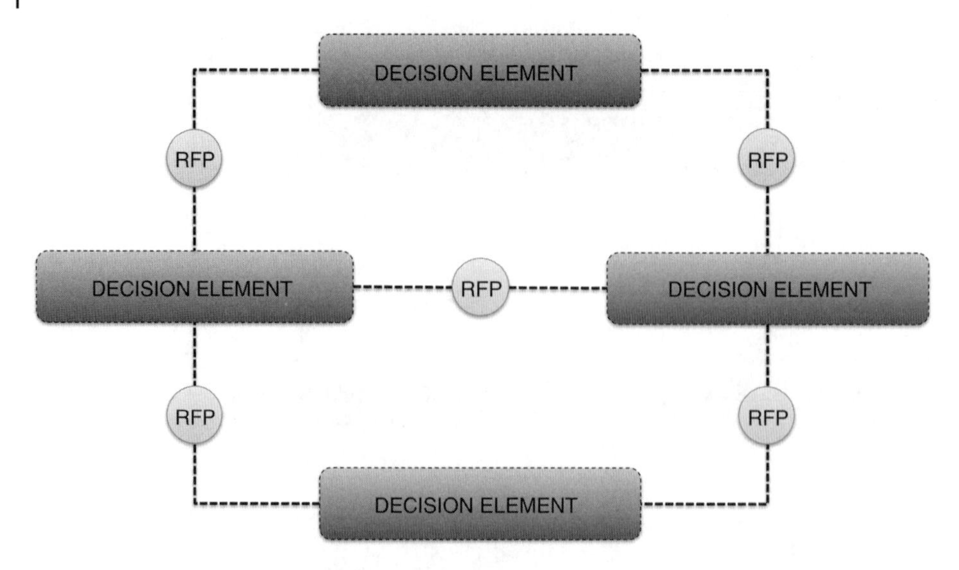

Figure 2.7 Reference points. Adapted from ETSI-GS-AFI-002 (2013).

limited by certain directions issued by some other entity. These signals are conveyed in the form of so-called characteristic information (CI) over specifically defined reference points, as show in Figure 2.7. This is where the KNP comes into the picture, as defined by Clark et al. (2007); it interweaves with the network level and adds an element of cognition, required especially when complex decisions need to be taken, to make it feasible for the networked system to perform tasks of a global scope, imposed, for instance, with the highest priority by an administrator or operator.[20] In parallel, certain local arrangements, such as the yet to be defined Autonomic Cooperative Behaviour (ACB), could be carried out seamlessly and without interruption, clearly as long as any of the policies in force are not being violated. As will be discussed, whether to exercise Autonomic Cooperative Behaviour at the node level or network level, or maybe assuming a broader scale, might depend on multiple characteristics and specific conditions (Wódczak, 2014).

Analysing the above-mentioned components, one could conclude that a working definition of autonomic networking could be approximated as an ability to align the ongoing operation of a networked system with the requirements arising not only from the collected monitoring data, but also from any externally or internally imposed policies, where all of those factors would be presumably changing over time in a continuous manner and at a varying pace (Wódczak, 2014). What is more, such processes should be expected to be observable at each level of abstraction, and only such an assumption would possibly lead to the ultimate impression that what distinct autonomic nodes of

20 Most obviously, all the inner workings of the Generic Autonomic Network Architecture were devised in such a way that the entire framework would build upon the concept of self-management without any specific demand for external intervention from an administrator or an operator. Yet, given the already highlighted convergence between the realms of autonomic computing and autonomic networking, as well as analysing the works of Cheng et al. (2008) and Jennings et al. (2007), one may conclude that any strict distinction between the notions of an administrator and an operator has already become rather artificial, and thus the term 'network operator' will be used throughout the remainder of this book.

the baseline[21] Generic Autonomic Network Architecture would express, in that sense, could be most naturally categorised as autonomic behaviour. As such, the notion of autonomic behaviour is tightly bound with the concept of an autonomic node which, being a core part of the Generic Autonomic Network Architecture, appears to be somewhat self-contained or even, theoretically, detachable, since the network level, where, in specific cases, the relevant global decisions may need to be taken, might function or exist only conceptually, as it is formed by purposely elevated network nodes, otherwise belonging to the node level (Wódczak, 2014). In fact, following the ETSI-GS-AFI-002 (2013) GS, autonomic behaviour is defined to be composed of sub-behaviours or subactions originating from DEs with the intention of achieving a prescribed objective, triggered either externally or internally, or possibly instantiated spontaneously.

2.3.2 Decision-Making Entities

As the conceptual background outlined so far is advanced even further, the transition from the original concept of autonomic computing to its current incarnation in the form of autonomic networking may start to become more conspicuous, mostly due to the accompanying translation of all the relevant nomenclatural and structural patterns. In fact, the initial concept of autonomic computing assumes that a system expressing such a capability should comprise numerous so-called autonomic elements (AEs) as the most comprehensive, yet not entirely atomic, constituents. As such, those AEs are expected not only to contain resources and deliver services, but also be able to manage their own behaviour and act pursuant to certain policies, imposed either internally, i.e. by other AEs, or externally, i.e. by a human operator. Thus, assuming continuous interaction, an autonomic system should be able, as indicated by Kephart and Chess (2003), to attain the level of 'social intelligence' inherent in a living ant colony.[22] Such a capability could not impose any particular notion of being cognitive, because even when imitated, the colony would never act as a single organism, but rather assume the form of a collection of components notifying one another and acting according to directions or rules.[23] In order to account for the instantiation of the pertinent architecture, first of all, the notion of legacy autonomic elements is presented in the light of the more up-to-date concept of DEs.[24] Based on this, the paradigm of an autonomic node is to be introduced as a

21 The notion of the baseline Generic Autonomic Network Architecture is introduced to accommodate its upcoming extension to the Autonomic Cooperative Networking Architectural Model (ACNAM), where the concept of the yet to be detailed autonomic node (ANO) will be upgraded to the notion of an Autonomic Cooperative Node (ACN). For additional context, in the light of the standardisation-related view of the Generic Autonomic Network Architecture to be exercised more towards the end of this book, the reader is also referred to Chaparadza et al. (2013).

22 In fact, the term 'social intelligence' inspired the author to employ the notion of autonomic intelligence in the title of this book, as a leveraged incarnation of autonomic computing, allowing it to be perceived as being more complementary to the concept of artificial intelligence yet elevated with a cooperative flavour.

23 Yet, such a comparison appears to both illustrate and reinforce the principle of autonomics, as the ants themselves could still be compared to the organs of a human body, interacting and exercising their behaviour under the umbrella of a 'single organism' of a much more virtual nature, yet retaining their standalone capabilities, so that the entire setup could adapt itself to the current situation, even though it would presumably lack any awareness, as indicated by Wódczak (2014).

24 Bearing in mind the generic background of decision-making entities (DMNs), apart from the already evolved full naming pattern of decision-making elements, decision elements are also referred to as autonomic decision-making elements (ADMEs).

high-level container thereof, maintaining a more virtual nature, in order to pave the way for the architectural considerations to be outlined later in the book.

In fact, the underlying question of the complexity of an autonomic system may become one of its inherently most critical characteristics in terms of guaranteeing properly accurate maintenance as understood through the aforementioned notion of self-management. Since such complexity might call for special operational tools, it was proposed that it should be tackled with and orchestrated by the application of the mathematical device of utility functions (UFs), as implied by Kephart and Das (2007), primarily devised to serve as a means of facilitating the task of preference specification, while spanning widely over the complementary domains of economic sciences (ESs) and artificial intelligence (AI). Clearly, the said utility functions allow for the specification of a multitude of parameters, based on which fully scrutinised actions could be taken by relevant DEs, as the original 'rational decisions' of 'automated agents' would be referred to in this context (Kephart and Das, 2007). Going further, as explained by Vassev and Hinchey (2010), a learned decision-making process certainly requires the acquisition and analysis of all information of relevance for the sake of knowledge collection and extraction. Similarly, the said parameters are claimed to directly translate into resource-level indicators or, even more straightforwardly, into the related metrics based on quality of service (QoS), intended not only to facilitate the formulation of the optimisation function but, most of all, identification of ways of addressing and solving the complexity issues, as outlined by Kephart and Das (2007). This calls immediately for a more detailed explanation of the workings of architectural designs based on autonomic computing.

Interesting though it may seem, and quite contrary to the follow-up developments, vastly capitalising on the core concept of autonomic computing, this legacy approach itself was proposed under the assumption that the distributed decision-making logic would be encapsulated within the aforementioned autonomic elements. Yet, as might have already become clear, the most recent designs, and the Generic Autonomic Network Architecture in particular, appear to rather expose the internals, thereby modifying the notion of an autonomic element, especially when perceived from the global perspective. In fact, as depicted in Figure 2.8, an autonomic element is normally composed of an autonomic manager (AM) and a managed element (ME), working in very close relation thanks to the existence of an autonomic control loop (ACL), which may work at a varying pace, as indicated by Yixin et al. (2005).[25] As the idea of autonomic managers has nowadays transitioned into the concept of decision elements, the functionality equivalent to a managed element is usually referred to as a managed entity (MEN), being orchestrated with the aid of a 'reformulated' hierarchical autonomic control loop.[26] Yet, it is crucial to note that the inherent feature of each autonomic manager, at the same time serving as its central and focal point, is the notion of knowledge, surrounded by the components of monitoring, analysis, planning, and

25 Referring to the cited publications on autonomic networking, the reader will frequently meet the notion of a 'control loop'. However, for better contrast with the yet to be introduced hierarchical autonomic control loops (HACLs), the naming pattern of 'autonomic control loops' has been assumed at this stage.

26 One should note that within the ETSI-GS-AFI-002 (2013) GS, both notions of a managed entity and a managed element tend to be used rather interchangeably. However, for consistency, the term 'managed entity' is preferred in this book.

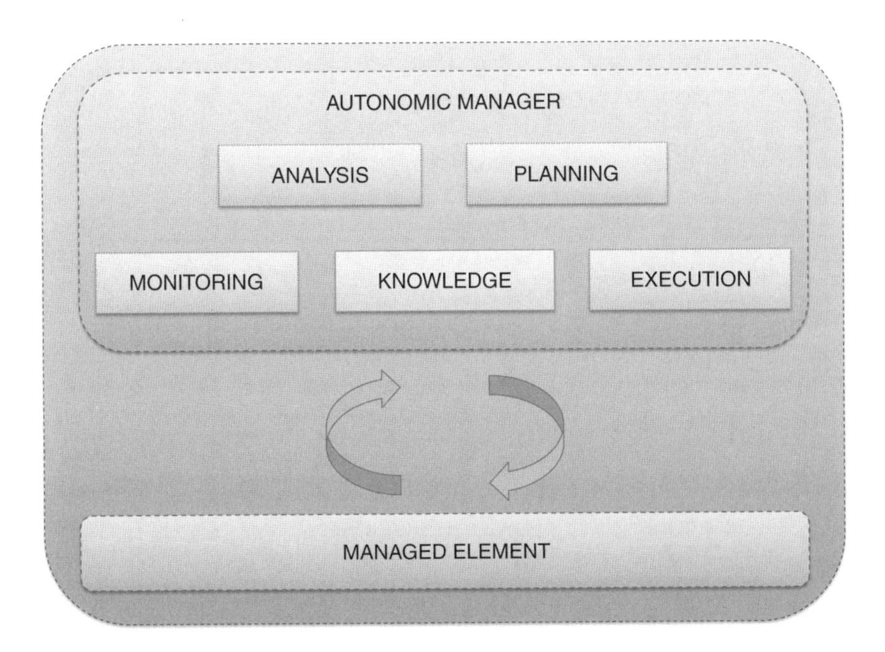

Figure 2.8 Autonomic element. Adapted from Kephart and Chess (2003).

execution.[27] This is, in fact, where the major difference between autonomic managers and decision elements becomes conspicuous, mostly due to the prevailing assumption that the encapsulation-like approach to distributed decision-making logic be evolved respectively.

Given such a context, moving, bottom-up, towards more abstract developments, a high-level depiction of the so-called autonomic node[28] is introduced, as shown in Figure 2.9, where the approach to the distributed decision-making process advocated by the Generic Autonomic Network Architecture is visible. In essence, such an autonomic node[29] is envisaged as composed of those three out of the aforementioned four levels of abstraction which may be characterised as exclusively internal when contrasted with the network level. The different nature of the latter results from the fact that, in a sense, it should rather be perceived as if it were an abstraction of an abstraction, especially since it is not physically attributable to any realistically quantifiable entity.

---

27 For additional details on each of the components listed, as well as useful context, the reader is referred, for example, to Kephart and Chess (2003).

28 A fully-fledged ANO incarnation will additionally include the network level as an inherent, if slightly detached, component.

29 In order to obtain additional insight into and comprehensive picture of what functionalities are covered by the autonomic node of the Generic Autonomic Network Architecture, the reader is referred directly to the ETSI-GS-AFI-002 (2013) GS. In particular, as outlined therein, the responsibility of the node main decision element is related to auto-configuration, security management, resilience and survivability, and fault management. Moving downwards, the areas orchestrated by the function level decision elements include, but are not limited to, monitoring, mobility management, QoS, data plane management, routing management, and service management. Finally, at the protocol level, control processes are exercised over usually external managed entities, such as, for instance, services and applications, monitoring tools, data plane protocols, intra-domain routing, inter-domain routing, and other protocols of relevance.

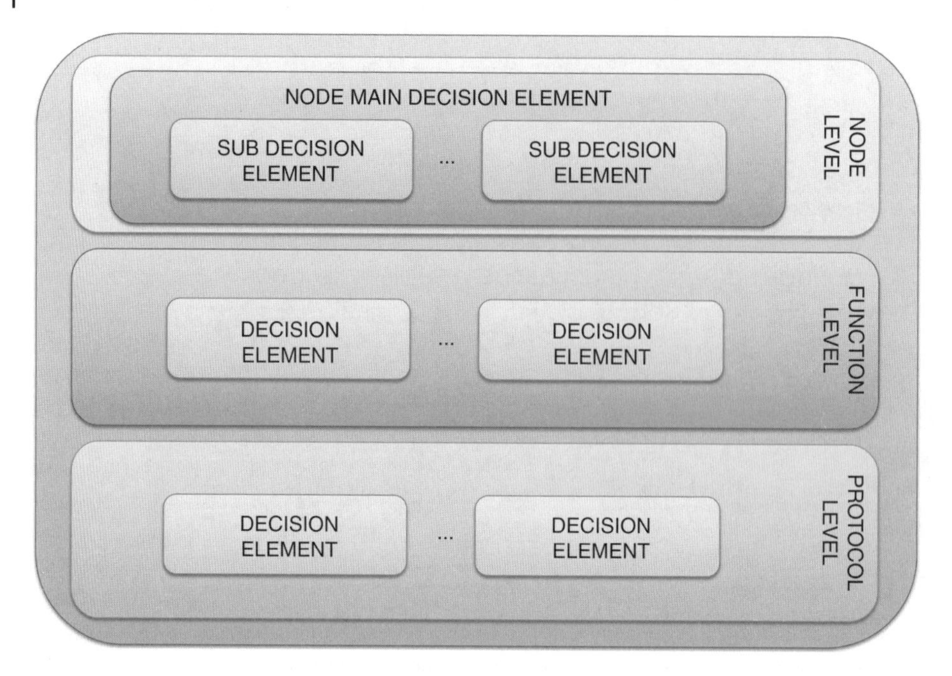

Figure 2.9 Simplified view of an autonomic node.

Looking more into such 'internal' levels of abstraction, one may discern immediately that decision elements should not be perceived as being atomic, since they might be further subdivided into sub-decision elements (SDEs). In general, however, since this type of subdivision is expected to be present at the node level, the question arises whether the node main decision element might be perceived as an aggregation of fully-fledged decision elements, with additional prefixing employed for linguistically justified nomenclatural consistency. One way or another, preceding a more detailed analysis of both the levels of abstraction and hierarchical autonomic control loops, as much as there are obvious differences between the conceptual construction of autonomic elements and decision elements, there are also sufficient similarities to make the ideological inheritance clearly conspicuous.

Keeping in mind the high-level structural depiction of an autonomic node, as well as the mutual interrelationship between decision elements, especially when perceived from the viewpoint implied by the framework of the levels of abstraction, it becomes most natural to generalise the information exchange between such decision elements, as advocated by the Generic Autonomic Network Architecture, where the already described characteristic information is conveyed over the said reference points.[30] In particular, on the one hand, it is clear that the related interfacing process is orchestrated

30 Looking at the 3GPP-TS-23.501 (2017) Technical Specification (TS), for example, one should note that, by comparison, the notion of a reference point may soon become obsolete as the 5G System standardisation has progressed even further, since the Reference Point Architecture (RPA) is currently being replaced with the more software-orientated Service Based Architecture (SBA). Although the background of 5G per se is not explicitly set within this book, partly because some of the concepts under investigation go well beyond the scope it is meant to cover, for certain more advanced topics of relevance the reader is referred, for example, to Ordonez-Lucena et al. (2017).

by the hierarchical autonomic control loops, where the policies of relevance are injected, either internally or externally, on the basis of inferences drawn from monitoring data. On the other hand, though, drawing a comparison between the legacy approach as envisaged by Kephart and Chess (2003) and the currently evolving designs, such as the Generic Autonomic Network Architecture itself, it transpires that the general distinction that apparently still exists between the role of an autonomic manager and a managed element could be further relaxed to become predominantly conceptual, when mapped onto the corresponding tandem formed by a decision element and a managed entity. In other words, especially when higher levels of abstraction are concerned, due to their mostly abstract nature, the managed entity may resemble the functionality and take the form of a decision element. Obviously, such a pattern may not fully hold at the lowest of the levels of abstraction, since the external protocols under control normally do not embody any intrinsic replica of the above-mentioned concepts.[31]

In this respect, before any specific data in the form of the above-mentioned characteristic information, encapsulated into proper messages, could be exchanged in a standardised manner over the said reference points, some internal preprocessing may need to be performed by the logic embedded in the corresponding decision elements. As such, the decision elements are supposed to undertake actions in accordance with the objectives imposed upon them, at the same time satisfying the assumption that the overall design remain within the flexibility frames of the Generic Autonomic Network Architecture. What is more, given the assumption of the existence of not only horizontal, i.e. peering or sibling, but also vertical, i.e. parental, relations, one may imagine a three-dimensional grid of decision elements attempting to fulfil the nontrivial task of maintaining the overall state of equilibrium of the autonomic system, while concurrently making the utmost effort to meet the aforementioned objectives. This may, once again, recall the biological analogy of the ant colony, thereby highlighting the ever returning question of the hardly tangible difference between the notion of being autonomic and the notion of being able to reason, just as if there had been any process of thinking in place. As already described, being separate organisms, ants would be said to move around together as a symbiotic group at most, this way conceptually becoming fairly alike to the similar interaction among human body organs, where the living processes could not be explicitly orchestrated with the aid of thinking, but instead result from the autonomic behaviour of such organs also forming a somewhat symbiotic arrangement.

Thus, the decision elements could be perceived more as configurable entities facilitating an object-orientated system design from the architectural perspective, while allowing for their functionality to be further split, shifted, or translated, which appears to remain fairly in line with the rationale behind autonomic system design on the one hand, yet certainly brings in the risk, or maybe rather the opportunity, of expanding the said functionality of being able to reason, so that AI-related approaches could be exercised equally well, on the other hand. Looking particularly from the perspective of

31 In fact, for the sake of clarity, one should note that the Open Shortest Path First (OSPF) routing protocol, as defined by Moy (1998), and therefore established well before the emergence of autonomic computing in the present understanding, not to mention autonomic networking, is claimed by Behringer et al. (2015) to 'exhibit properties of self-management and can thus be considered autonomic', although Meriem et al. (2016) highlight that 'such autonomic-like feature in OSPF is not cognitive'. For additional context in this respect the reader may also refer to Retvari et al. (2011).

the lowest-level decision elements it becomes obvious that they need to communicate with their respective managed entities over well-defined and preferably standardised interfaces which may create a much more rigid, limited, and 'hard-coded' environment, following the nomenclature of Kephart and Chess (2003). However, should a slightly broader view be assumed, it could naturally transpire that, regardless of the evolution stage of a particular instantiation of the concept of autonomic computing, both the original design and the most recent developments thereof, such as the Generic Autonomic Network Architecture, tend to be rather strongly based on the assumption that the decision elements need to manage not only their internal behaviour, but also external interactions (Wódczak, 2014). Such interactions mostly translate into the exchange of well-defined and well-structured signals or messages with other decision elements, once again putting a special emphasis on the inclusion of the external world, both in the microscopic and macroscopic context, as originally advocated by Kephart and Chess (2003).

2.3.3 Abstraction Levels and Control Loops

The levels of abstraction originating from the Generic Autonomic Network Architecture could not be reflected upon exhaustively should they not be put against their corresponding hierarchical autonomic control loops. In general, following what has already been signalled, one may distinguish between four disjoint levels of abstraction from the top network level, through the node level and the function level, down to the protocol level. Apart from their respective hierarchical autonomic control loops, each level is characterised by its inherent decision elements, thanks to which the process of interaction between them takes place in both a top-down and bottom-up manner. In particular, such interaction is performed with the aid of vertical reference points (VRPs), i.e. over hierarchical autonomic control loops between hierarchically-orientated decision elements in the vertical dimension, as well as through horizontal reference points (HRPs), directly between their peer or sibling counterparts in the horizontal dimension. Thus, each decision element may retain its exclusive responsibilities without becoming subject to any restrictions where the allowed interaction among all of them or their respective subsets is concerned (Wódczak, 2014). However, such a design might turn out to become highly complex and, therefore, have substantial implications on the overall stability and scalability of the autonomic system in question. The above issue may be addressed through the adoption of a specific approach to the definition of the role and scope of the hierarchical autonomic control loops, as outlined in the ETSI-GS-AFI-002 (2013) GS.

In fact, analysed either in the top-down or bottom-up direction, the multi-tier arrangement of the internals of the Generic Autonomic Network Architecture appears to create an otherwise highly evocative impression that the entire setup would be managed in a substantially hierarchical manner. Unsurprisingly, the higher the level of abstraction, the broader the scope of certain decision elements and their respective hierarchical autonomic control loops becomes; in particular, the scope of control of any of such decision element spans recursively over the numerous lower-level decision elements to be overseen, as well as over the respective hierarchical autonomic control loops to be orchestrated. What is more, it is necessary to highlight that, generally, the higher the level of abstraction, the lower the pace at which a given hierarchical

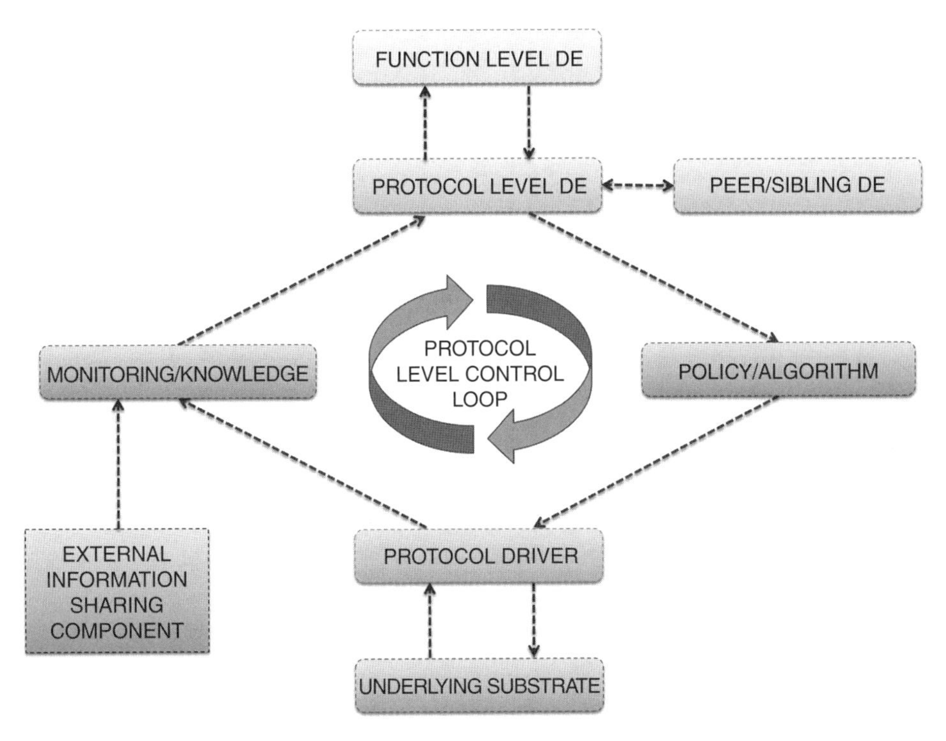

Figure 2.10 Protocol level hierarchical autonomic control loop. Adapted from ETSI-GS-AFI-002 (2013).

autonomic control loop should be running (Wódczak et al., 2011). Thus, the autonomic system may be expected to demonstrate scalability, as it should always allow for the incorporation of 'parent' decision elements at a specific level of abstraction in order to orchestrate the behaviour of their 'child' decision elements. One should also note, however, that in the case of a decision element at the protocol level, there would be no level of abstraction below, so that such a decision element would normally orchestrate the external resource of a managed entity, usually with a relevant protocol.[32] Complementary to scalability would remain the said pace a given hierarchical autonomic control loop be running at which should change with the level of abstraction to guarantee a sufficient degree of stability.

In the following, all the levels of abstraction are to be characterised in reverse order on the basis of their respective hierarchical autonomic control loops and in accordance with the ETSI-GS-AFI-002 (2013) GS. Commencing from the bottom protocol level, as depicted in Figure 2.10, the relevant hierarchical autonomic control loop is orchestrated by the protocol level decision element which interfaces with a managed entity in the form of a software driver supervising a specific protocol, most likely belonging to the Open Systems Interconnection (OSI) or Transmission Control Protocol/Internet Protocol (TCP/IP) stack, as outlined by Wódczak et al. (2011). The particular protocol

32 In particular, such a protocol could be equipped with internal hierarchical autonomic control loops of its own, or rather autonomic control loops (ACLs), because, as explained in the next note, the concept of Generic Autonomic Network Architecture does not advocate the introduction of protocol-intrinsic hierarchical autonomic control loops, mainly for complexity reasons.

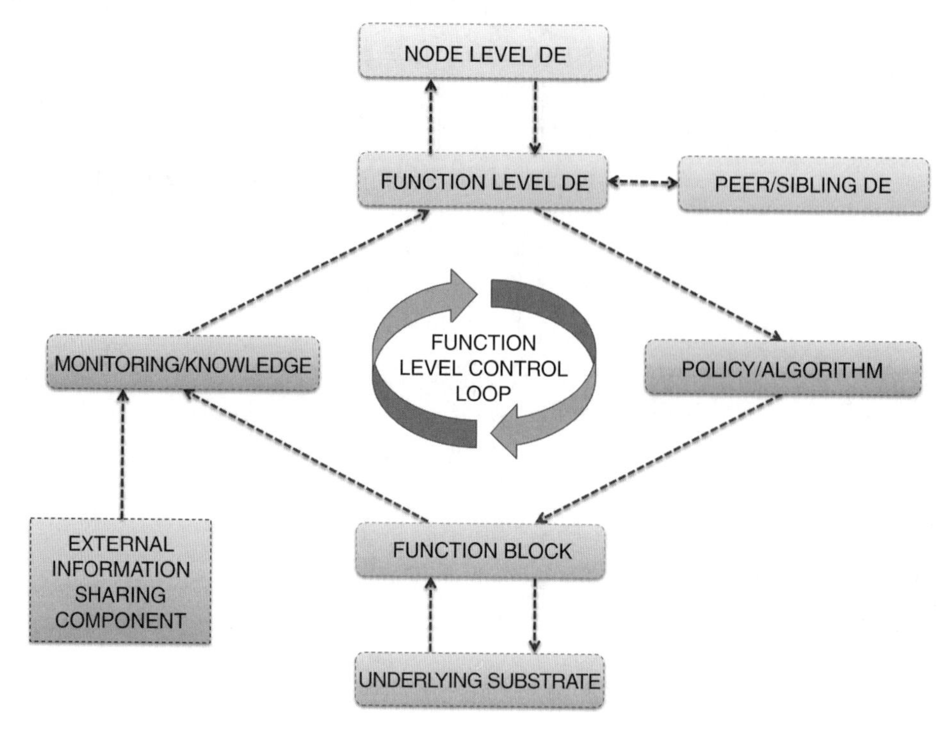

Figure 2.11 Function level hierarchical autonomic control loop. Adapted from ETSI-GS-AFI-002 (2013).

might take either a modular or monolithic shape and, potentially, even be equipped with its own intrinsic autonomic control loops (ACLs)[33] not originating from the Generic Autonomic Network Architecture design, particularly since, according to previous note, the deployment of hierarchical autonomic control loops within protocols may not be recommended to keep them simple and driven directly by the Generic Autonomic Network Architecture.[34] Moving upwards, immediately above the protocol level resides the function level, as outlined in Figure 2.11, mostly encompassing the so-called 'abstract network layer functions', and not really the protocols themselves (ETSI-GS-AFI-002, 2013). Such a function may incorporate, for instance, the general functionality of routing or mobility management. Thus, the role of the function level consists in the proper configuration of specific protocols with the aid of hierarchical autonomic control loops intended to facilitate the process of event monitoring and tracking.

One next comes across the node level, being the highest physical entity of this sort, as the network level to follow is more of a conceptual, though still practical, nature. As such, being located right above the function level, the node level contains the logic

33 As previously mentioned, the term 'autonomic control loop' is purposely used instead of 'hierarchical autonomic control loop' in order to underline that a non-GANA internal design is referred to, even though, at a certain level of generality, the two could be perceived as if they were equivalent.
34 In fact, as summarised by Wódczak (2014), it is claimed that protocol-internal autonomic control loops could unnecessarily put the Generic Autonomic Network Architecture at the risk of stability problems given the already high complexity of its current incarnation.

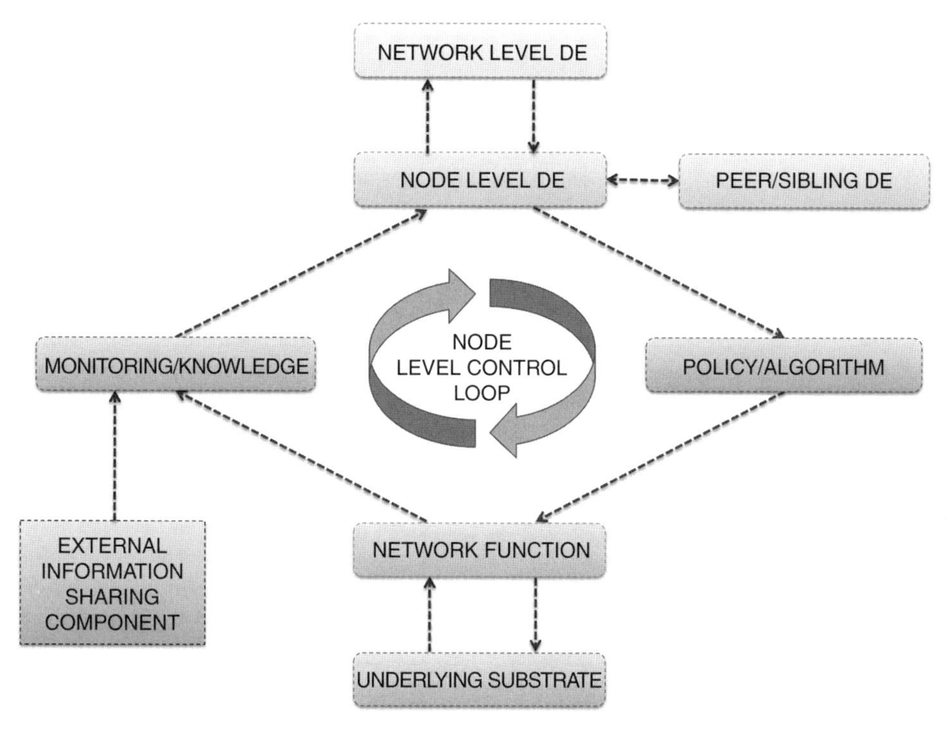

Figure 2.12 Node level hierarchical autonomic control loop. Adapted from ETSI-GS-AFI-002 (2013).

responsible for the autonomic node as a whole. Thus, as outlined in Figure 2.12, not only is the node level decision element granted direct access to the requirements exposed by the function level decision elements, but it may access all the information pertaining to the internal priorities of the node level itself, as advocated by the ETSI-GS-AFI-002 (2013) GS. What is more, because it takes its directions immediately from the network level decision element, it may be perceived as a kind of a hub responsible for the entire autonomic node. However, since a corresponding physical network device cannot be claimed to exist, at least not in the form of a single piece of hardware, the network level should be perceived as being formed by some purposely elevated autonomic nodes, characterised by extended capabilities.[35] Since only a single network level decision element should exist, as presented in Figure 2.13, it could become rather complicated because of its elevated nature. For this reason, it would be recommended that the core functionality of this decision element should follow a modular pattern, so that a conceptually centralised design could capitalise on being physically detachable. Consequently, a fully distributed solution could be implemented, where there would be multiple network level decision elements to make decisions of a global scope in a cooperative manner (Wódczak, 2014).

35 In order to account for such a notion, especially in the case of mobile ad hoc networks (MANETs), the capability in question could be translated directly into enhanced durability requiring that the autonomic node be equipped with a more capacious energy source.

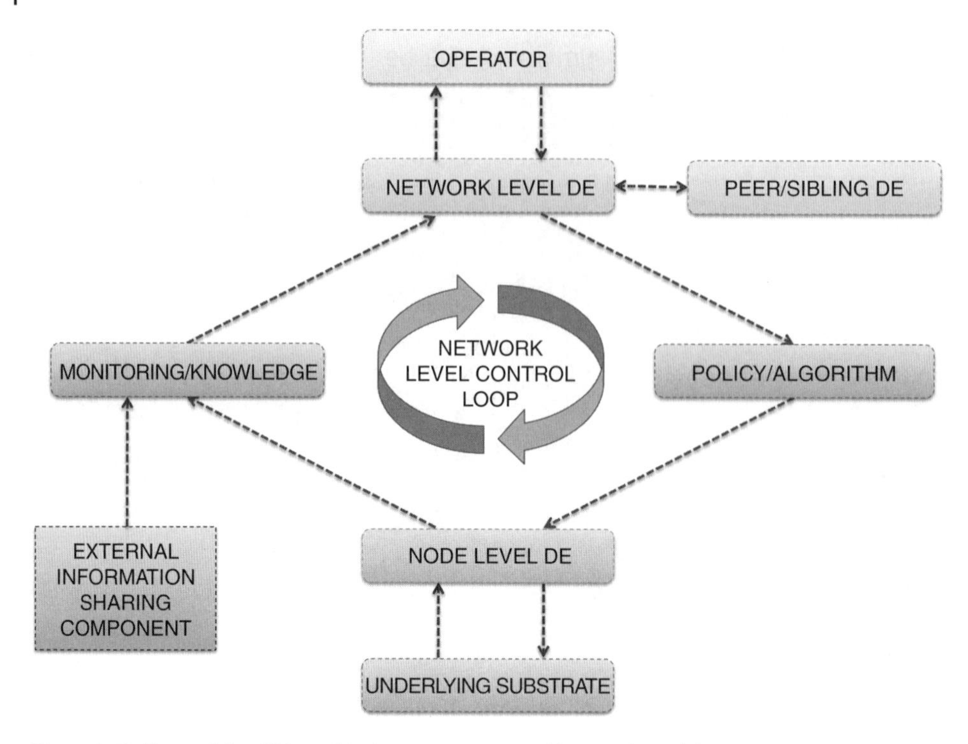

Figure 2.13 Network level hierarchical autonomic control loop. Adapted from ETSI-GS-AFI-002 (2013).

2.4 Synergetic Cooperative Approach

2.4.1 Vertical Technological Pillars

Given the background of the introduction of the workings of autonomic computing per se, as well as with the complementary discussion of the rationale behind its evolved incarnation in the form of the Generic Autonomic Network Architecture, the time has come to provide a comprehensively detailed, yet still correspondingly high-level, outline of the synergetic approach to autonomically driven cooperative networked system design to be advocated in this book. In this respect, first of all, the major technological pillars,[36] conceptually assumed to follow a vertically orientated pattern, will be introduced in this subsection, to be later complemented with newly proposed horizontally orientated architectural extensions, so that it is possible to present a fully-fledged and incrementally arranged theoretical blueprint of the target design later in the book. Focusing on the Vertical Technological Pillars (VTPs) in the first place, one could, in fact, enumerate four major components of relevance, the first three of which,

36 One should note that since from the general linguistic perspective 'pillars' are bound to be vertically orientated for clearly semantically justified reasons, the overall structure composed of layers and normally presented in the bottom-up direction, as exemplified by the Open Systems Interconnection (OSI) Reference Model (RM), will need to be visualised horizontally. Apart from linguistic reasons, this decision has been made by the author mainly to stress that the entire design is built on such legacy pillars with the addition of new vertically positioned levels of abstraction brought into the picture by the Generic Autonomic Network Architecture. This approach will be further discussed in the following subsection.

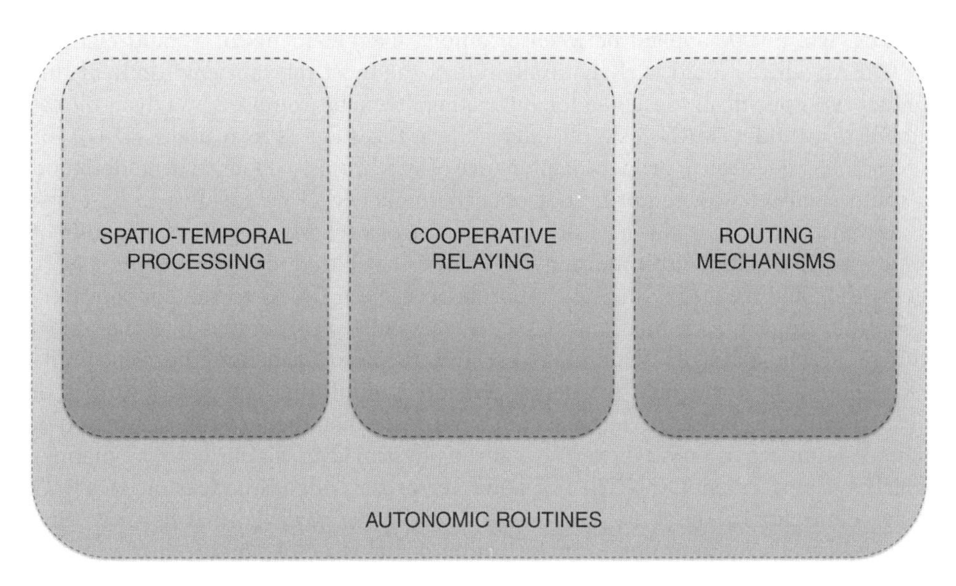

Figure 2.14 Vertical Technological Pillars.

spatio-temporal processing (STP), cooperative relaying (COR), and routing mechanisms (RMEs), are going to be placed altogether against the background of the fourth one, the so-called Autonomic Routines (ARs). The structure created in this way is intended to prepare the ground for the further analysis and introduction of the concepts to be gradually proposed, creating clear boundaries delimiting their scope and area of application. At the same time, the novel aspects will be introduced as the plot develops, including the Horizontal Architectural Extensions (HAEs) and the incremental conceptual outline.

As shown in Figure 2.14, the leftmost pillar of STP is described first to reflect the composition of the Open Systems Interconnection Reference Model, assuming a bottom-up traversal of the specific layers. Belonging to the physical layer (PHY), STP provides the very desirable capability of radio channel orthogonalisation, allowing for immediate improvement of wireless transmission performance, especially in terms of data transfer reliability. In fact, the name of spatio-temporal processing has not been chosen accidentally, as the specific technique, space-time block coding (STBC), employed in the analysis throughout this book, is claimed to be much more of a modulation-driven than a coding-based approach (Alamouti, 1998). This is so since, by contrast with space-time trellis coding (STTC), no coding gain (CG) may be achieved in the former case, yet the highly relevant diversity gain (DG), to be discussed in more detail at a later stage, is still observable. Space-time block coding has been chosen as the most representative approach to STP given the nature of the overall concept under discussion. It can obviously be replaced with other, more complicated, techniques offering even better performance, yet given the overall complexity of the ultimate Autonomic Cooperative Networking Architectural Model, it became of prime importance to keep all the building blocks reasonably balanced, still allowing for their being advanced, should this turn out necessary.

Most importantly, however, irrespective of the approach to STP employed, the utmost emphasis will be laid on the realism of the proposed solution, especially given the fact that STBC will not be applied between or among fixed elements of a multi-element array (MEA), which would fairly naturally reflect the most relevant environment this technique was specifically designed for, but, rather, it will be exercised by a dynamic set of mobile relay nodes (MRNs), simply referred to at this stage as relay nodes (RNs). For this reason, the question of proper maintenance of synchronism needs to be understood in a comprehensive way, requiring not only the higher-level perspective to be taken into account, where it is the network layer protocol orchestrating the behaviour of specific relay nodes, but also requiring the focus to be shifted to the link layer (LNK), where orthogonal frequency-division multiple access (OFDMA) to the cooperatively shared radio transmission medium may be successfully incorporated into the global picture of the whole design (Wódczak, 2012). Thus the rationale behind the cyclic prefix (CPX) inherent in orthogonal frequency-division multiplexing (OFDM) becomes of critical relevance, as it may perfectly serve its purpose when the fine-tuning portion of synchronisation provisioning is exercised at the physical layer, similarly to the manner in which it is performed in the case of Digital Terrestrial Video Broadcasting (DVB-T) systems (Gonzalez-Bayon et al., 2010). One should note, however, that despite being mentioned here, a more profound investigation of the issue of synchronism remains beyond the scope of this book.

Moving forward to the pillar of cooperative relaying immediately shifts the overall context of the discussion to the link layer, where also the medium access control (MAC) sublayer is located, as explained, for example, by Tanenbaum and Wetherall (2011). In fact, the MAC-related region of interest has already been touched upon, although it happened in a rather indirect way: OFDMA, mentioned in the previous paragraph, undoubtedly qualifies as belonging to its set of inherent protocols. On the other hand, however, it clearly contrasts with OFDM since the latter belongs directly to the physical layer given its pure characteristics of a modulation-type technique. In the light of such slight differences in terms of nomenclature and their reciprocally significant discrepancies in technical meaning, there is no wonder that in spite of stemming from STBC, the concept of distributed space-time block coding (DSTBC), which is yet to play a key role in the conceived solution, should be assigned to the link layer, too. Such an allocation is conditioned by the fact that DSTBC, as introduced by Laneman and Wornell (2003), is already more of a mapping of the legacy STBC onto a networked setup, yet without involving any network layer routines per se. This is so, even though it is fairly usual to perceive DSTBC as a cooperative relaying protocol, while the notion of a protocol should be understood in a different way in this context, since its semantic field leans more towards a typical link layer mode of operation.

Even though DSTBC may be perceived to be the key element of the overall design, it is necessary to align it with the concept of virtual antenna arrays (VAAs), as outlined by Dohler et al. (2003), since usually the latter[37] will be referred to throughout the chapters to follow, while, in fact, the former will be meant. The reason for such a duality appears to be conditioned by a couple of factors. On the non-technical side, the introduction of the initial versions of both approaches appears to have coincided in time, at least to some extent, thereby making any rigidly definite distinction between them in terms

37 In fact, the elevated version of VAAs, in the form of autonomic cooperative sets (ACSs), is referred to indirectly here. This is done to avoid any unnecessary confusion, as the concept of ACSs immediately calls for the notion of Autonomic Cooperative Behaviour still to be described more extensively.

of naming somewhat artificial, if not confusing or inadequate. On the technical side, VAAs are often claimed not to be limited to employing just the technique of STBC, thus elevating the DSTBC approach to the status of a special mode of operation, which, at the same time, seemingly creates the most desirable compromise in this respect, also fully respected throughout this book. One should note, however, that despite the said mode of operation, the relay nodes supposed to form a VAA need to employ one of the yet to be discussed cooperative relaying protocols, commonly known under the names of amplify-and-forward (AF), decode-and-forward (DF), and decode-and-reencode (DR), to ensure that the data conveyed in a wireless manner from a given source node (SN) to the relevant destination node (DN) will be properly relayed.

Last, but not least, among the three major pillars are the routing mechanisms so necessary for orchestrating the system behaviour from the yet more elevated viewpoint of the network layer. In fact, the cooperative relaying protocols mentioned in the previous paragraph are predominantly part of the link layer since, as much as they can cover local interactions among adjacent relay nodes, for pragmatically justified reasons they are not designed to scale up too extensively, and, therefore, are also not ready to serve communication between disjoint cooperative groups. One should note, however, that by no means should such a characteristic be perceived as a deficiency, since any related limitation is clearly imposed by the scope of operation attributable to the link layer, thereby allowing for the proper functioning of such protocols on a small scale, yet being far away from what is meant to be delivered by a fully-fledged routing scheme. This is, in fact, where the routing mechanisms come into the global picture of the solution to be presented; they are, additionally, carefully chosen to address the requirements imposed by a setup originating from and following the nature of mobile ad hoc networks. In particular, the Optimised Link State Routing (OLSR) protocol is going to be employed, not only because of the fact that it belongs to the so-called proactive class, which is characterised by continuous and periodic dissemination and collection of control data, but especially thanks to its innate mechanism of multi-point relay (MPR) station selection heuristics found by the author to perfectly integrate with VAA-driven cooperation (Wódczak, 2007).

As such, the above-mentioned MPR station selection heuristics allows for an immediate upgrade of the workings, and therefore also a related improvement in performance and reliability, of the underlaying cooperative relaying protocol operating at the link layer, chiefly because of the fact that additional information of a much wider scope becomes instantly available, allowing for much more accurate preselection of the relay nodes to be assigned to their most relevant VAAs. At a later stage, as the plot develops even more comprehensively, it is expected to become highly conspicuous that the enhancement of accuracy may be obtained virtually at adjustably diminishable additional cost in terms of the incurred control overhead of the respectively modified Optimised Link State Routing protocol, as well as with the possibility of maintaining, in most cases, full backward compatibility thereof. The network layer would seem to be the highest extent to which one could traverse in an attempt at providing the fullest attainable cooperative networked system automation, should it not be for the slightly separately positioned fourth pillar, the so-called Autonomic Routines. As such, this pillar will not be addressed too extensively at this stage, not only because a significant part of this chapter has already been devoted to the description of its workings, but especially due to the fact that its advantages will gradually unfold with every paragraph to follow. It will become especially visible in next subsection, where the Horizontal Architectural Extensions will be presented in the context of the VTPs.

In order to summarise this part, as well as provide a proper background for the upcoming developments, certain high-level pieces of information will be delivered in the form of a conclusion, so that a comprehensively consistent picture of the overall design may be fully outlined. In essence, even though the idea of the Autonomic Cooperative Networking Architectural Model derives directly from the conceptual assumptions behind the discovery of the legacy autonomic computing, the Horizontal Architectural Extensions will expand upon the most up-to-date instantiation of autonomic computing in the form of the Generic Autonomic Network Architecture, where all the workings of the former are well adjusted to the nowadays prevailing approaches in this respect. For this reason, the entire design will be somewhat driven by the emerging interrelationship between the layered structure following the well-known pattern of the Open Systems Interconnection Reference Model and the levels of abstraction inherent in the Generic Autonomic Network Architecture. As a result, all the carefully and specifically drafted decision elements belonging to the said levels of abstraction will be allowed to orchestrate the behaviour of their pertinent protocol-like entities allocated to the most relevant region of the said Open Systems Interconnection Reference Model. Most obviously, such a manner of operation will be tightly monitored and overseen thanks to the deployment of the already introduced hierarchical autonomic control loops. Consequently, an additional dimension will be incorporated into the workings of both the above-mentioned developments, so that it will become possible to further advance the state of the art in this area of science.

2.4.2 Horizontal Architectural Extensions

Given the background and context of the VTPs, the time has come to present the Horizontal Architectural Extensions (HAEs) along with the related design challenges in order to pave the ground for the ultimate incremental conceptual outline. In fact, as already mentioned, the naming pattern adopted in this respect by the author has not been chosen accidentally, but with clearly identified objectives that go well beyond pure linguistics. To some extent at the expense of the Open Systems Interconnection Reference Model, normally orientated in the bottom-up direction, it has become possible to expand the concept of the Autonomic Cooperative Networking Architectural Model by building on top of each of the VTPs, even though their modified left-to-right arrangement shown in Figure 2.14 might not appear to be immediately and fully understandable for reasons stemming from conceptual convictions. Thanks to such an approach, though, it will be possible to use two-dimensional drawings for the sake of expressing the rationale behind the three-dimensional architectural advancements to be presented across this book. This will be achieved not only by applying the conceptual expansion of the pillars in the upward direction, where normally other layers belonging to the Open Systems Interconnection Reference Model would be drawn, but also through exercising the possibility of spanning the said extensions, as inspired by the workings of the Generic Autonomic Network Architecture, over the adjacent technological pillars and beyond.

While this approach, consisting in the expansion of the concept in the bottom-up direction upon the vertically orientated technological pillars as explained above, is

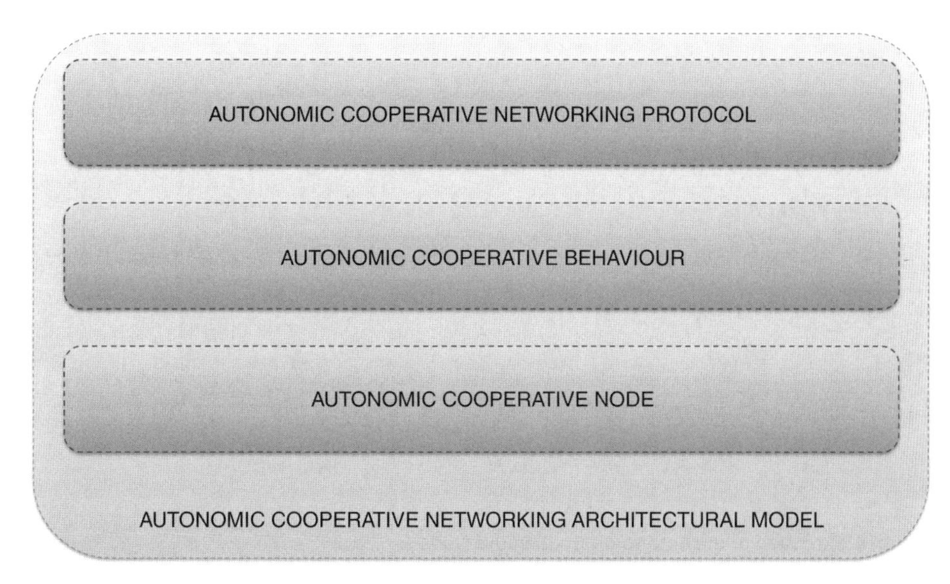

Figure 2.15 Horizontal Architectural Extensions.

necessary for the sake of making the layers of the Open Systems Interconnection Reference Model and the levels of the Generic Autonomic Network Architecture mutually orthogonal, the possibility of spanning the Horizontal Architectural Extensions over the adjacent pillars and beyond becomes critically necessary because of the need to accommodate all the related cross-layer enhancements. Thus, assuming the perspective presented in Figure 2.15, one may enumerate four extensions arranged in an incremental order: the first three, the Autonomic Cooperative Node (ACN), Autonomic Cooperative Behaviour (ACB), and Autonomic Cooperative Networking Protocol (ACNP), are presented as belonging to one group; the fourth, in the form of the Autonomic Cooperative Networking Architectural Model, is chosen to provide the synergetic background spanning the former three. It is little wonder that both the VTPs and the HAEs contain the same number of elements arranged in a fairly similar manner, i.e. three of them grouped together in each case with the fourth taking an elevated position. While such elevated elements of both classifications appear to glue together without doubt, it will turn out shortly that the others overlap thematically, too. This results not only from the assumed orthogonality, but also from building the extensions on top of their respective pillars.

In order to shed more light on the aforementioned extensions, and to emphasise the potential complexity of the related transition in terms of providing a comprehensive mapping, each of the cases will be discussed on a rather high level, before a much more in-depth analysis is to be carried out in the following chapters. In fact, looking at the first pair of components, encompassing both STP and ACN, as depicted in Figure 2.16, one may discern immediately that while the former component appears to be rather strictly narrowed down when the semantic perspective is concerned, the latter seems much more capacious, especially since its ultimate definition has not yet been put forward. Interestingly, even though both components appear to be nouns, STP falls into the category of a gerund, while the ACN is more of a proper noun,

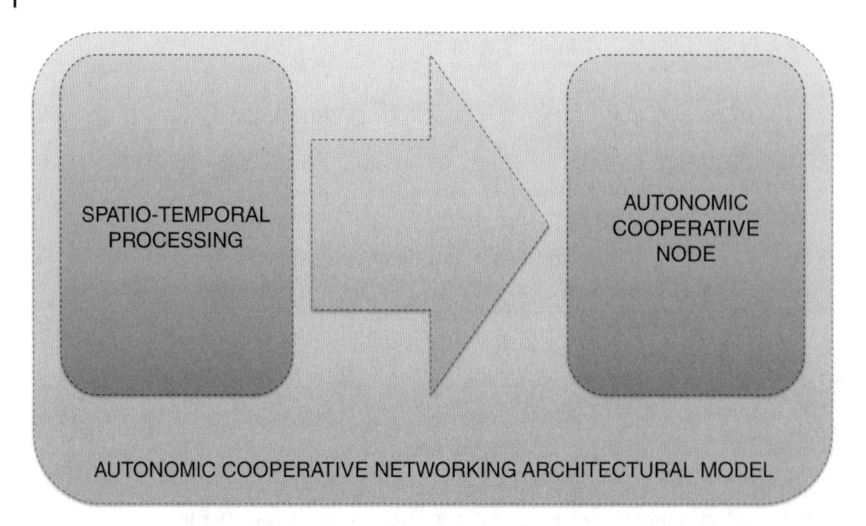

Figure 2.16 Extension to the Autonomic Cooperative Node.

making the explanatory process much more demanding at the level of nomenclature. The explanation of such a discrepancy is in the assumed flexibility allowing the architectural extensions to span over adjacent technological pillars, which is definitely true in the analysed example. In other words, in the case of the technological pillars it became fairly straightforward to distinguish between STBC and DSTBC by simply assigning them to the physical layer and link layer, respectively. At the same time, the Autonomic Cooperative Node is assumed to cover both those areas, and even go well beyond them.

In fact, looking more broadly, one could claim that certain parts of STP seem to fall more into the category of Autonomic Cooperative Behaviour, as shown in Figure 2.17,

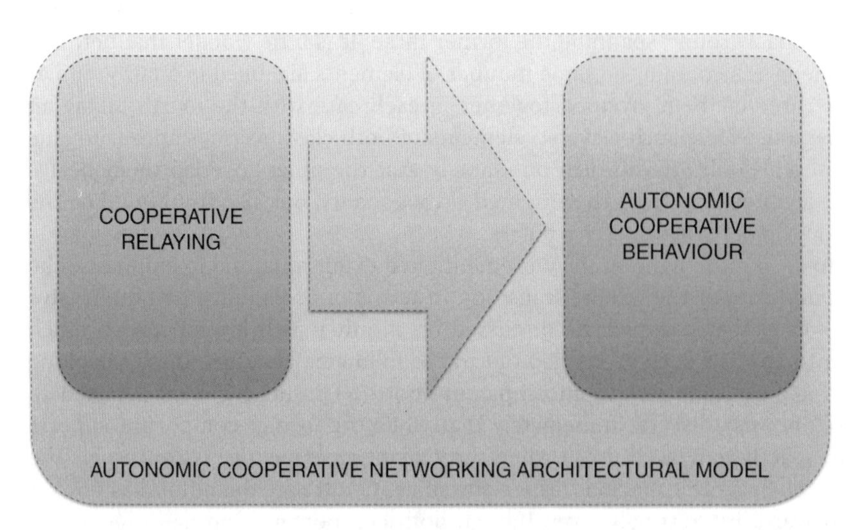

Figure 2.17 Extension to the Autonomic Cooperative Behaviour.

where the originally expected transition would apparently originate from cooperative relaying. In order to account for such a situation, it could be helpful to assume that the notion of an Autonomic Cooperative Node would rather stem from a legacy relay node exposing an inclination towards Autonomic Cooperative Behaviour by being characterised by a set of accompanying cooperative capabilities manifesting themselves through the ability to become part of a VAA performing the operation of distributed space-time block coding. In other words, the transition in question from STP to ACN would need to be defined through the inclusion of cooperative relaying, while the more obvious one from cooperative relaying to Autonomic Cooperative Behaviour could be further justified with the aid of STP. Moreover, the Autonomic Cooperative Node could not even be limited to spanning just the bottom two adjacent pillars; as will be shown, it will rather remain in close relation to all of them, while, at the same time, it will play its most fundamental role as an element of a VAA in order to exercise cooperative relaying. This is exactly why all the transitions under investigation have been purposely presented in separate figures, as what could initially seem to be a fairly straightforward operation of simple mapping immediately highlights the challenges ahead of the overall design to be described.

To unintentionally complicate things even further, yet still leaving all the details to be revealed at the appropriate time, the reader should note that the already discussed Autonomic Cooperative Behaviour would not be sufficiently comprehensive, should it not encompass, at least to some extent, the multi-point relay station selection heuristics of the Optimised Link State Routing protocol, introduced previously with the VTPs. Such an assumption becomes interestingly important, as it implies that certain parts of the Optimised Link State Routing protocol are to be redefined when the Autonomic Cooperative Networking Architectural Model is outlined, especially because, according to Figure 2.18, it might be expected that the multi-point relay station selection heuristics should remain a key part of the Autonomic Cooperative Networking Protocol. Most obviously, even if such a modification is advocated for architectural reasons, the

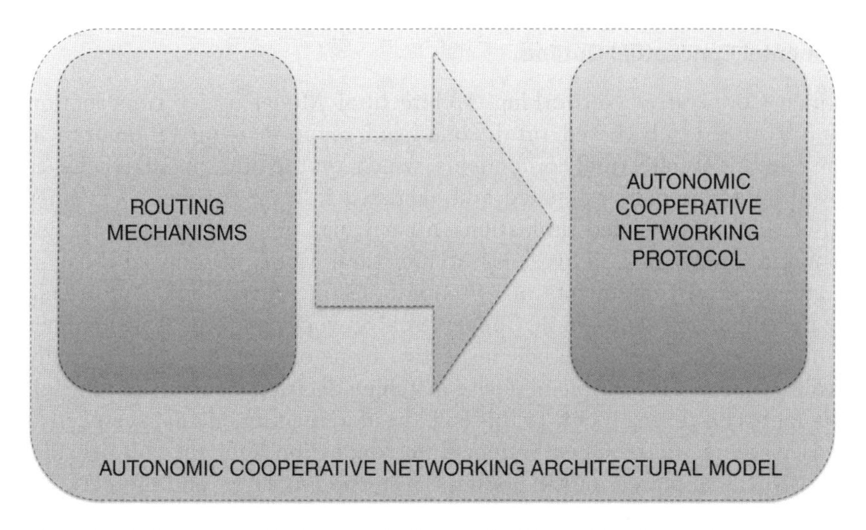

Figure 2.18 Extension to the Autonomic Cooperative Networking Protocol.

overall functionality of the modified Optimised Link State Routing protocol should not become inferior when compared to its original version. Quite the contrary: the modular approach of the proposed type may only improve the overall system stability and scalability, especially given the fact that additional developments stemming from and conditioned on the Generic Autonomic Network Architecture are to be incorporated, for example appropriately designed decision elements along with their respective hierarchical autonomic control loops, to provide a fully-fledged and self-sustainable solution in the form of the Autonomic Cooperative Networking Architectural Model.

Finally, the last transition is expected to pertain to the relation between the Autonomic Routines and the Autonomic Cooperative Networking Architectural Model, where the focus is much more shifted to the extraction of the relevant workings of the Generic Autonomic Network Architecture and to the provision of newly designed decision elements entering into mutual relations in both the vertical and horizontal directions, predominantly over their respective hierarchical autonomic control loops. This transition appears to be the most complicated, given the number of elements involved, but since at this stage of design not too much relevant information has been disclosed for practical reasons, it is not really possible to present a similarly high-level discussion of the sort exercised for the other extensions. Taking into account the background nature of the Autonomic Cooperative Networking Architectural Model, one may foresee that its elements will undoubtedly be present in virtually every aspect of the eventual design, including not only all of the components of Autonomic Cooperative Node, Autonomic Cooperative Behaviour, and Autonomic Cooperative Networking Protocol, but also the architectural overlay formed by itself. In such a context, based on the ground-setting VTPs, as well as taking into account the HAEs, the ultimate incremental conceptual outline will be provided in the next subsection. In this way the first, introductory, part of this book will be considered to have been completed and the focus will move to a discussion of the specific components of the target design.

2.4.3 Incremental Conceptual Outline

The Autonomic Cooperative Networking Architectural Model under construction in this book is assumed to be based, on the one hand, on an incremental theoretical design of the relevant architectural components, which, on the other hand, whenever applicable, will be additionally evaluated and validated in an experimental manner with the aid of computer-assisted simulation. In particular, given the general frames of the Autonomic Cooperative Networking Architectural Model introduced thus far, the crucial architectural components of Autonomic Cooperative Node, Autonomic Cooperative Behaviour, and Autonomic Cooperative Networking Protocol are going to be defined in the sense of outlining their respective scope of operation and mutually dependent roles, so that it is possible to devise matching decision elements to take full responsibility for orchestrating them. To this end, the aforementioned dual perspective will be exercised where, on the Open Systems Interconnection Reference Model side, special emphasis will be placed on the physical layer, the link layer, and the network layer, while on the Generic Autonomic Network Architecture side similarly increased attention will be paid to the protocol level, the function level, the node level, and the

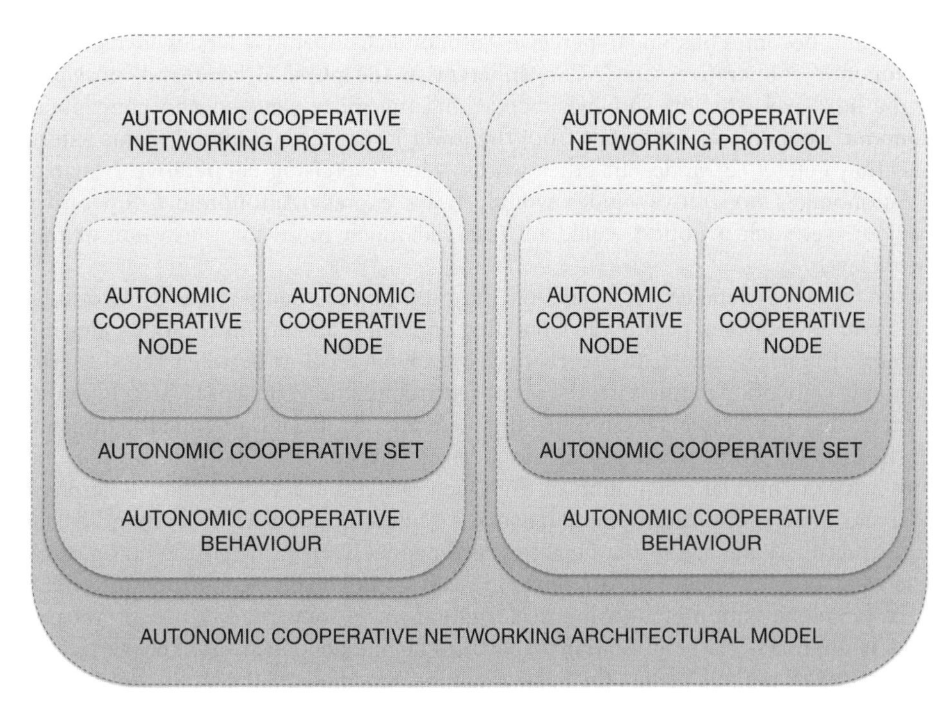

Figure 2.19 Hierarchical relations among autonomic entities.

network level. In this way the previously made assumption will be satisfied, according to which the analysis and design effort has been arranged in accordance with two orthogonal collocated axes in the form of the VTPs and HAEs. At the same time, additional architectural components will be incorporated into the blueprint in order to progress further towards making it even more concrete.

In fact, looking at the more detailed outline provided in Figure 2.19, one may discern a number of hierarchical relations driving the Autonomic Cooperative Networking Architectural Model, purposely presented in a way that does not involve any notion of the Open Systems Interconnection Reference Model. In particular, the overall drawing should really be analysed in the third dimension by placing special emphasis on the component of depth, chosen to be represented in the bottom-up direction in such a way that the smallest architectural entities, visible as Autonomic Cooperative Nodes, should be perceived as if they were located closer to the spectator, while the biggest one, manifesting itself as the Autonomic Cooperative Networking Architectural Model, would play the role of an umbrella-like overlay, binding all the other entities tightly together.[38] Going this way, one may conclude immediately that as much as in the previous subsection it was necessary to account for Autonomic Cooperative Behaviour rather indirectly, with the aid of the notion of VAAs stemming from the

38 One should note that such a representation will, at least for the time being, be perceived as a reference for further discussion; the very vital question of the precedence between the Autonomic Cooperative Networking Protocol and the Autonomic Cooperative Behaviour will be addressed towards the end of the book.

VTPs, it now becomes possible to perceive Autonomic Cooperative Nodes as members of autonomic cooperative sets (ACSs) inherent in the physical layer and, possibly, also the link layer, thereby somewhat detaching the more elevated and conceptual Autonomic Cooperative Behaviour from the lower layers of the Open Systems Interconnection Reference Model. In other words, when exercising cooperative relaying, the Autonomic Cooperative Nodes would always express Autonomic Cooperative Behaviour, yet such a notion would be reserved much more for the needs of the network layer.

Before a more detailed overview of the Autonomic Cooperative Networking Architectural Model can be provided, building on what is already known of its general workings, it appears crucial to sketch the conceptual relation between the Generic Autonomic Network Architecture, the proposed architectural concept it directly stems from, and the Open Systems Interconnection Reference Model it is intended to interact with. In fact, following what has already been described, and what can be seen in Figure 2.20, the mutual orientation of the Open Systems Interconnection Reference Model and the Generic Autonomic Network Architecture tends to be perpendicular, if not, using the language of mathematics, orthogonal. In fact, given the complexity of the interaction in question, any analogy with orthogonality could possibly hold true, as far as the notion of a geometrical 'right angle' is concerned, yet by no means should it approach the connotation of 'statistical independence'. As the vertically orientated layers are put against the horizontally arranged levels, the additional notion

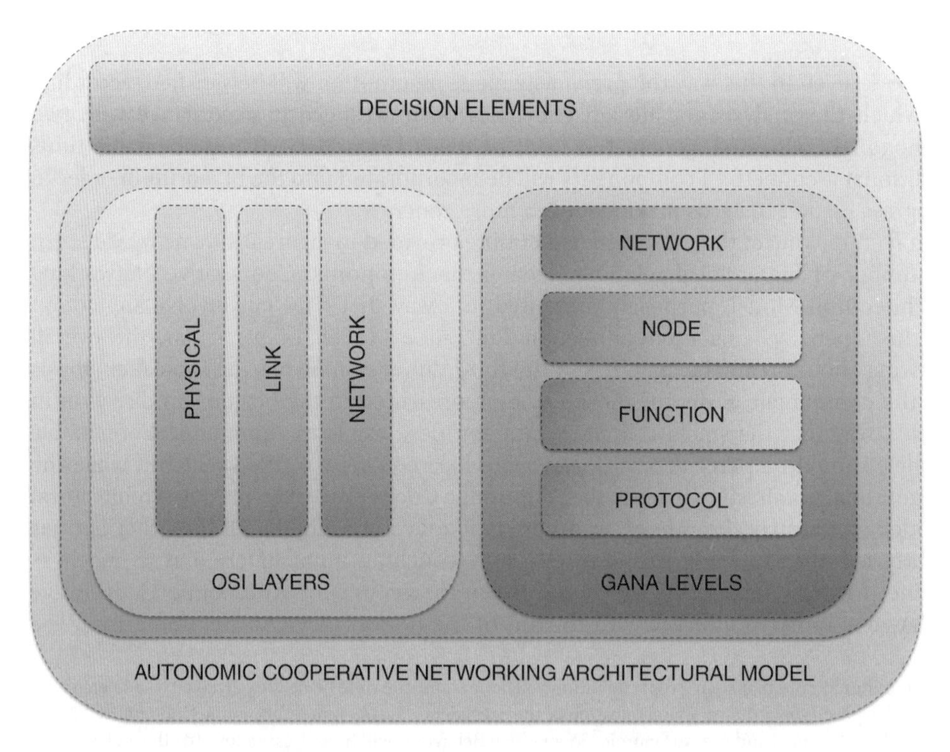

Figure 2.20 Correspondence between vertical layers and horizontal levels.

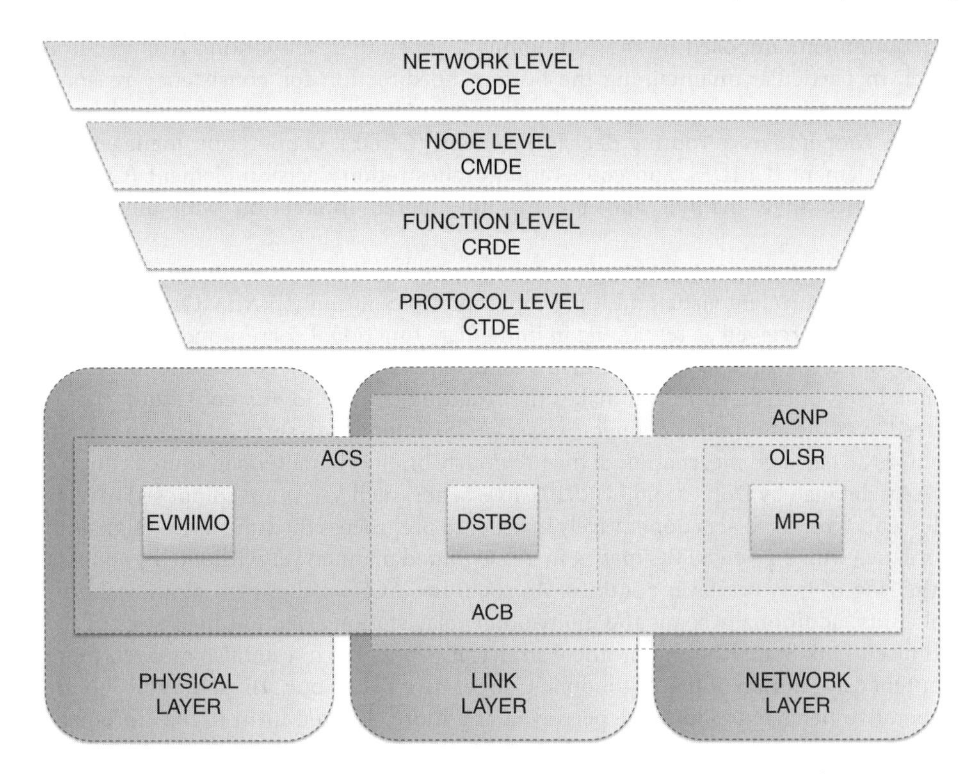

Figure 2.21 Overview of the Autonomic Cooperative Networking Architectural Model.

of the related decision elements is brought into the overall picture with the intention of highlighting their supervisory nature vested in them for the sake of becoming responsible for overseeing the entire setup under the umbrella of the Autonomic Cooperative Networking Architectural Model. However, for design-related reasons, the architectural entities stemming from various levels will need to be directly mapped onto specific layers, while the arrangement of decision elements, forming a reversed trapezoid, will occupy the top-most part thereof.

Such an arrangement is presented in Figure 2.21, where the aforementioned generalities of the Open Systems Interconnection Reference Model and the Generic Autonomic Network Architecture are replaced with the much more specific workings of the target Autonomic Cooperative Networking Architectural Model. To this end a fragmentary, vertically rearranged, protocol stack is employed consisting of the three lowest layers of the Open Systems Interconnection Reference Model. Even though such a reduction could make the resulting stack, in certain sense, appear to lean towards the one advocated by the TCP/IP protocol suite, as described by Tanenbaum and Wetherall (2011), it should by no means be perceived in such a way. What is more, keeping the above context in mind and changing the viewpoint slightly towards the decision-making logic located on top of the overall arrangement in question, defined directly on the grounds of the architectural principles advocated by the Generic Autonomic Network Architecture, one may discern that each of the four levels of abstraction contains the corresponding decision elements reflecting the requirements imposed by the Autonomic Cooperative Networking Architectural Model. In particular, maintaining the bottom-up direction for consistency reasons, one may enumerate the following entities: cooperative transmission decision element (CTDE), cooperative re-routing decision element (CRDE), cooperation management decision element (CMDE), and cooperation orchestration decision element (CODE), all positioned in a manner allowing for unhindered interaction with any entity available.[39]

In fact, commencing the description with the physical layer, the yet to be introduced entity of the equivalent virtual multiple-input multiple-output (EVMIMO) radio channel should be perceived as remaining in immediate functional correspondence with its related counterpart responsible for distributed space-time block coding and located at the link layer. Such an approach makes the resulting tandem jointly constitute what is referred to as the autonomic cooperative set (ACS), intended to serve, at least to certain extent, as an elevated incarnation of the previously briefly characterised source concept of VAAs, detailed by Dohler and Li (2010). In essence, such VAAs are composed of relay nodes whose role is to act cooperatively in order to preprocess the transmitted signals in accordance with a given STBC matrix in a distributed manner, yet without any support on the side of network layer routines. As such, the ACS, in the form of an architectural entity, additionally spans the multi-point relay station selection heuristics of the Optimised Link State Routing protocol so that it is possible to establish an even more comprehensive notion of the Autonomic Cooperative Behaviour. Thus, the Autonomic Cooperative Behaviour should be perceived as a more elevated form of the aforementioned ACS, mostly because it is predestined to combine the pertinent routines of all three layers of the Open Systems Interconnection Reference Model. What follows is the Autonomic Cooperative Networking Protocol, which builds upon the Optimised Link State Routing protocol, yet extracts certain parts to instantiate the ultimate notion of Autonomic Cooperative Behaviour.

Last, but not least, come the above-mentioned decision elements stemming directly from the Generic Autonomic Network Architecture and fully meeting the design principles it prescribes, while having been defined exclusively for the needs of the proposed Autonomic Cooperative Networking Architectural Model, quite similarly to the various other architectural components introduced in their evolved forms under the umbrella of the Open Systems Interconnection Reference Model. Taking into account that there are three layers of the Open Systems Interconnection Reference Model involved, as opposed to the four levels of the Generic Autonomic Network Architecture intended to interact with them, one may most naturally expect that there is not a one-to-one correspondence in this respect. In particular, in the bottom-up direction typically followed in this book, first comes the CTDE, to be responsible for the establishment of DSTBC with the involvement of both the physical layer and the link layer. Immediately to follow is the CRDE, responsible for the management of the cooperative re-routing (CRR) functionality. Next appears the CMDE, chiefly visible at the network layer and accountable for the coordination of the related cooperative transmissions and noncooperative transmissions, respectively. Finally, the globally distributed Autonomic

39 For obvious reasons, not every interaction between the decision elements listed and the architectural entities may be possible or reasonable. Moreover, such interactions will be outlined more explicitly as the overall design is advanced.

Cooperative Behaviour is to be overseen by the CODE, covering, and often going well beyond, the network layer.

2.5 Conclusion

In this chapter the rationale behind the concept of autonomic computing was introduced, and its convergence with modern networked systems was emphasised. Extensive conceptual analysis was undertaken to provide a comprehensive overview of the current status and role of autonomics. In particular, the general vision and the state of the art in the field of autonomic computing was scrutinised from the perspective of technological adaptation of the mechanisms responsible for the functioning of the human autonomic nervous system. As such, this involved analysis of the key aspects of self-configuration, self-optimisation, self-healing, and self-protection, altogether forming the notion of self-management, and was extended to cover the relevant architectural assumptions and variations, complemented with insight into the overlapping nature of autonomics and agent systems. Based on this, it was possible to address the question of convergence between autonomic computing and autonomic networking, and, thus, lay the groundwork for the discussion of the role of self-awareness, with the ultimate aim of introducing the rationale behind the Autonomic Cooperative Networking Architectural Model. In order to make this possible, an investigation of the state-of-the-art concept of the Generic Autonomic Network Architecture was first carried out, with special attention being paid to the explanation of the role of decision elements and hierarchical autonomic control loops, together with their respective levels of abstraction, presented in an incremental bottom-up order, i.e. starting from the lowest protocol level, through the function level and node level, up to the network level.

The scope of the Autonomic Cooperative Networking Architectural Model was then further investigated with the introduction of the Vertical Technological Pillars and the Horizontal Architectural Extensions. In fact, making the layers of the Open Systems Interconnection Reference Model perpendicular to the levels of the Generic Autonomic Network Architecture allowed the identification of the key architectural challenges to be addressed by the Autonomic Cooperative Networking Architectural Model. To this end, an incremental conceptual outline was presented involving the key architectural components in the form of the Autonomic Cooperative Node, Autonomic Cooperative Behaviour, and Autonomic Cooperative Networking Protocol, as well as the major decision elements of relevance. In particular, first, the protocol level cooperative transmission decision element was presented, with its responsibility for virtual multiple-input multiple-output channel based and distributed space-time block coding enabled cooperative relaying. Then, the function level cooperative re-routing decision element was deployed, with responsibility for triggering cooperative re-routing to ensure transmission resiliency. Moving forward, the node level cooperation management decision element was introduced to facilitate the integration between cooperative relaying and routing mechanisms. Finally, the network level cooperation orchestration decision element was presented to be accountable for comprehensive oversight of the overall system. All in all, a high-level blueprint of the Autonomic Cooperative Networking Architectural Model was presented to be further advanced in the chapters to follow.

References

3GPP-TS-23.501 2017 3rd Generation Partnership Project; Technical Specification Group Services and System Aspects; System Architecture for the 5G System; Stage 2 (Release 15).

Alamouti S 1998 A simple transmit diversity technique for wireless communications. *IEEE Journal on Selected Areas in Communications* **16**(8), 1451–1458.

Alippi C, Fantacci R, Marabissi D, and Roveri M 2016 A cloud to the ground: The new frontier of intelligent and autonomous networks of things. *IEEE Communications Magazine* **54**(12), 14–20.

Beck F, Chrisment I, Droms R, and Festor O 2010 Autonomic renumbering in the future internet. *IEEE Communications Magazine* **48**(7).

Behringer M, Pritikin M, Bjarnason S, Clemm A, Carpenter B, Jiang S, and Ciavaglia L 2015 *Autonomic Networking: Definitions and Design Goals*. RFC7575.

Brazier FMT, Kephart JO, Van Dyke Parunak H, and Huhns MN 2009 Agents and service-oriented computing for autonomic computing: A research agenda. *IEEE Internet Computing* **13**(3), 82–87.

Chaparadza R 2008 Requirements for a Generic Autonomic Network Architecture (GANA), suitable for standardizable autonomic behaviour specifications of decision-making elements (DMEs) for diverse networking environments. In *International Engineering Consortium (IEC) Annual Review of Communications*.

Chaparadza R, Meriem TB, Radier B, Szott S, Wódczak M, Prakash A, Ding J, Mihailovic A, and Soulhi S 2013 Implementation guide for the ETSI AFI GANA model: A standardized reference model for autonomic networking, cognitive networking and self-management. *Fifth IEEE International Workshop on Management of Emerging Networks and Services (IEEE MENS 2013)*, IEEE GLOBECOM 2013, Atlanta, Georgia, USA.

Chaparadza R, Papavassiliou S, Kastrinogiannis T, Vigoureux M, Dotaro E, Davy A, Quinn K, Wódczak M, and Toth A 2009 Creating a viable evolution path towards self-managing future internet via a standardizable reference model for autonomic network engineering. *Future Internet Assembly (FIA 2009)*, Prague, Czech Republic.

Cheng Y, Leon-Garcia A, and Foster I 2008 Toward an autonomic service management framework: A holistic vision of SOA, AON, and autonomic computing. *IEEE Communications Magazine* **46**(5), 138–146.

Chiti F, Fantacci R, Loreti M, and Pugliese R 2016 Context-aware wireless mobile autonomic computing and communications: Research trends and emerging applications. *IEEE Wireless Communications* **23**(2), 86–92.

Clark D, Partridge C, Ramming J, and Wroclawski J 2007 A knowledge plane for the internet. *Proceedings of ACM SIGCOMM 2003*.

Dohler M and Li Y 2010 *Cooperative Communications: Hardware, Channel & PHY*. Wiley.

Dohler M, Gkelias A, and Aghvami H 2003 2-hop distributed MIMO communication system. *IEE Electronics Letters* **39**(18) 1350–1351.

ETSI-GS-AFI-002 2013 Autonomic network engineering for the self-managing Future Internet (AFI); Generic Autonomic Network Architecture (An Architectural Reference Model for Autonomic Networking, Cognitive Networking and Self-Management). *ETSI Group Specification*.

Gadsby A 2003 *Longman Dictionary of Contemporary English*. Pearson Education Limited.

Gonzalez-Bayon J, Fernandez-Herrero A, and Carreras C 2010 Improved schemes for tracking residual frequency offset in DVB-T systems. *IEEE Transactions on Consumer Electronics* **56**(2).

Greenberg A, Hjalmtysson G, Maltz D, Myers A, Rexford J, Xie G, Yan H, Zhan J, and Zhang H 2005 A clean slate 4D approach to network control and management. *ACM SIGCOMM Computer Communication Review* **35**(5), 41–54.

Hinchey MG and Sterritt R 2006 Self-managing software. *IEEE Computer* **39**(2), 107–109.

Jennings B, van der Meer S, Balasubramaniam S, Botvich D, Foghlu M, Donnelly W, and Strassner J 2007 Towards autonomic management of communications networks. *IEEE Communications Magazine* **45**(10), 112–121.

Jennings N, Sycara K, and Wooldridge M 1998 A roadmap of agent research and development. *Journal of Autonomous Agents and Multi-Agent Systems* **1**(1), 7–38.

Kephart JO and Chess DM 2003 The vision of autonomic computing. *IEEE Computer* **36**(1), 41–50.

Kephart JO and Das R 2007 Achieving self-management via utility functions. *IEEE Internet Computing* **11**(1), 40–48.

Laneman JN and Wornell GW 2003 Distributed space-time-coded protocols for exploiting cooperative diversity in wireless networks. *IEEE Transactions on Information Theory* **49**(10), 2415–2425.

Liakopoulos A, Zafeiropoulos A, Polyrakis A, Grammatikou M, Gonzalez J, Wódczak M, and Chaparadza R 2008 Monitoring issues for autonomic networks: The EFIPSANS vision. *European Workshop on Mechanisms for the Future Internet*.

Mainsah E 2002 Autonomic computing: The next era of computing. *Electronics and Communication Engineering Journal* **14**(1), 2–3.

Meriem TB, Chaparadza R, Radier B, Soulhi S, Lozano-Lopez JA, and Prakash A 2016 *GANA: Generic Autonomic Networking Architecture*. ETSI White Paper.

Movahedi Z, Ayari M, Langar R, and Pujolle G 2012 A survey of autonomic network architectures and evaluation criteria. *IEEE Communications Surveys and Tutorials* **14**(2), 464–490.

Moy J 1998 *OSPF Version 2*. RFC2328.

Ordonez-Lucena J, Ameigeiras P, Lopez D, Ramos-Munoz J, Lorca J, and Folgueira J 2017 Network slicing for 5G with SDN/NFV: Concepts, architectures, and challenges. *IEEE Communications Magazine* **55**(5).

Paulson LD 2002 Computer system, heal thyself. *IEEE Computer* **35**(8), 20–22.

Phan CV 2009 Formal aspects of self-* in autonomic networked computing systems. In *Autonomic Computing and Networking*, ed. Denko MK, Yang LT, and Zhang Y, Springer Science+Business Media, LLC, pp. 381–410.

Retvari G, Nemeth F, Prakash A, Chaparadza R, Hokelek I, Fecko M, Wódczak M, and Vidalenc B 2011 A guideline for realizing the vision of autonomic networking: Implementing self-adaptive routing on top of OSPF. In *Formal and Practical Aspects of Autonomic Computing and Networking: Specification, Development, and Verification*, ed. Cong-Vinh P, IGI Global.

Sinclair J 1997 *Collins COBUILD English Dictionary*. HarperCollins Publishers.

Smirnov M, Tiemann J, Chaparadza R, Rebahi Y, Papavassiliou S, Karyotis V, Pouli V, Merekoulias V, Gulyas A, Heszberger Z, Retvari, G. Nemeth F, Wódczak M, Kaldanis V, Markopoulos A, Karantonis G, Davy A, Quinn K, Cheng S, Li Y, Jin Y, Gong X, Cui Y, Hu B, Shi Y, Wang W, Liakopoulos A, Zafeiropoulos J, Lopez J, Munoz J, Vigoreux M,

Berde B, Cleary D, and Toth A 2009 Demystifying self-awareness of autonomic systems. *ICT Mobile Summit*, Santander, Spain.

Tanenbaum A and Wetherall D 2011 *Computer Networks*. Prentice Hall.

Truszkowski WF, Hinchey MG, Rash JL, and Rouff CA 2006 Autonomous and autonomic systems: A paradigm for future space exploration missions. *IEEE Transactions on Systems, Man, and Cybernetics Part C (IEEE Transactions on Human-Machine Systems)* **36**(3), 279–291.

Vassev E and Hinchey M 2010 The challenge of developing autonomic systems. *IEEE Computer* **43**(12), 93–96.

Wódczak M 2007 Extended REACT: Routing information enhanced algorithm for cooperative transmission. *16th IST Mobile & Wireless Communications Summit 2007*, Budapest, Hungary.

Wódczak M 2011a Convergence aspects of autonomic cooperative networking. *IEEE Fifth International Conference on Next Generation Mobile Applications, Services and Technologies*, Cardiff, Wales, UK.

Wódczak M 2011b Resilience aspects of autonomic cooperative communications in the context of cloud networking. *IEEE First Symposium on Network Cloud Computing and Applications*, Toulouse, France.

Wódczak M 2012 *Autonomic Cooperative Networking*. Springer.

Wódczak M 2014 *Autonomic Computing Enabled Cooperative Networked Design*. Springer.

Wódczak M, Meriem TB, Radier B, Chaparadza R, Quinn K, Kielthy J, Lee B, Ciavaglia L, Tsagkaris K, Szott S, Zafeiropoulos A, Liakopoulos A, Kousaridas A, and Duault M 2011 Standardizing a reference model and autonomic network architectures for the self-managing future internet. *IEEE Network* **25**(6), 50–56.

Yixin D, Hellerstein JL, Parekh S, Griffith R, Kaiser GE, and Phung D 2005 A control theory foundation for self-managing computing systems. *IEEE Journal on Selected Areas in Communications* **23**(12), 2213–2222.

3

Protocol Level Spatio-Temporal Processing

3.1 Introduction

Following the rationale behind the autonomically driven cooperative design established in the previous chapter, where a specific umbrella was created with the intention to encompass all the architectural considerations to be outlined across the entire book, the time has come to commence the rollout thereof with the foundations related to protocol level spatio-temporal processing. To this end, first of all, the emphasis is laid on the bottom-most multiple-input multiple-output channel conditioned developments, to provide a good understanding of their workings, and especially to put them in the correct context before they can be properly integrated into the Autonomic Cooperative Networking Architectural Model. Thus, the diversity-rooted origins of the approach based on multiple-input multiple-output technology are discussed, so that it is possible to clearly justify the role of and necessity for the later deployment of spatio-temporal processing enabled techniques on top of it. Before the focus is shifted, however, the question of radio channel virtualisation is addressed, where the singular-value decomposition theorem is explained in order to introduce the notion of an equivalent virtual multiple-input multiple-output radio channel to be deployable among Autonomic Cooperative Nodes. The radio channel capacity is then brought into the bigger picture of the whole analysis to account for its linear scaling with the number of generic transmitters or generic receivers. What is more, an external model for radio channel coefficient calculation, to be employed in the next chapter, is summarised, and the difference between coding gain and diversity gain is addressed for clarity.

The focus then moves towards space-time coding techniques, with the intention of accounting for their superiority over the above-mentioned diversity techniques, especially the possibility of shifting the complexity related to the deployment of multiple antennae on, usually relatively small, mobile user terminals to a base station or an access point, and to pave the way for their later elevation to the level of networked configurations, where the transition towards the concept of distributed space-time block coding shall prevail. In this respect, first of all the most baseline approach to space-time coding, in the form of orthogonal block coded designs, is outlined, with a special emphasis on space-time block coding, where not only the method of operation behind this technology is explained, but the question of its being perceived more as a modulation rather than a coding technique is touched upon. Then the plot is extended even further in this respect, to outline the derivation of the decoding metrics for a selected set of space-time block coding matrices with the aim of building on it in introducing the

Autonomic Intelligence Evolved Cooperative Networking, First Edition. Michał Wódczak.

concept of the equivalent distributed space-time block encoder in the next chapter. This derivation is performed not only to make the related processes of coding and decoding more transparent, but also to clarify certain inconsistencies the author came across in the source materials. Last, but not least, an extension towards space-time trellis coding is presented, where the expected additional coding gain becomes obtainable.

Once all the aforementioned technological aspects, key to the overall Autonomic Cooperative Networking Architectural Model, have been outlined, their mutual relation with the protocol level related control logic is discussed for architectural integration purposes. In this way a specific pattern is created to be followed throughout the entire book to outline an incremental advancement of the proposed design. In particular, this part opens with the introduction of the notion of an Autonomic Cooperative Node being, de facto, one of the major building blocks when the workings of the proposed concept are concerned. On this occasion, not only is the relation between autonomics and cooperation discussed in detail, but the internal structure of such an Autonomic Cooperative Node is scrutinised from the perspective of the Generic Autonomic Network Architecture, so that it is possible to outline the adopted interaction model over the Open Systems Interconnection Reference Model protocol stack. Then, the cooperative transmission decision element is brought into the analysis as being characterised as belonging to the protocol level, while being mostly responsible for the interaction with the routines of the physical layer. Given such a context, not only is the role and notion of the concept of a protocol discussed, but certain adaptive logic is presented where the relevant code matrices are switched on the basis of the radio channel parameters. Finally, all the pertinent architectural integration aspects are outlined, with a particular approach to the graphical representation thereof, and the way is prepared for further extensions.

3.2 Multiple-Input Multiple-Output Channel

3.2.1 Diversity-Rooted Origins

From the scope of this book and the high-level design presented thus far, one could apparently attempt to say, probably without too extensive an exaggeration, that diversity laid the foundations of most of the aspects of the latter, if not all of them, even if it happened more or less directly. In fact, this claim may seem well justified, as not only did this notion generally add new dimensions to the overall perception of multi-path propagation,[1] but its evolution inevitably led to the invention of key technologies so crucial to further advancement of the proposed concept, just to mention its influence on spatio-temporal processing (STP) or cooperative relaying, as respectively summarised by Phan et al. (2013) and Liu and Lin (2015). In fact, the rationale behind the introduction of the related gain obtainable thanks to such diversity is deeply rooted in the assumption that multiple replicas of the transmitted signal, as observed by the receiver, convey constructively redundant information, while the fading they shall be subject to may be rather poorly correlated. This way, it could become highly unlikely that all such

1 What is, most generally, meant here is the mental and technological transition from the troublesome legacy multi-path propagation to the modern mostly desirable orthogonalised approaches, to be introduced on the basis of the considerations outlined in this section.

replicas would encounter a deep fade at the same time, which increases the probability of proper reception. Since there exist numerous approaches in this respect, the most generic categorisation of the diversity techniques of relevance seems to pertain to the domain of their application, and thus it is possible to distinguish generally between temporal diversity (TLD), frequential diversity (FLD), and spatial diversity (SLD) (Wódczak, 2014).[2] In this context, the shortlisted techniques will be analysed, and their related developments will be discussed in the light of the upcoming developments.

Looking at the first of these categories, one may immediately identify that, as originally explained by Jankiraman (2004), the role of temporal diversity consists in the transmission of multiple replicas of a signal using disjoint time slots where, according to Vucetic and Yuan (2003), it is mandatory that a separation should exist between the time slots such that its size would be equal at least to the coherence time (CT) of the channel, defined by Goldsmith (2005) as the time during which the autocorrelation function of the channel impulse response is approximately nonzero. Apparently, as such an approach may cause decoding delays, it is said to be mostly suitable for fast fading environments, where the aforementioned coherence time is relatively insignificant. Moving forward to frequential diversity, one may see that this approach assumes, in turn, the exploitation of different frequencies for the very similar task of transmitting replicas of the original signal. As for temporal diversity, such frequencies need to be properly separated, since different parts of the spectrum undergo independent fades (Jankiraman, 2004). To reflect this situation, the relevant separation is denoted by the complementary term of coherence bandwidth (CB), which is defined by Goldsmith (2005) as the frequency range across which the entire signal bandwidth is highly correlated. In other words, also following Vucetic and Yuan (2003), should the fading statistics for different frequencies appear to be essentially uncorrelated, a frequency separation of the order of several times the channel coherence bandwidth would be required for proper operation.

By contrast to the above categories, spatial diversity does not really induce redundancy either in the temporal or frequential domain, as detailed by Vucetic and Yuan (2003); in its case, the constituents of a multi-element array (MEA) are separated by a few wavelengths to guarantee that replicas of the transmitted signal do not become correlated; hence, this type of separation is usually referred to as coherence distance (CD), to follow Jankiraman (2004). What is more, according to both Vucetic and Yuan (2003) and Jankiraman (2004), this category may be additionally extended to encompass two representative approaches to exercising radio transmission diversity in the spatial domain, known as polar diversity (PLD) and angular diversity (ALD).[3] In particular, in the case of polar diversity the signals, characterised by being polarised either horizontally or vertically, are transmitted and received by two sets of differently polarised antennae in order to ensure that there is no correlation between such signals, even if the antennae belonging to such sets are not guaranteed to be separated by a distance corresponding to a few wavelengths. As for angular diversity, such an approach is applicable given carrier frequencies of the order of several gigahertz, which

2 In general, in the literature, most often the following naming pattern is followed: time diversity, frequency diversity, and space diversity. Yet, for linguistically justified reasons, the author prefers to use adjectival forms.

3 As in the previous note, the following naming pattern is usually applied in the literature: polarisation diversity and angle diversity. Once again, for linguistically justified reasons, the author prefers to refer to these with adjectival forms.

are characteristic of environments featuring rich scattering in the spatial domain. Thus, it would suffice to use two highly directional receiving antennae, each facing in a different direction, to fully exploit spatial diversity (Vucetic and Yuan, 2003).

However, given the major theme of this book, it becomes pragmatically justified to provide additional insight into the category of spatial diversity, especially since space-time coding (STC) techniques will be addressed shortly. In essence, depending on whether the MEA is located at the transmitter or at the receiver, two major subcategories, known as reception diversity (RND) and transmission diversity (TND), may be distinguished.[4] In fact, as explained by Vucetic and Yuan (2003), one of the least sophisticated approaches to reception diversity is selection combining (SC), outlined in Figure 3.1, where a given signal is preferred should it be characterised by the highest value of the instantaneous signal-to-noise ratio (SNR), which obviously requires that all the diversity branches be monitored simultaneously. Therefore, the suboptimal solution of switched combining (SWC) or scanning diversity (SCD) was introduced, where that diversity branch remains selected which is able to maintain the SNR above a specified threshold. Moreover, there is the linear method of maximal ratio combining (MRC),[5] as depicted in Figure 3.2, where distinct diversity branches are weighted with the use of α_i coefficients and then submitted to the superposition module. In this case, the α_i coefficient may be written as

$$\alpha_i = A_i e^{-j\varphi_i}, \tag{3.1}$$

where A_i is the amplitude and φ_i is the phase of the received signal r_i observed at antenna i, so that the overall received signal r_i may be expressed as

$$r = \sum_{i=1}^{M} \alpha_i r_i. \tag{3.2}$$

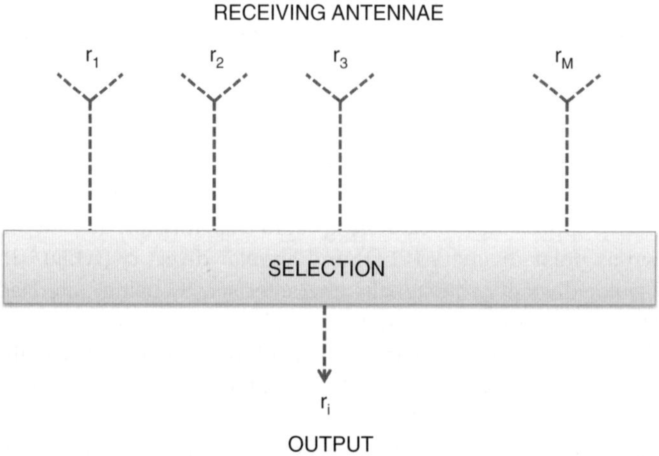

RECEIVING ANTENNAE

r_1 r_2 r_3 r_M

SELECTION

r_i

OUTPUT

Figure 3.1 Selection combining. Adapted from Vucetic and Yuan (2003).

4 Generally, in the literature, these techniques are referred to as receive diversity and transmit diversity; however, also for linguistically justified reasons, the author prefers to refer to these using nominal forms, in turn.

5 Also known as maximal ratio receive combining (MRRC) (Alamouti, 1998). One should note that the original naming pattern is maintained in this footnote, even though, following the previous commentary, the form of "maximal ratio reception combining" would be preferred by the author.

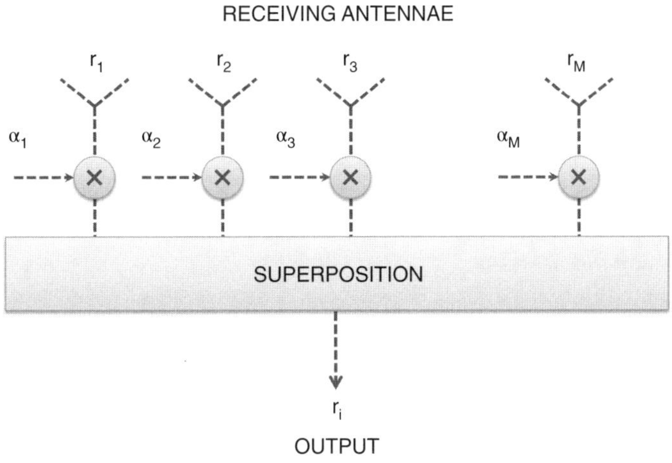

Figure 3.2 Maximal ratio combining. Adapted from Vucetic and Yuan (2003).

As additionally analysed, for example by Vucetic and Yuan (2003), maximal ratio combining is an optimum combining method from the perspective of the maximisation of the SNR at the output.

One should note, however, that there is also a suboptimal version of maximal ratio combining that is characterised by the lack of the necessity for the estimation of the fading amplitude for each diversity branch. Quite the contrary: it is designed to presume that the amplitudes A_i are all equal to 1 – hence its name, equal gain combining (EGC). While intuitively the performance of such an approach would be expected to decrease, the concept prevails in terms of reduced implementation complexity. Nonetheless, from a practical perspective, the implementation of reception diversity should be considered either on the side of a base station (BS) or an access point (AP), since it could obviously become troublesome, despite all the technological advancements, to equip user terminals (UTs), in the form of mobile phones, with more than two antennae and a battery capacious enough to support separate radio frequency (RF) chains (Wódczak, 2012a). In fact, this is why the major emphasis was placed even more on the case of transmission diversity, despite the fact that it would be more difficult to apply, as indicated by Vucetic and Yuan (2003). Most of all, keeping in mind the assumption of the location of the MEA on the side of a base station or an access point, once the signals it has transmitted have arrived at the receiver, they would be clearly mixed, at least in the spatial domain, so that additional processing at both the transmitter and the receiver would need to be employed, obviously depending on the applied communication technology, in order to facilitate successful completion of the whole transmission process.[6]

In fact, with the emerging multiple-input multiple-output (MIMO) technology promising a substantial increase in capacity from the very outset, research in the field of transmission diversity was clearly fostered, resulting in a number of schemes. A classic example, in the form of delayed transmission diversity (DTD), as discussed,

6 One of the most natural issues to be brought up in this respect would be that unless it should be fed back from the receiver, the transmitter could lack any instantaneous information describing the parameters of the radio channel. It turns out, however, that approaches such as space-time block coding do not require any channel state information on the side of the transmitter at all, as indicated, for example, by Jankiraman (2004).

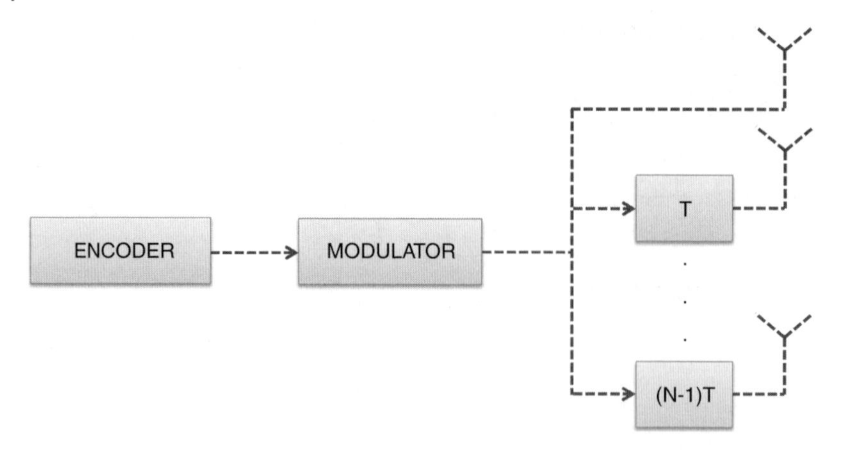

Figure 3.3 Delay transmit diversity. Adapted from Larsson and Stoica (2003b).

for instance, by Larsson and Stoica (2003b), is presented in Figure 3.3. Here, the replicas of the transmitted signal are purposely delayed so that the receiver may observe them as if the original copy of the signal were distorted by typical multi-path propagation and thus apply a maximum likelihood sequence estimator (MLSE) or minimum mean square error (MMSE) equaliser in order to obtain the diversity gain (DG). At this stage, it appears to be no accident that transmission diversity was chosen to be presented last, even though logically it could appear before reception diversity. Clearly, the reason for such an approach lies in the fact that the discussion related to transmission diversity is going to be further extended into the analysis of the related spatio-temporal processing, so necessary for the later introduction of distributed space-time block coding enabled and virtual antenna array driven cooperative relaying.[7] Before such an extension may be possible, however, first of all the rationale behind the multiple-input multiple-output radio channel needs be discussed; in particular, the possibility of its equivalent virtualisation[8] will be taken into account in the light of the ultimate goal of having it transposed into a networked setup of cooperative relay nodes.

3.2.2 Radio Channel Virtualisation

The question of a virtual multiple-input multiple-output (VMIMO) channel is especially vital in the case of cooperative relaying, where the notion of a fixed MEA is translated into a dynamic set of relay nodes forming a virtual antenna array (VAA) to exercise the operation of distributed space-time block coding. Given the fact that such a virtual multiple-input multiple-output radio channel is formed by relay nodes,[9] it may in general be perceived as sufficiently, if not entirely, uncorrelated, which, in general, translates into the applicability of the theoretical gains attainable by legacy

7 One should note that either a fixed or a mobile scenario could be considered in this respect, as outlined by Doppler *et al.* (2009) or Wódczak (2007), respectively.
8 Consequently the notion of an equivalent virtual multiple-input multiple-output channel (EVMIMO) is going to be introduced.
9 Even though, for the time being, the concept of relay nodes is going to be used, chiefly for consistency reasons, it will be soon elevated to the notion of Autonomic Cooperative Nodes, as it was already discussed in the previous chapter.

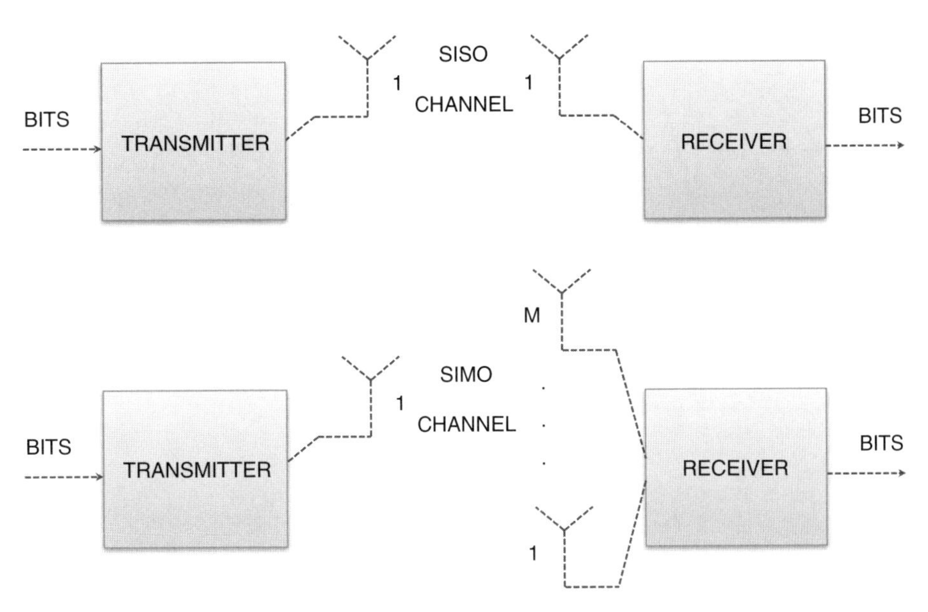

Figure 3.4 Single-input single-output and single-input multiple-output systems.

systems employing the classic MIMO approach, as long as a sufficiently high level of synchronism is guaranteed. One should note, however, that before the emergence of systems capable of exploiting transmission diversity in the spatial domain, radio transmission was generally performed in the temporal and frequential domain, over single-input single-output (SISO) or single-input multiple-output (SIMO) channels, as outlined in Figure 3.4. Then, once it turned out that additional information may, easily and efficiently, be conveyed over the third, spatial, dimension, the global scientific trends in this respect were immediately shifted towards multiple-input single-output (MISO) and MIMO-based technologies, as presented in Figure 3.5. In fact, from the mathematical perspective, the wireless multiple-input multiple-output[10] radio channel in question is usually defined by a channel matrix (CM), denoted by $H_{N \times M}$, where the coefficients $h_{i,j}$ indicate the gains of the radio links formed between each transmitting antenna i $(1 \leq i \leq N)$ and each receiving antenna j $(1 \leq j \leq M)$:

$$H_{N \times M} = \begin{bmatrix} h_{1,1} & h_{1,2} & \cdots & h_{1,M} \\ h_{2,1} & h_{2,2} & \cdots & h_{2,M} \\ \vdots & \vdots & \ddots & \vdots \\ h_{N,1} & h_{N,2} & \cdots & h_{N,M} \end{bmatrix}. \tag{3.3}$$

In order to fully quantify the potential behind such a channel representation, it suffices to make a comparison with the extraordinary milestone work of Shannon (1948), which, although it does not cover the most recent advancements in this respect, provides a thorough analysis of a SISO channel with additive white Gaussian noise (AWGN), for the sake of expressing a single user data rate bound, as described by the following formula,

10 For a wider context of adaptive MIMO transmission techniques, the reader is referred to Chae *et al.* (2010).

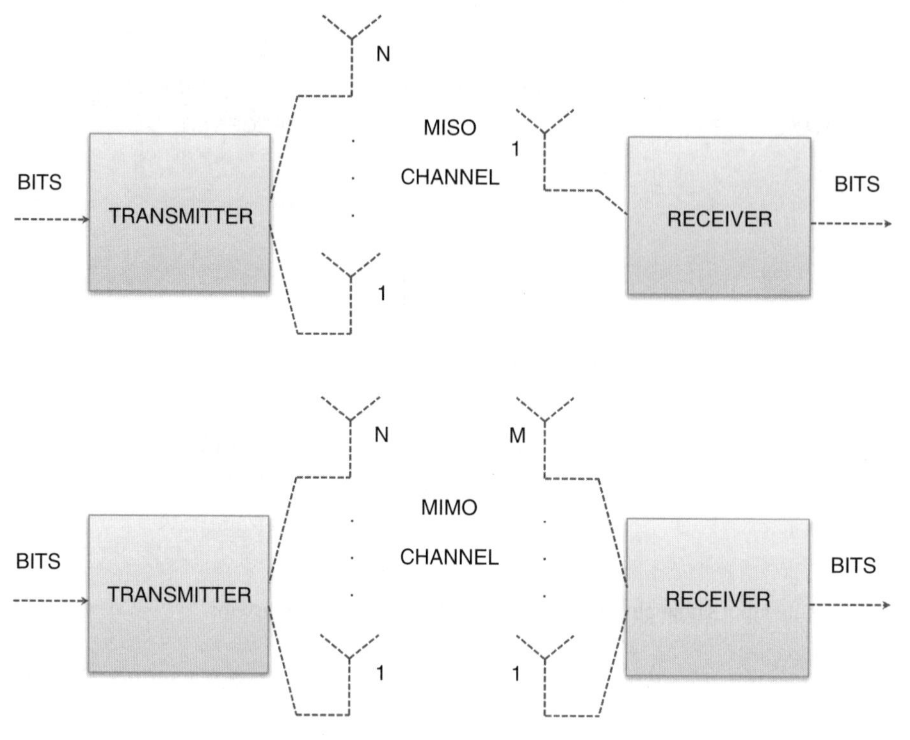

Figure 3.5 Multiple-input single-output and multiple-input multiple-output systems.

where $|h|^2$ represents the channel gain (CHG):

$$C = B\log_2\left(1 + \frac{P_{\mathrm{T}}}{\sigma^2}|h|^2\right). \tag{3.4}$$

The major strength of Equation 3.4 stems from the fact that it combines the channel capacity C with the related channel bandwidth B and the SNR, expressed as the ratio of the transmitted signal power P_{T} to the noise power σ^2. Fairly obviously, it would seem that the most straightforward method of increasing such a capacity, and, therefore, the attainable data rate bound, would consist in broadening the bandwidth of the radio channel (Wódczak, 2012a). Yet, setting aside technological considerations, given the fact that bandwidth is a commercially controlled product, such an approach would be too inefficient and costly, and consequently not justifiable. Going further, one could also potentially consider increasing the SNR; however, since such an approach would immediately translate into the need to increase the power of the transmitted signal, it would naturally enlarge the level of both the co-channel interference (CCI) and inter-channel interference (ICI). What is more, due to the logarithmic relation between both the parameters, the effectively attainable channel capacity gain (CCG) would be less significant when compared to the first case (Wódczak, 2012a).

As a result, there is no denying that the above issue would need to be resolved in some other way, and the situation changes immediately when MIMO[11] technology is employed as outlined, for example, by Telatar (1999). In general, as long as the number of both the transmitting N and receiving M antennae remains equal to 1, any throughput gain may amount merely to about 1 bit per Hz for each 3 dB increase in the SNR, as clearly explained by Foschini and Gans (1998). Should MEAs of size N be employed at both ends of the radio channel, the achievable capacity would scale linearly with N, as proven by Lozano et al. (2001). Consequently, once again referring to the work of Foschini and Gans (1998), it appears feasible to achieve almost as many as N bits per Hz, and, in order to verify such a statement, one may follow, for example, Vucetic and Yuan (2003), where the singular-value decomposition (SVD) theorem is applied, according to which the previously introduced channel matrix $H_{N \times M}$ (Equation 3.3) may be presented in the following rewritten form:

$$H = UDV^{\mathrm{H}}. \tag{3.5}$$

In this case, D is a nonnegative and diagonal matrix of size $M \times N$, while U and V are unitary matrices of size $M \times M$ and $N \times N$, respectively, where ^{H} denotes a Hermitian transposition. Clearly, the advantage of the SVD theorem lies in the fact that all the above-mentioned matrices are characterised by having nonzero elements solely on their main diagonals. What it implies mathematically is that $UU^{\mathrm{H}} = I_M$ and $VV^{\mathrm{H}} = I_N$, where I_M is the identity matrix of size $M \times M$, and I_N the identity matrix of size $N \times N$ (Vucetic and Yuan, 2003).

Similarly, the diagonal entries of D are equal to the nonnegative square roots of the eigenvalues of the matrix HH^{H} and are denoted by λ:

$$HH^{\mathrm{H}}y = \lambda y, \quad y \neq 0, \tag{3.6}$$

where y is an eigenvector of size $M \times 1$ associated with λ. Based on the above, one might conclude most correctly that it is possible to think of an equivalent virtual multiple-input multiple-output (EVMIMO) channel comprising solely k uncoupled parallel subchannels, with k being understood as the rank of the channel matrix H, and, thereby, presumably equal to at most the minimum of N and M. This situation is outlined in Figure 3.6, where the concept of generic transmitters (GTs) and generic receivers (GRs) is presented, as originally introduced by Wódczak (2014).[12] The generic transmitters and generic receivers will be denoted by GT_X and GR_Y, respectively, where $1 \leq X \leq N$ and $1 \leq Y \leq M$, or $1 \leq X \leq M$ and $1 \leq Y \leq N$, depending on whether $N < M$ or $N > M$. Most generally, the said generic transmitters and generic receivers could be assumed to either take the form of the legacy MEA, or, as in the analysed case, to reflect the networked design, become perceived as relay nodes. Should the latter be the case, certain clarification might be necessary to avoid any unintended inconsistency, possibly resulting from the fact that MEAs are also covered. In other words, should the

11 In the light of the following derivation, the reader is also referred to the latest advancement of relevance in the form of massive multiple-input multiple-output channel (MMIMO), as outlined, for instance, by Bjornson *et al.* (2016) or Carvalho *et al.* (2017).

12 Relaxing the assumptions even further, it is necessary to keep in mind that either a GT or a GR may be instantiated by a single element of a fixed or mobile MEA, as well as by a single fixed or mobile antenna, where the fixed and mobile configurations should be understood as being mounted on a cellular mast or a relay node, respectively.

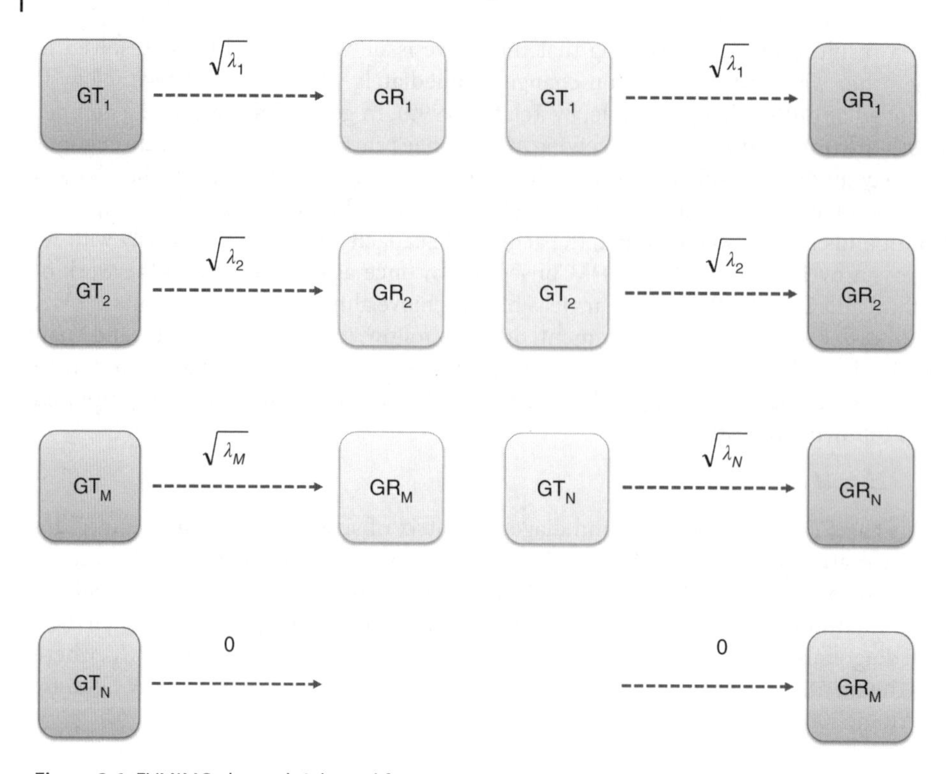

Figure 3.6 EVMIMO channel. Adapted from Vucetic and Yuan (2003).

said relay nodes be homogeneous, and thus equipped with the same number of antennae, then N would always be greater than M, as long as a virtual antenna array became instantiated.[13]

3.2.3 Capacity, Modelling, and Gains

Attempting to visualise the attainable capacity of the equivalent virtual multiple-input multiple-output radio channel and following the example of Vucetic and Yuan (2003), yet assuming the most representative case where $N = M$, so that the linear relation between the channel capacity and the number of generic transmitters or generic receivers could be emphasised, one could express the formula of channel capacity as follows, given that the det() function provides the determinant of a matrix:

$$C = B\log_2 \det\left[I_N + \frac{P_T}{N\sigma^2}Q\right]. \tag{3.7}$$

Here, Q would take the value of either HH^H, for $N > M$, or $H^H H$, for $N \leq M$. Bearing in mind that the generic transmitters and generic receivers would be virtually interconnected exclusively over the aforementioned orthogonal parallel subchannels, the

13 The reader should note, however, that once relay nodes have been elevated to Autonomic Cooperative Nodes and VAAs have been transposed to take the form of autonomic cooperative sets, the notion of VMIMO technology, as described in this subsection, will become one of the major constituents of what will be called Autonomic Cooperative Behaviour.

respective channel matrix H could be expressed as:

$$H = \sqrt{N} I_N, \tag{3.8}$$

where $\sqrt{N}$ denotes a scaling factor related to power normalisation, so that, following Vucetic and Yuan (2003), the capacity C of the equivalent virtual multiple-input multiple-output radio channel could be rewritten as

$$C = B \log_2 \det \left[I_N + \frac{N P_\mathrm{T}}{N \sigma^2} I_N \right], \tag{3.9}$$

which, thanks to the existence of the unitary matrix I_N in both the components of the logarithmic function, allowing for a direct translation into a diagonal matrix with the relevant expression on its main diagonal, may be rewritten, using diag() function to represent the diagonal matrix, as

$$C = B \log_2 \det \left[\mathrm{diag} \left(1 + \frac{P_\mathrm{T}}{\sigma^2} \right) \right]. \tag{3.10}$$

Based on the properties of the determinant of such a matrix, which is equal to the product of the elements on its main diagonal, as well as applying basic logarithmic operations, one may further rewrite the equation in the form

$$C = N\, B \log_2 \left(1 + \frac{P_\mathrm{T}}{\sigma^2} \right). \tag{3.11}$$

Given the above, it immediately becomes clear that the capacity of the equivalent virtual multiple-input multiple-output radio channel may scale linearly with the number of generic transmitters or generic receivers. In fact, the normalised form of the achievable channel capacity is illustrated in Figure 3.7, where an equal number N, ranging from 1 to 8, of generic transmitters and generic receivers is assumed. In general, such a potential may appear exploitable in two somewhat disjoint yet overlapping areas, following Molisch and Win (2004). Namely, on the one hand it is possible to create a highly effective diversity scheme to increase the robustness of the system against impairments induced by the radio channel, while on the other hand there is the option of transmitting multiple data streams in parallel to increase the system throughput (Wódczak, 2012a). Essentially, it all boils down to the characteristics of the underlying radio channel and, therefore, a properly adjusted mathematical model is required. For this reason, the approach to modelling cited below, as originally proposed by Zheng and Xiao (2003), is brought into the global picture of this book after having been thoroughly verified, in particular because it will be referred to for illustrative simulation purposes related to parts of the overall Autonomic Cooperative Networking Architectural Model. Thus, mutually comparable results are expected to be provided, which, given the complexity of the simulated components, may highly capitalise on the moderate computational complexity of this model.

In particular, the model proposed by Zheng and Xiao (2003) has been chosen since it may be directly used for generating multiple uncorrelated fading waveforms for frequency-selective fading channels and diversity-combining scenarios, and especially for multiple-input multiple-output radio channels. As such, the model is capable of producing radio channel coefficients fully satisfying the reference Rayleigh distribution through the application of an elegantly straightforward sum-of-sinusoids statistical

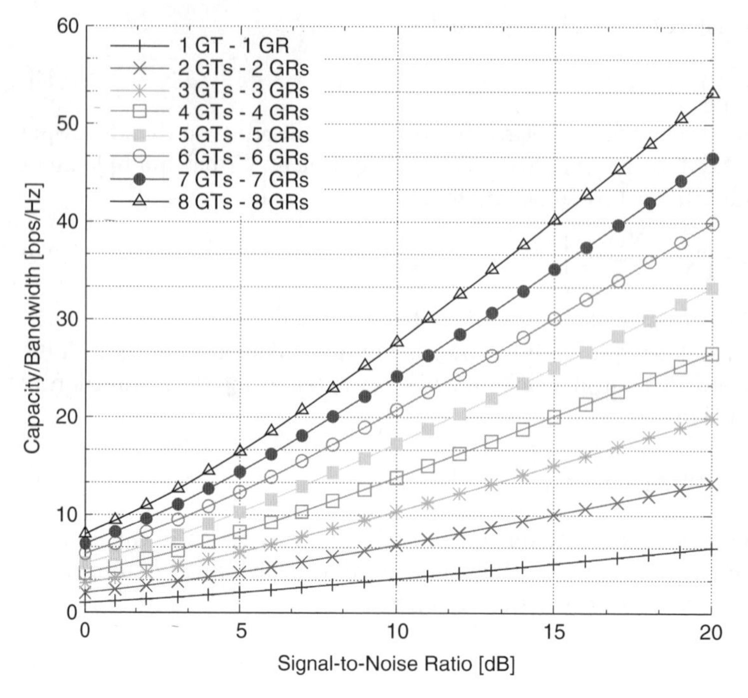

Figure 3.7 Capacity of VMIMO channel for $N = M$.

approach (Zheng and Xiao, 2003). The major advantage of the model in question stems from the fact that the autocorrelation and cross-correlation functions of the quadrature components, as well as the autocorrelation function of the complex envelope, all match the desired reference ones even when the number of sinusoids is expressed as a single digit number. As verified by the author, it suffices to use as few as five sinusoids to obtain a virtually ideal distribution. In particular, following Zheng and Xiao (2003), the normalised fading process of such a statistical model may be written as

$$X(t) = X_c(t) + jX_s(t), \tag{3.12}$$

where $X_c(t)$ is defined by

$$X_c(t) = \frac{2}{\sqrt{K}} \sum_{k=1}^{K} \cos(\varphi_k) \cos(\omega_d t \cos \alpha_k + \phi) \tag{3.13}$$

and $X_s(t)$ is defined by

$$X_s(t) = \frac{2}{\sqrt{K}} \sum_{k=1}^{K} \sin(\varphi_k) \cos(\omega_d t \cos \alpha_k + \phi), \tag{3.14}$$

while α_k is

$$\alpha_k = \frac{2\pi k - \pi + \theta}{4K}, \quad k = 1, 2, \ldots, K, \tag{3.15}$$

and the random variables θ, ϕ, and φ_k are all statistically independent and uniformly distributed over the range $[-\pi, \pi)$ for all values of k. The respective probability density

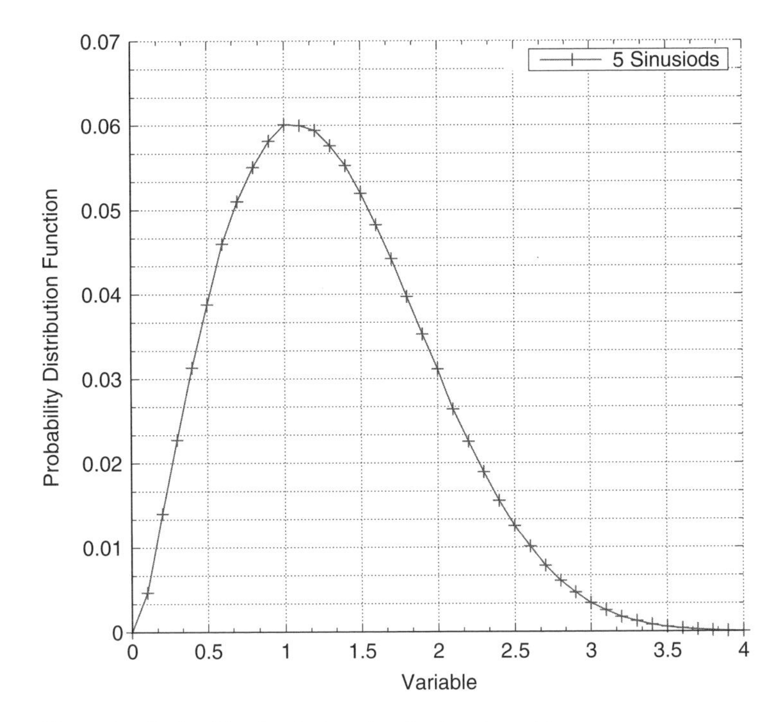

Figure 3.8 Probability density function for five sinusoids.

function (PDF) for the apparently optimum number of five sinusoids is presented in Figure 3.8, to confirm the convergence efficiency.[14]

Last, but not least, keeping in mind the diversity-rooted origins of spatio-temporal processing and the general characteristics of the multiple-input multiple-output radio channel, it becomes necessary to introduce the notion of diversity gain (DG). In fact, diversity techniques are, in general, useful in terms of alleviating the impairments introduced by wireless communications translating themselves into signal quality deterioration related to the fading effect. As discussed by Jankiraman (2004), the larger the number of independent fading branches or paths, or simply the receiving antennae, the bigger the so-called diversity order (DO). Moving forward, following Larsson and Stoica (2003b), for example, one should note that once maximum likelihood detection (MLD) or maximal ratio combining is assumed, the average error probability in a region of high SNR values may be expressed as

$$P\,(e) \sim G_{\mathrm{c}}(SNR)^{-G_{\mathrm{d}}}, \tag{3.16}$$

where G_{c} denotes the coding gain, according to Larsson and Stoica (2003b) also referred to as coding advantage (CA), which is provided by a block or convolutional coding scheme in the temporal domain, whereas G_{d} represents directly the aforementioned diversity order. In particular, should the $P(e)$ curve be plotted as a function of SNR on a log–log scale, then G_{c} would determine the horizontal position of this curve while G_{d} would correspond to its slope (Molisch and Win, 2004). What is more, thanks to the diversity order, the said diversity gain would be observable; it is defined as the gain

14 For additional illustration and justification in this respect the reader is also referred to Figures A.1, A.2, A.3, A.4, and A.5, where the PDFs for 2, 3, 4, 10, and 30 sinusoids, respectively, are presented.

provided by the spatial diversity across channels on the transmitter side, the receiver side, or both, allowing compensation for the fading effect (Jankiraman, 2004).

3.3 Space-Time Coding Techniques

3.3.1 Orthogonal Block-Coded Designs

Given the background established thus far with the introduction of the transmission medium in the form of the equivalent virtual multiple-input multiple-output radio channel along with all the related commentary on diversity, virtualisation, capacity, modelling, and gains, the time has come for the next step in advancing the analysis by shifting the focus onto the approach known as space-time block coding (STBC). In fact, not without sound justification was its extended version, distributed space-time block coding (DSTBC), highlighted in the opening chapter under the umbrella of spatio-temporal processing as one of the major technological pillars of the entire Autonomic Cooperative Networking Architectural Model. Regardless of its pivotal role to the entire design, one of the key advantages of space-time block coding consists in the fact that, as indicated by Jankiraman (2004), no channel state information (CSI) is required on the side of the transmitter, which, along with its orthogonality-driven design as proposed by Alamouti (1998), makes it one of the biggest technological advancements in the realm of modern wireless communications. In fact, looking at Figure 3.9, one may easily discern that such an STBC-enabled system shall perfectly match the previously discussed characteristics of the multiple-input multiple-output channel, as it is intended to exploit a transmitting MEA of size N and a receiving MEA of size M. In this context, the workings of space-time block coding are outlined below, along with example code matrices of further interest.

In general, among other spatio-temporal processing techniques of potential relevance, which could be successfully employed for the pre-processing of the signals to be transmitted in such a way that they would become more robust to the impairments typically induced by the radio channel, there is the said space-time block coding; it is characterised (Alamouti, 1998) by the ability to offer the said diversity gain, yet it cannot offer any inherent option of introducing additional coding gain. Despite this, the value of space-time block coding should never be diminished, especially if outer coding is employed. What appears to be most interesting, however, is that, despite its commonly

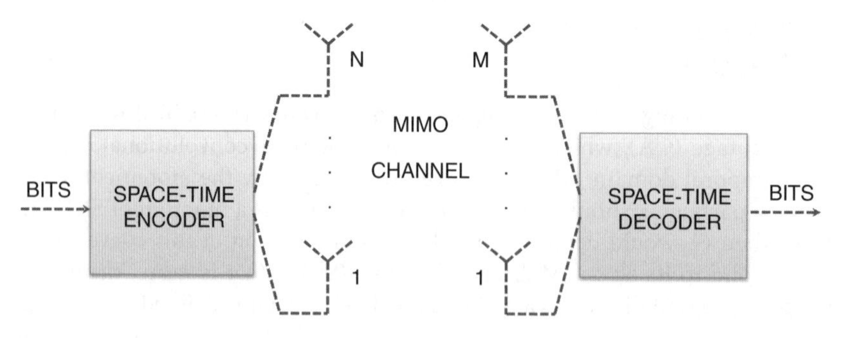

Figure 3.9 Diagram of space-time coded system.

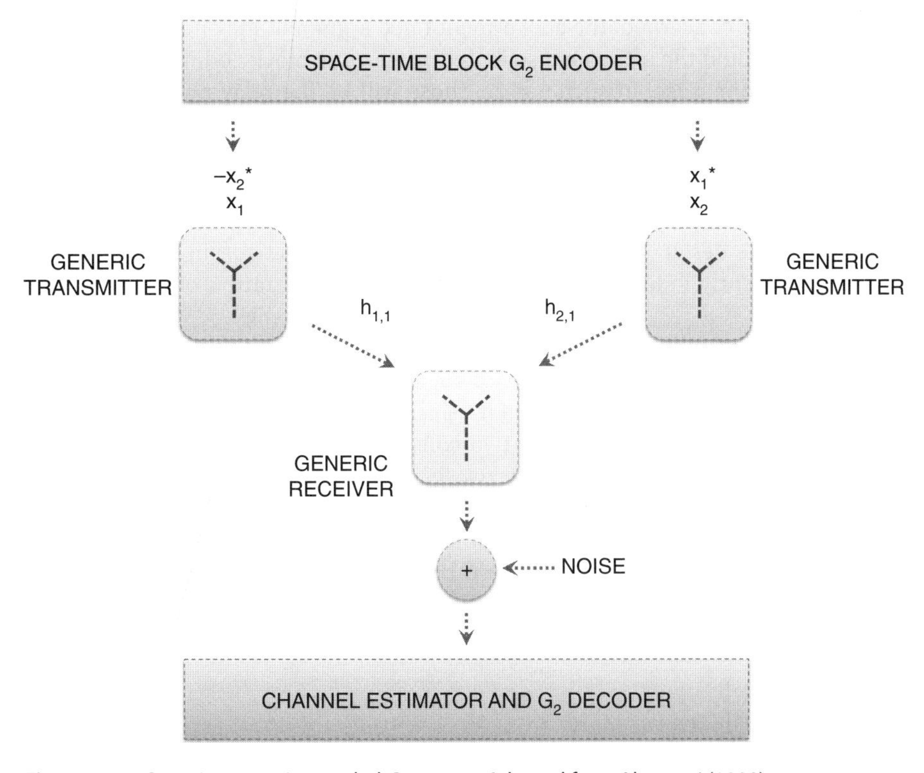

Figure 3.10 Generic space-time coded G_2 system. Adapted from Alamouti (1998).

accepted name, in general the concept of space-time block coding should rather be perceived as a modulation and not an explicit coding technique. This is so, as the rationale behind its design was specifically directed towards the provision of additional diversity in both the spatial and temporal domains, should a wireless communications system employ MEAs with the objective of increasing its transmission capabilities. One should note, however, that the notion of MEAs is brought up solely for consistency reasons and, as was the case for the multiple-input multiple-output radio channel eventually being upgraded to its counterpart in the form of the equivalent virtual multiple-input multiple-output radio channel, MEAs are immediately replaced with generic transmitters and generic receivers, as outlined in Figure 3.10, in order to pave the way for the ultimate transition from space-time block coding into its networked concept of distributed space-time block coding.

However, apart from all the advantages of space-time block coding signalled thus far, no sooner does the major selling point behind this technology become clear than the comparison with legacy approaches to reception diversity is made. It then immediately transpires that space-time block coding allows for the complexity related to the deployment of multiple antennae on small mobile UTs to be shifted to a BS or an AP. Such an approach is highly advantageous because the complexity of the UTs can be reduced, and the cost of installing a single MEA solely on the BS or AP side appears less substantial, while the spacing among the elements of such an MEA may no longer be physically limited. As much as this capability could be appealing when more classic use cases are concerned, the above-mentioned argument may no longer remain fully convincing should the ultimate distributed space-time block coding be taken into account, especially because, in such a case, those will be the relay nodes to become preselected to form a virtual antenna array expected to behave in a way resembling a dynamic MEA. One should note, however, that a use case of this type already seems to go well beyond the assumptions inherent in the classic version of STBC, where MEAs would normally be employed. The technological upgrade results from the fact that the approach to the design of the Autonomic Cooperative Networking Architectural Model fairly frequently induces certain changes to legacy operation of the leveraged technologies.

In order to discuss the manner in which transmission based on a complex orthogonal design[15] may operate, the initial version of such an approach is first introduced as invented by Alamouti (1998), who defined it using the G_2 code matrix (3.17). While its operation follows the precise pattern of two consecutive time slots, the G_2 code was designed for a system employing strictly two generic transmitters and an unlimited number of generic receivers. In particular, moving into its workings, and following the pattern of the G_2 matrix, during the first time slot both the x_1 and x_2 symbols are transmitted by the first and second generic transmitters, respectively, while during the second time slot both the $-x_2^*$ and x_1^* symbols are transmitted in exactly the same manner by the same generic transmitters.

$$G_2 = \begin{bmatrix} x_1 & x_2 \\ -x_2^* & x_1^* \end{bmatrix}. \tag{3.17}$$

One should note, however, that there exist more complicated space-time block coding matrices, just to mention the generalised complex orthogonal designs (GCODs), which were additionally and separately invented and proposed by Tarokh et al., 1999a,b. Such enhanced designs include the G_3 (Equation 3.18), G_4 (Equation 3.19), H_3 (Equation 3.20), and H_4 (Equation 3.21) code matrices, which are applicable when sets of generic transmitters larger than two are to be employed. What is more, it is also necessary to keep in mind that there is a trade-off between the robustness of these codes and their code rates (CDRs) being strictly related to the number of generic transmitters. In fact, the code rate is equal to 1 only in the case of the G_2 code; although more reliable, the other codes offer worse code rates, equal to $\frac{1}{2}$ for the G_3 and G_4 matrices and $\frac{3}{4}$ for H_3 and H_4.

$$G_3 = \begin{bmatrix} x_1 & x_2 & x_3 \\ -x_2 & x_1 & -x_4 \\ -x_3 & x_4 & x_1 \\ -x_4 & -x_3 & x_2 \\ x_1^* & x_2^* & x_3^* \\ -x_2^* & x_1^* & -x_4^* \\ -x_3^* & x_4^* & x_1^* \\ -x_4^* & -x_3^* & x_2^* \end{bmatrix}, \tag{3.18}$$

15 As explicitly defined by Tarokh et al. (1999a).

$$G_4 = \begin{bmatrix} x_1 & x_2 & x_3 & x_4 \\ -x_2 & x_1 & -x_4 & x_3 \\ -x_3 & x_4 & x_1 & -x_2 \\ -x_4 & -x_3 & x_2 & x_1 \\ x_1^* & x_2^* & x_3^* & x_4^* \\ -x_2^* & x_1^* & -x_4^* & x_3^* \\ -x_3^* & x_4^* & x_1^* & -x_2^* \\ -x_4^* & -x_3^* & x_2^* & x_1^* \end{bmatrix}, \tag{3.19}$$

$$H_3 = \begin{bmatrix} x_1 & x_2 & \frac{x_3}{\sqrt{2}} \\ -x_2^* & x_1^* & \frac{x_3}{\sqrt{2}} \\ \frac{x_3^*}{\sqrt{2}} & \frac{x_3^*}{\sqrt{2}} & \frac{(-x_1 - x_1^* + x_2 - x_2^*)}{\sqrt{2}} \\ \frac{x_3^*}{\sqrt{2}} & -\frac{x_3^*}{\sqrt{2}} & \frac{(x_2 + x_2^* + x_1 - x_1^*)}{\sqrt{2}} \end{bmatrix}, \tag{3.20}$$

$$H_4 = \begin{bmatrix} x_1 & x_2 & \frac{x_3}{\sqrt{2}} & \frac{x_3}{\sqrt{2}} \\ -x_2^* & x_1^* & \frac{x_3}{\sqrt{2}} & -\frac{x_3}{\sqrt{2}} \\ \frac{x_3^*}{\sqrt{2}} & \frac{x_3^*}{\sqrt{2}} & \frac{(-x_1 - x_1^* + x_2 - x_2^*)}{\sqrt{2}} & \frac{(-x_2 - x_2^* + x_1 - x_1^*)}{\sqrt{2}} \\ \frac{x_3^*}{\sqrt{2}} & -\frac{x_3^*}{\sqrt{2}} & \frac{(x_2 + x_2^* + x_1 - x_1^*)}{\sqrt{2}} & -\frac{(x_1 + x_1^* + x_2 - x_2^*)}{\sqrt{2}} \end{bmatrix}. \tag{3.21}$$

Moreover, with the passage of time there have been many approaches to space-time block coding proposed in pursuit of achieving the often disjoint, at least when generalised complex orthogonal designs are concerned, objectives of increasing the number of generic transmitters and the related code rate at the same time. In particular, attempting to complement the above classic code matrices by analysing the further milestone developments in this area, one may, in the first place, come across the work of Su and Xia (2003), where a couple of generalised complex orthogonal designs capable of using five and six generic transmitters and characterised by code rates of $\frac{7}{11}$ and $\frac{3}{5}$, respectively, were proposed. Then, a full-rate GCOD was proposed by Jewel and Rahman (2009) which could employ four generic transmitters, and this was complemented by another full-rate GCOD proposed by Murthya and Gowrib (2012) which, in turn, could accommodate as many as eight generic transmitters. Interesting though it may seem, should the criterion of orthogonality be relaxed even slightly, one may not only discover quasi-orthogonal (QO) full-rate and full-diversity designs proposed by Jung *et al.* (2008) and capable of employing any odd number of generic transmitters, but many other concepts of relevance displaying novelty with regard to different aspects of spatio-temporal processing. However, any further classification of these is beyond the scope of this book; the above-mentioned code matrices will be referred to at a later stage.

3.3.2 Derivation of Decoding Metrics

Even though, as it has already been highlighted, there apparently exist a variety of relevant spatio-temporal processing techniques, fairly thoroughly scrutinised in the literature, the focus from now on in this book will be deliberately limited to space-time block coding. Such an assumption is being made despite the fact that one could contemplate using any other relevant approach to spatio-temporal processing as a replacement. Given the objective and theme of this book, such a replacement should not be expected to incur any substantial changes to the most crucial architectural assumptions governing the place and role of the major components of the presented system design. In this respect, more precisely, the code matrices presented thus far, which fulfil the generalised complex orthogonal design criterion to be outlined in detail below, will be employed for further analyses related to the comprehensive depiction of the workings of the Autonomic Cooperative Networking Architectural Model. Given such an approach, it is necessary to provide a high-level overview of the decoding process, mostly for quick reference reasons. However, in scrutinising the literature covering the topic of space-time block coding the author gained the impression that there may exist some inconsistency in the mathematical forms of the decoding metrics, manifested in certain apparently unintentional typographical errors. In order to resolve any doubts potentially arising from the above observation, all the pertinent metrics been recalculated by the author, and they are presented with a commentary intended to indicate the said inconsistencies, when applicable.

In particular, moving to the reception side of the system, it might be the case that a generic receiver is equipped with an MEA to obtain an even better performance. Should this be the case, the radio signal received by receiving antenna j could be expressed as (Alamouti, 1998)

$$r_t^j = \sum_{i=1}^{N} h_{i,j} s_t^i + \eta_t^j, \tag{3.22}$$

where $h_{i,j}$ denotes the channel coefficient previously defined for the multiple-input multiple-output channel matrix (Equation 3.3), s_t^i represents the symbol transmitted by the antenna of the generic transmitter[16] i, and the noise samples η_t^j are modelled by a complex Gaussian process of zero mean and $N_0/2$ variance per dimension. This is, in fact, where the condition of GCOD comes into the global picture of space-time block coding, being the main condition under which the operation of decoding may be successfully performed. Following the work of Larsson and Stoica (2003a), for example, the condition of orthogonality may be defined as

$$G_N G_N^{\mathrm{H}} = \left(\sum_{i=1}^{N} |x_i|^2 \right) I_N, \tag{3.23}$$

16 The analysis could be further advanced by allowing each GT to be equipped with more than one transmitting antenna, thereby forming an MEA. For the sake of clarity, however, it will be assumed from now on that, contrary to GRs, a single antenna is used by each GT.

where N is the number of generic transmitters and I_N is the $N \times N$ identity matrix. The process of decoding is based on MLD, aiming to minimise the decision metric as defined by Tarokh et al. (1999a):

$$z = \sum_{t=1}^{L} \sum_{j=1}^{M} \left| r_t^j - \sum_{i=1}^{N} h_{i,j} s_t^i \right|^2 . \tag{3.24}$$

This metric can easily be derived on the basis of the theory outlined, for instance, by Goldsmith (2005). In order to account for such a mathematical formula, one should note that its most obvious interpretation boils down to the fact that, for a given code, those potentially transmitted symbols are chosen which minimise this metric. Unfortunately, its use in the above form would render the calculation process computationally inefficient.

Therefore, to address this issue, a set of relevantly modified metrics was introduced by Tarokh et al. (1999b). In particular, for the G_2 code matrix (Equation 3.17), the metric in Equation 3.24 may be rewritten as

$$\sum_{j=1}^{M} \left(\left| r_1^j - h_{1,j} s_1 - h_{2,j} s_2 \right|^2 + \left| r_2^j + h_{1,j} s_2^* - h_{2,j} s_1^* \right|^2 \right) . \tag{3.25}$$

While MLD aims to find the minimum value of this metric for all possible combinations of the variables s_1 and s_2, given the attempt at avoiding the aforementioned computational inefficiency, the above expression may be expanded so that the terms independent of the said variables can be deleted. In other words, the problem of the minimisation of this metric becomes equivalent to minimising the following expression:

$$\begin{aligned} &|s_1|^2 \sum_{j=1}^{M} \sum_{i=1}^{2} |h_{i,j}|^2 - \sum_{j=1}^{M} \left[r_1^j h_{1,j}^* s_1^* + (r_1^j)^* h_{1,j} s_1 + r_2^j h_{2,j}^* s_1 + (r_2^j)^* h_{2,j} s_1^* \right] \\ &+ |s_2|^2 \sum_{j=1}^{M} \sum_{i=1}^{2} |h_{i,j}|^2 - \sum_{j=1}^{M} \left[r_1^j h_{2,j}^* s_2^* + (r_1^j)^* h_{2,j} s_2 - r_2^j h_{1,j}^* s_2 - (r_2^j)^* h_{1,j} s_2^* \right] . \end{aligned} \tag{3.26}$$

Such a formula, in turn, may be clearly perceived as being composed of two components, the former being exclusively a function of the variable s_1, while the latter being exclusively a function of the variable s_2. As a result, one may conclude that the total value of this metric is a minimum when each of these components is a minimum. Consequently, following Tarokh et al. (1999b), the metric for the G_2 code can be expressed as

$$\left| \sum_{j=1}^{m} R_j - s \right|^2 + \left(-1 + \sum_{j=1}^{m} \sum_{i=1}^{2} |h_{i,j}|^2 \right) |s|^2, \tag{3.27}$$

where for $s = s_1$ the signal R_j received by antenna j is given by

$$R_j = r_1^j h_{1,j}^* + (r_2^j)^* h_{2,j} \tag{3.28}$$

and for $s = s_2$ the signal R_j received by antenna j is given by

$$R_j = r_1^j h_{2,j}^* - (r_2^j)^* h_{1,j}. \tag{3.29}$$

Similarly, the metric for the G_3 code may be written as

$$\left|\sum_{j=1}^{m} R_j - s\right|^2 + \left(-1 + 2\sum_{j=1}^{m}\sum_{i=1}^{3}|h_{i,j}|^2\right)|s|^2, \tag{3.30}$$

where for $s = s_1$ the signal R_j received by antenna j is given by

$$R_j = r_1^j h_{1,j}^* + r_2^j h_{2,j}^* + r_3^j h_{3,j}^* + (r_5^j)^* h_{1,j} + (r_6^j)^* h_{2,j} + (r_7^j)^* h_{3,j}, \tag{3.31}$$

for $s = s_2$ the signal R_j received by antenna j is given by

$$R_j = r_1^j h_{2,j}^* - r_2^j h_{1,j}^* + r_4^j h_{3,j}^* + (r_5^j)^* h_{2,j} - (r_6^j)^* h_{1,j} + (r_8^j)^* h_{3,j}, \tag{3.32}$$

for $s = s_3$ the signal R_j received by antenna j is given by

$$R_j = r_1^j h_{3,j}^* - r_3^j h_{1,j}^* - r_4^j h_{2,j}^* + (r_5^j)^* h_{3,j} - (r_7^j)^* h_{1,j} - (r_8^j)^* h_{2,j}, \tag{3.33}$$

and for $s = s_4$ the signal R_j received by antenna j is given by

$$R_j = -r_2^j h_{3,j}^* + r_3^j h_{2,j}^* - r_4^j h_{1,j}^* - (r_6^j)^* h_{3,j} + (r_7^j)^* h_{2,j} - (r_8^j)^* h_{1,j}. \tag{3.34}$$

Similarly again, the metric for the G_4 code can be expressed as

$$\left|\sum_{j=1}^{m} R_j - s\right|^2 + \left(-1 + 2\sum_{j=1}^{m}\sum_{i=1}^{4}|h_{i,j}|^2\right)|s|^2, \tag{3.35}$$

where for $s = s_1$ the signal R_j received by antenna j is given by

$$R_j = r_1^j h_{1,j}^* + r_2^j h_{2,j}^* + r_3^j h_{3,j}^* + r_4^j h_{4,j}^* + (r_5^j)^* h_{1,j} + (r_6^j)^* h_{2,j} + (r_7^j)^* h_{3,j} + (r_8^j)^* h_{4,j}, \tag{3.36}$$

for $s = s_2$ the signal R_j received by antenna j is given by

$$R_j = r_1^j h_{2,j}^* - r_2^j h_{1,j}^* - r_3^j h_{4,j}^* + r_4^j h_{3,j}^* + (r_5^j)^* h_{2,j} - (r_6^j)^* h_{1,j} - (r_7^j)^* h_{4,j} + (r_8^j)^* h_{3,j}, \tag{3.37}$$

for $s = s_3$ the signal R_j received by antenna j is given by

$$R_j = r_1^j h_{3,j}^* + r_2^j h_{4,j}^* - r_3^j h_{1,j}^* - r_4^j h_{2,j}^* + (r_5^j)^* h_{3,j} + (r_6^j)^* h_{4,j} - (r_7^j)^* h_{1,j} - (r_8^j)^* h_{2,j}, \tag{3.38}$$

and for $s = s_4$ the signal R_j received by antenna j is given by[17]

$$R_j = r_1^j h_{4,j}^* - r_2^j h_{3,j}^* + r_3^j h_{2,j}^* - r_4^j h_{1,j}^* + (r_5^j)^* h_{4,j} - (r_6^j)^* h_{3,j} + (r_7^j)^* h_{2,j} - (r_8^j)^* h_{1,j}. \tag{3.39}$$

---

17 During the derivation of this metric the author noticed that the formula given by Tarokh et al. (1999b) is erroneous, as is the one given on p. 107 of Vucetic and Yuan (2003). Such an error does not exist in the work of Jankiraman (2004); however, one should note that all the G_4 metrics provided therein also contain a typographical error and are written as if they were pertaining to a code matrix composed of three columns (index i on pp. 87–88), while this code was apparently designed for four GTs.

The metric for the H_3 code can similarly be written as

$$\left|\sum_{j=1}^{m} R_j - s\right|^2 + \left(-1 + \sum_{j=1}^{m}\sum_{i=1}^{3} |h_{i,j}|^2\right)|s|^2, \tag{3.40}$$

where for $s = s_1$ the signal R_j received by antenna j is given by

$$R_j = r_1^j h_{1,j}^* + (r_2^j)^* h_{2,j} - \frac{r_3^j h_{3,j}^*}{2} + \frac{r_4^j h_{3,j}^*}{2} - \frac{(r_3^j)^* h_{3,j}}{2} - \frac{(r_4^j)^* h_{3,j}}{2}, \tag{3.41}$$

for $s = s_2$ the signal R_j received by antenna j is given by

$$R_j = r_1^j h_{2,j}^* - (r_2^j)^* h_{1,j} + \frac{r_3^j h_{3,j}^*}{2} + \frac{r_4^j h_{3,j}^*}{2} - \frac{(r_3^j)^* h_{3,j}}{2} + \frac{(r_4^j)^* h_{3,j}}{2}, \tag{3.42}$$

and for $s = s_3$ the signal R_j received by antenna j is given by

$$R_j = \frac{r_1^j h_{3,j}^*}{\sqrt{2}} + \frac{r_2^j h_{3,j}^*}{\sqrt{2}} + \frac{(r_3^j)^* h_{1,j}}{\sqrt{2}} + \frac{(r_3^j)^* h_{2,j}}{\sqrt{2}} + \frac{(r_4^j)^* h_{1,j}}{\sqrt{2}} - \frac{(r_4^j)^* h_{2,j}}{\sqrt{2}}. \tag{3.43}$$

Finally, the metric for the H_4 code can be written as

$$\left|\sum_{j=1}^{m} R_j - s\right|^2 + \left(-1 + \sum_{j=1}^{m}\sum_{i=1}^{3} |h_{i,j}|^2\right)|s|^2, \tag{3.44}$$

where for $s = s_1$ the signal R_j received by antenna j is given by

$$R_j = r_1^j h_{1,j}^* + (r_2^j)^* h_{2,j} - \frac{r_3^j (h_{3,j}^* - h_{4,j}^*)}{2} + \frac{r_4^j (h_{3,j}^* - h_{4,j}^*)}{2} - \frac{(r_3^j)^* (h_{3,j} + h_{4,j})}{2} - \frac{(r_4^j)^* (h_{3,j} + h_{4,j})}{2}, \tag{3.45}$$

for $s = s_2$ the signal R_j received by antenna j is given by

$$R_j = r_1^j h_{2,j}^* - (r_2^j)^* h_{1,j} + \frac{r_3^j (h_{3,j}^* - h_{4,j}^*)}{2} + \frac{r_4^j (h_{3,j}^* - h_{4,j}^*)}{2} - \frac{(r_3^j)^* (h_{3,j} + h_{4,j})}{2} + \frac{(r_4^j)^* (h_{3,j} + h_{4,j})}{2}, \tag{3.46}$$

and for $s = s_3$ the signal R_j received by antenna j is given by

$$R_j = \frac{r_1^j (h_{3,j}^* + h_{4,j}^*)}{\sqrt{2}} + \frac{r_2^j (h_{3,j}^* - h_{4,j}^*)}{\sqrt{2}} + \frac{(r_3^j)^* (h_{1,j} + h_{2,j})}{\sqrt{2}} + \frac{(r_4^j)^* (h_{1,j} - h_{2,j})}{\sqrt{2}}. \tag{3.47}$$

3.3.3 Trellis-Coded Approach

In the quest for further improvement, two different paths were followed, the first of them, to be yet revisited, in the form of a trellis coded modulation (TCM) driven, concatenated version of space-time block coding, as proposed by Alamouti et al. (1998), and, the second one, assuming an all-in-one solution of space-time trellis coding

(STTC), as introduced in the publications of Tarokh et al. (1998, 1999c). Commencing the analysis with the latter approach, when compared to space-time block coding, mostly perceived as a modulation technique, space-time trellis coding intends to capitalise on additional relations not only between specific sequences transmitted by distinct Generic Transmitters, but also between the symbols being the constituents of such sequences, so that apart from the discussed diversity gain, additional coding gain may be observed, too. Most naturally, the trellis-coded nature of this approach should be perceived as being inherently embedded into the structure of the space-time-driven arrangement itself. Quite the opposite occurs when the former concatenated solution is considered: despite a certain level of integration, especially on the side of the very outer TCM, the underlying inner space-time block coding still seems, undeniably, detachable. Even though the level of internal integration clearly differs between the two trellis-coded approaches, the former will be especially emphasised once the latter has been outlined in more detail, chiefly because, as additionally elaborated by Gong and Letaief (2002), this type of concatenation may prove to outperform even an STTC-based design, making it especially interesting given the major theme of this book.

Following Tarokh et al. (1998), the base code trellis exploiting the phase-shift keying (PSK) modulation, in the form of 4-PSK, is presented in Figure 3.11. Examining the structure of such a code, one should note that the order of the numbers placed to the left of the corresponding trellis diagram should be interpreted in a specific way: the most significant digit represents the information on the current state, while the least significant one corresponds not only to the input itself, but particularly to the next state, to which the transition is to be made. In other words, consecutive pairs of bits arriving at the input of the respective encoder would determine the transition from the current state to the following one, where two symbols would be relayed to the generic transmitters in such a way that the first generic transmitter would transmit the channel symbol informing about the current state, while the second generic transmitter would transmit the channel symbol corresponding to the next state. The process of decoding, in turn, is based on the widely recognised algorithm by Viterbi (1967), the logic of which revolves around the fact that each transition on the trellis diagram is assigned a metric

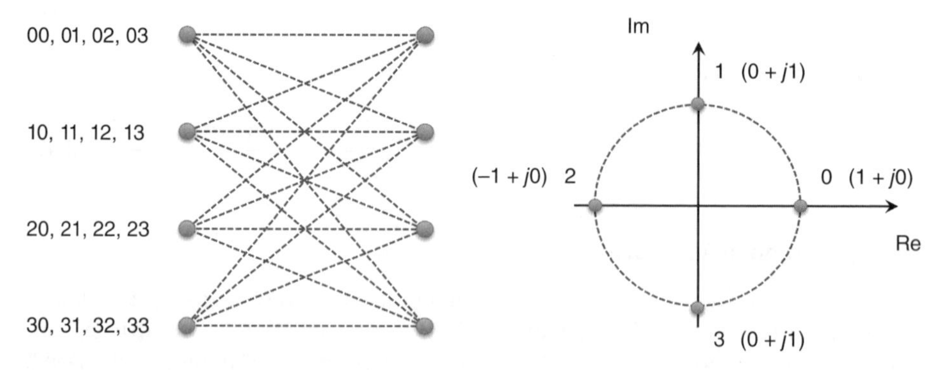

Figure 3.11 Base code for 4-PSK modulation. Adapted from Tarokh et al. (1998).

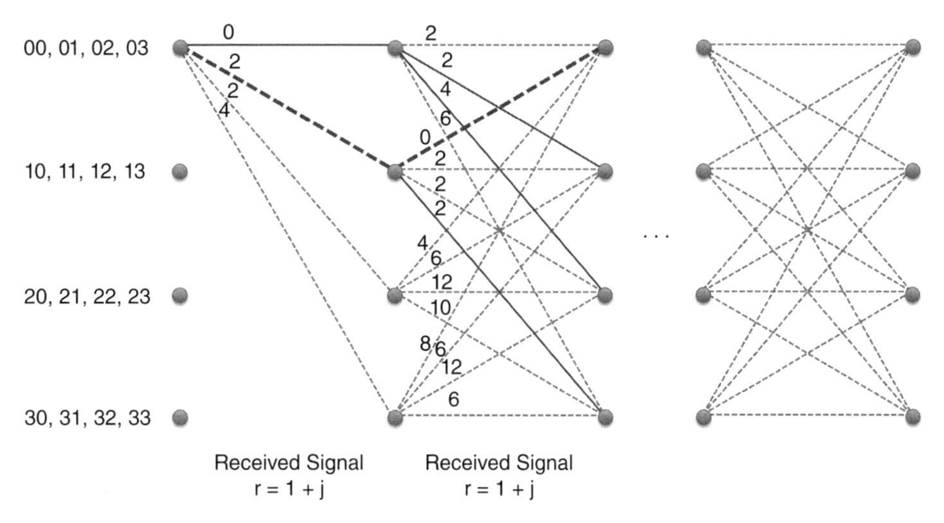

Figure 3.12 Example of STTC decoding process.

value, in this case calculated according to the formula devised by Tarokh et al. (1998):

$$w_{x,y} = \sum_{j=1}^{M} \left| r_t^j - \sum_{i=1}^{N} h_{i,j} s_t^i \right|^2, \tag{3.48}$$

where x and y denote the consecutive trellis states between which a given transition is taking place. Assuming no channel impairments, the decoding procedure of an example input sequence of $\{1, 0, 3, 1, 2, 2\}$ fed to the space-time code trellis illustrated in Figure 3.11 would be carried out in accordance with Figure 3.12 (Wódczak, 2012a).

In particular, assume the encoder is initially in the zero state, and at each moment t it can transition from the current state to the next one in the subsequent moment $t + 1$, given one of the input values of $\{0, 1, 2, 3\}$. Thus, the input sequence could be translated into the passage of the symbol pairs $\{01, 10, 03, 31, 12, 22\}$ to the generic transmitters. In such a context, keeping in mind the assumed nomenclature, the first generic transmitter would transmit the signals corresponding to the symbol sequence $\{0, 1, 0, 3, 1, 2\}$, while, concurrently, the second generic transmitter would transmit the signals corresponding to the symbol sequence $\{1, 0, 3, 1, 2, 2\}$. More precisely, this would mean that the respective modulated sequences of $\{1, j, 1, -j, j, -1\}$ and $\{j, 1, -j, j, -1, -1\}$ would be observed at the output of the modulator. In other words, during the first modulation interval the (s_0, s_1) signal pair, $(1 + j0, 0 + j1)$, would be transmitted, so that, referring to Equation 3.22, the received signal could be written as $r_1^1 = 1 + j$. Consequently, according to Equation 3.48, the following metrics would be calculated for their corresponding transitions: $w_{0,0} = 2$, $w_{0,1} = 0$, $w_{0,2} = 4$, $w_{0,3} = 2$. Then, the same procedure would be performed during the second modulation interval, where the signal pair $(s_0, s_1) = (0 + j1, 1 + j0)$ would be transmitted. The received signal could be expressed as $r_2^1 = j + 1$, while the metrics would take the values of: $w_{0,0} = 2$, $w_{0,1} = 0$, $w_{0,2} = 2$, $w_{0,3} = 4$, $w_{1,0} = 0$, $w_{1,1} = 2$, $w_{1,2} = 2$, $w_{1,3} = 2$, $w_{2,0} = 4$, $w_{2,1} = 8$, $w_{2,2} = 10$, $w_{2,3} = 2$, $w_{3,0} = 2$, $w_{3,1} = 0$, $w_{3,2} = 4$, $w_{3,3} = 2$. As this process continues, a number of possible paths could lead to the same trellis state.

Essentially, given the rationale behind the aforementioned Viterbi algorithm, should more than one path cross the same trellis state at a given modulation interval, proceeding according to a certain rule referred to below, the path characterised by the lowest cumulative metric would need to be chosen. Looking at the illustrative example above, one may discern that apparently there would be as many as four candidate paths to be taken into account, and, if the selection decision were to be taken at this (normally too early) stage, then it would be the path in bold in Figure 3.12 that would be selected. However, one should note immediately that usually the analysis of such a trellis diagram should proceed as far as from three to five times the constraint length of the convolutional code for the sake of guaranteeing that the most reliable decisions are taken (Wesołowski, 2002). Given the related performance results depicted in Figure 3.13, there also exist more advanced codes in this respect, for example those proposed by Tarokh et al. (1998) and shown in Figures 3.14 and 3.15. While both these code trellis graphs are based on the same number of eight states and are generally designed for the PSK modulation scheme, the former follows the 4-PSK constellation pattern, while the latter reflects the application of the 8-PSK one. In this context, the focus is now going to shift slightly from space-time trellis coding in order to touch upon the alternative approach consisting in the employment of a TCM-driven, concatenated version of space-time block coding, where the additional operation of interleaving is exploited for better performance, as outlined, for example, by Gong and Letaief (2002).

As such, on the transmitting side the system exploits a TCM encoder responsible for the generation of a sequence of complex symbols, the consecutive vector of which, once interleaved, is submitted to the space-time block encoder (STBE) to perform, in

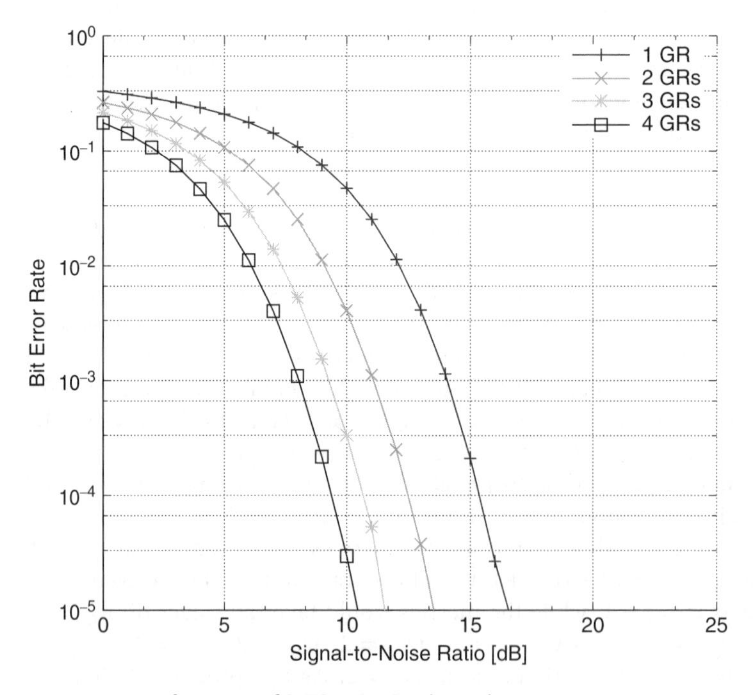

Figure 3.13 Performance of STTC in AWGN channel.

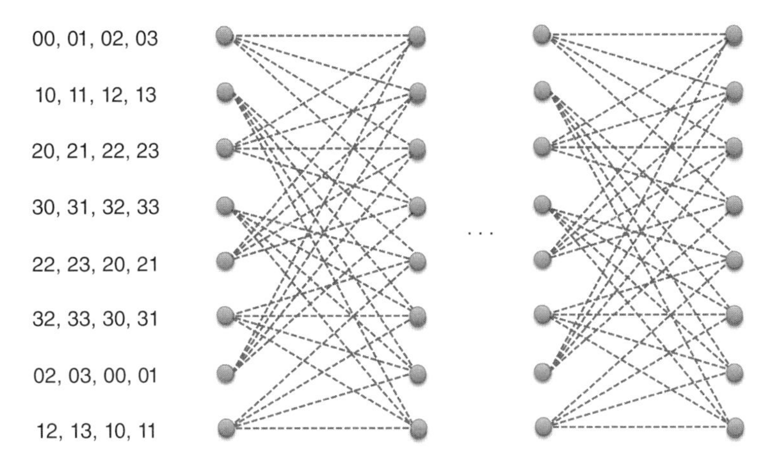

Figure 3.14 Example of STTC with 4-PSK modulation. Adapted from Tarokh et al. (1998).

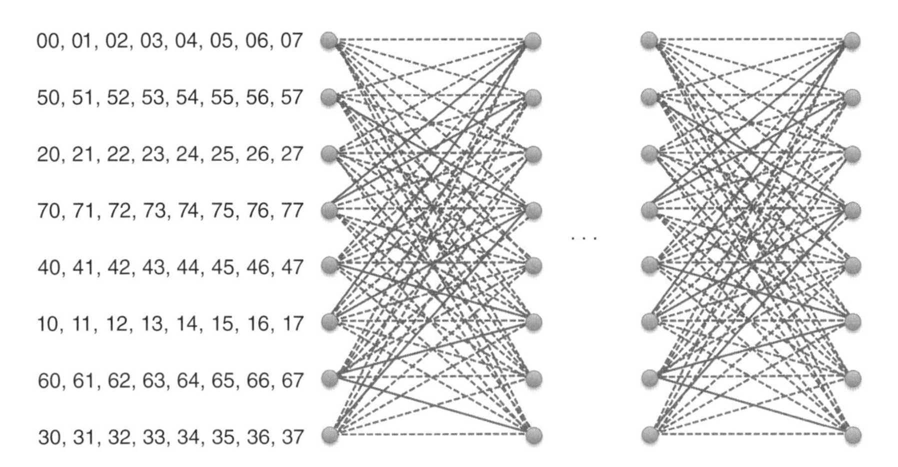

Figure 3.15 Example of STTC with 8-PSK modulation. Adapted from Tarokh et al. (1998).

the previously explained manner, the rather modulation-related operation outlined in Figure 3.16. Where the receiving side is concerned, the entire chain of processing modules is generally reversed by first involving the space-time block decoder (STBD) so that its output sequence, after having been deinterleaved, may be submitted to the TCM decoder, where the familiar operation of Viterbi decoding takes place, as indicated by Gong and Letaief (2002). Apparently, in order to be able to really capitalise on such a communication setup, one would need to establish a set of relevant criteria defining the proper construction of appropriate trellis codes, so that it could operate under the assumption of a specific fading process conditioned by the characteristics of the radio channel (Gong and Letaief, 2002). Should the reader be interested in additional details, the cited material is suggested as further reference since, despite the unquestionable advantages of the summarised trellis-based approaches, the major focus in this book will remain on space-time block coding, to be additionally advanced with the introduction of the equivalent distributed space-time block encoder (EDSTBE)

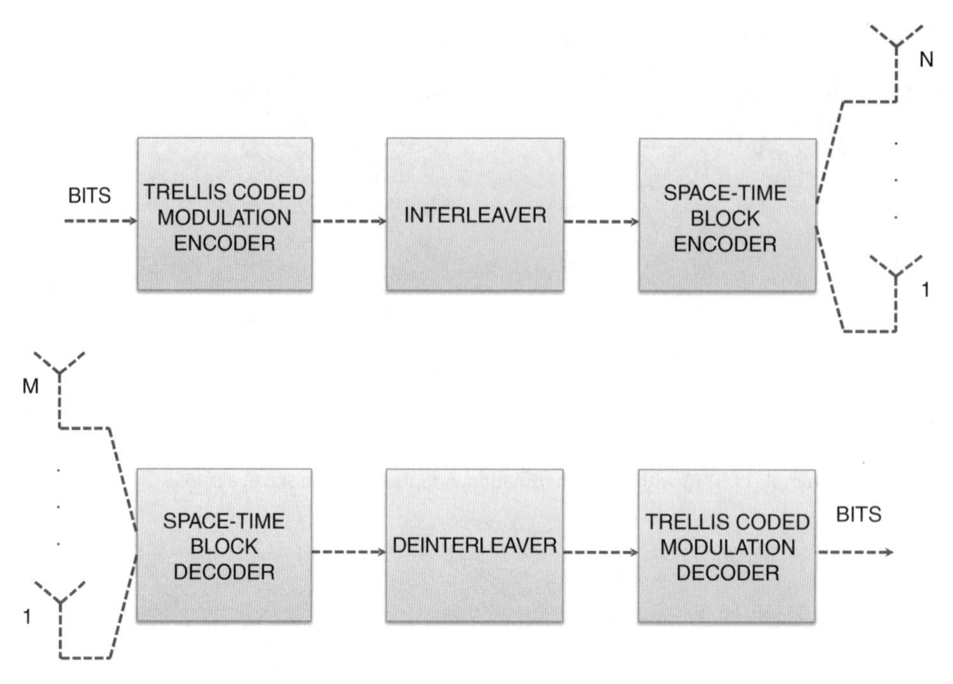

Figure 3.16 Concatenated STBC with TCM. Adapted from Gong and Letaief (2002).

as the plot unfolds more towards networked designs relating to cooperative relaying, after special emphasis has been placed on the protocol level architectural extensions.

3.4 Protocol Level Overlay Logic

3.4.1 Autonomic Cooperative Node

Given the initial overview of the high-level incarnation of the Autonomic Cooperative Networking Architectural Model in the opening chapter and the analysis of the relevant spatio-temporal processing techniques introduced in this chapter, the time has come to commence the incremental and detailed presentation of the proposed architectural design in a bottom-up manner, beginning with the workings of both the vertically orientated physical layer, inherent in the Open Systems Interconnection Reference Model, and the horizontally positioned protocol level, embedded in the Generic Autonomic Network Architecture. In fact, once the description has been advanced, it will immediately become more than apparent that the already implied lack of correspondence between the layers and levels is highly factual, causing the conceptual overhead required to maintain consistency between the incrementally expanding design stages. In particular, the Autonomic Cooperative Node (ACN), to be scrutinised in more detail below, appears to serve as a highly representative example of such a situation, simply because it extends way beyond the protocol level. Consequently, it has been assumed by the author that whenever a fairly monolithic concept is to be referred to, its entire architectural structure will be presented in a general form at the introductory stage, even though some of its parts may not yet seem directly relevant. Thus, once the relation between autonomics and cooperation has been described, the concept of an Autonomic Cooperative Node will be outlined in the above-mentioned manner so that it may be complemented with the adopted interaction model over the protocol stack.

In the light of the above introduction, chiefly for the sake of reflecting the necessity of paving the ground for further developments, yet still before the rationale behind the Autonomic Cooperative Node may be introduced, the question of mutual relation between autonomics and cooperation is to be addressed, assuming certain dose of abstraction. On the one hand, the inherent feature of biologically inspired autonomics may be directly translated into the capacity of such a networked system to perform self-management through the orchestration of its behaviour, presumably without any configuration, optimisation, healing, or protection-related intervention from a network operator (Bicocchi and Zambonelli, 2007; Nakano, 2011). On the other hand, however, the, somewhat, similarly inherent, yet maybe a bit less exposed, idea of instantiating cooperation among the network nodes of such a system appears to imply that cooperation might add constructively to an increase of overall system resiliency in many different dimensions, as justified by Wódczak (2014). Keeping in mind the above context, where the whole networked system was termed autonomic, there is no reason why the concept of a cooperation-enabled node could not be integrated into the bigger picture of the same. Looking at the legacy autonomic node of the Generic Autonomic Network Architecture, however, the highly relevant question arises of how the flavour of being cooperative should be integrated into its logic, so that its self-managing nature could not be perceived as posing any preclusion in this respect.

Nonetheless, there is little wonder that, at least at first sight, should both the terms of autonomics and cooperation be subject to a joint treatment, they could be, somewhat, perceived as if they were stemming from two disjoint perspectives of being mutually exclusive. To be able to account for such a duality, one could consider two fairly contradictory connotations, especially should it be possible to imagine that a preselected set of cooperative network nodes might form a unified entity. On the one hand, this entity could be understood as an atomic embodiment of all those network nodes in the sense that there should not exist any contradiction between any two or more of its members. On the other hand, however, should the assumption of atomicity not fully hold, as might be the case in reality, the member network nodes could equally well express egoistic tendencies. Most obviously, such tendencies would not necessarily translate immediately into any intentional or deliberate actions, but, given the rationale behind the functioning of an autonomic system, they could be expected to rather result from a specific design, possibly not being able to handle all the potential behaviour stemming from dissociated policies, imposed either internally or externally. In general, such a connotation shall seem more inherent in artificial-intelligence-orientated approaches, where the system would be expected to reason and take, in some sense, 'learned' decisions by itself. Even though the notion of autonomic intelligence also assumes a certain flavour of cognition, it appears that the primary metaphor of the human autonomic nervous system should take precedence, thereby exposing the drive to achieve a commonly constructive state of global equilibrium.

As such, the instantiation of the Autonomic Cooperative Node, being one of the core functionalities of the entire Autonomic Cooperative Networking Architectural Model, definitely requires certain extensions to be proposed on top of the workings of

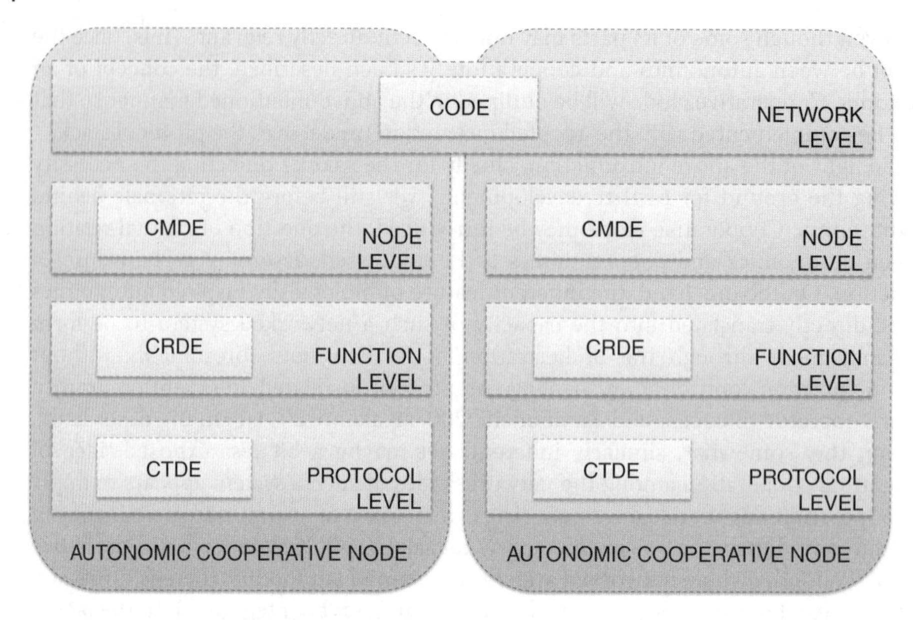

Figure 3.17 ACNs from a level-driven orthogonal perspective.

the previously introduced elements of the Generic Autonomic Network Architecture. One such extension, Autonomic Cooperative Behaviour (ACB), becomes of special interest at this stage of the analysis because of its becoming directly responsible for enforcing the upgrade of the legacy autonomic node to the Autonomic Cooperative Node in question. In order to create a fully comprehensive context, however, one may need to go back to the foundations of the Generic Autonomic Network Architecture itself, where the ability to perform the key task of auto-discovery, possibly spanning multiple levels of abstraction, as only generally touched upon by Meriem et al. (2016), appears to be critically instrumental in incorporating the notion of cooperation into the global picture of autonomics. It is claimed to be so, since the performance of such a task is predominantly related to the possibility of answering the question of whether the target Autonomic Cooperative Node should operate in its fully-fledged and newly designed version, as it is yet to be outlined, or maybe it should fall back to, at least from the perspective of this book, the already outdated autonomic node related mode of operation due to backwards compatibility justified reasons.

Moving into the details, the proposed conceptual outline of an Autonomic Cooperative Node is depicted in Figure 3.17,[18] where a set of comprehensively drafted and mutually aligned decision elements of relevance is presented.[19] Not only should these decision elements be perceived as instrumental in the functioning of the entire Autonomic Cooperative Networking Architectural Model, but the existence and scope of the previously introduced levels of abstraction, inherent in the Generic Autonomic

18 Such a construction may also be referred to as an autonomic cooperative set (ACS), though one should note that in general the size of the latter is by no means limited to two Autonomic Cooperative Nodes.
19 One should note that the presented concepts will become relevant not only to future internet (FI) related developments (Wódczak, 2010), but also to emergency communications networks (ECNs) (Wódczak, 2012b) and vehicular ad hoc networks (VANETs) (Li *et al.* 2012).

Network Architecture, should remain virtually unaltered (Wódczak, 2012a). In particular, perceiving the entire setup from the usually assumed bottom-up perspective, first comes the protocol level where the cooperative transmission decision element (CTDE) is deployed with the intention of combining the equivalent virtual multiple-input multiple-output radio channel inherent in the physical layer with the distributed space-time block coding belonging to the link layer, so that the above-mentioned ACSs may be created, where the instantiation of cooperative transmission is assumed to be taking place. Next, in the upward direction, is the cooperative re-routing decision element (CRDE) of the function level, responsible mainly for providing additional dependability. This is achieved through interaction with two other decision elements of relevance in the form of the interrelated resilience and survivability decision element (RSDE) and fault management decision element (FMDE), both being able to jointly identify the root causes and symptoms to infer that a physical or logical system failure may be imminent (Wódczak, 2014). Consequently, the pertinent cooperative re-routing procedure may be triggered well ahead of any accordingly foreseeable failure to ensure that overall service continuity is properly guaranteed.

Progressing further towards the node level, one may clearly discern expansion of complementary decision-making logic with special focus on the incorporation of the management of a cross-layer driven interaction between the said distributed space-time block coding of the link layer and the multi-point relay (MPR) station selection heuristics of the Optimised Link State Routing (OLSR) protocol residing at the network layer for the needs of guaranteeing that the Autonomic Cooperative Nodes of interest may become properly assigned to their most respective autonomic cooperative sets. Last, but not least, comes the top-most cooperation orchestration decision element (CODE) located at the network level, responsible for overseeing the ACB-related interactions under the umbrella of the Autonomic Cooperative Networking Architectural Model. Given the above context of the presented entities, one should keep in mind that such an overlay stemming from the Generic Autonomic Network Architecture should be perceived as perpendicular to the protocol stack based on the Open Systems Interconnection Reference Model. It is deemed to be so since the role of said overlay should consist in interacting with and steering various protocols of the pertinent stack, while communication between or among various Autonomic Cooperative Nodes would still need to follow the typical encapsulation process performed over that stack, as generally outlined in Figure 3.18, in the light of Wódczak (2012a).

Following the above assumptions, as well as taking into account the nature of wireless communications, should a given source node (SN) decide to transmit data packets towards a chosen destination node (DN), possibly located two hops away, such a transmission would most obviously be heard not only by the intermediate Autonomic Cooperative Nodes, presumably intending to instantiate Autonomic Cooperative Behaviour, but also by the other neighbours residing within the same range.[20] To reduce protocol overhead, those Autonomic Cooperative Nodes not preselected to cooperate should silently discard any such received packets. The relevant operation would already be performed at the link layer, taking the responsibility of encoding

20 In general, both the SN and the DN may also be perceived as Autonomic Cooperative Nodes. In the context of this example, however, they would work rather in a standalone mode, not necessarily displaying their cooperative capabilities.

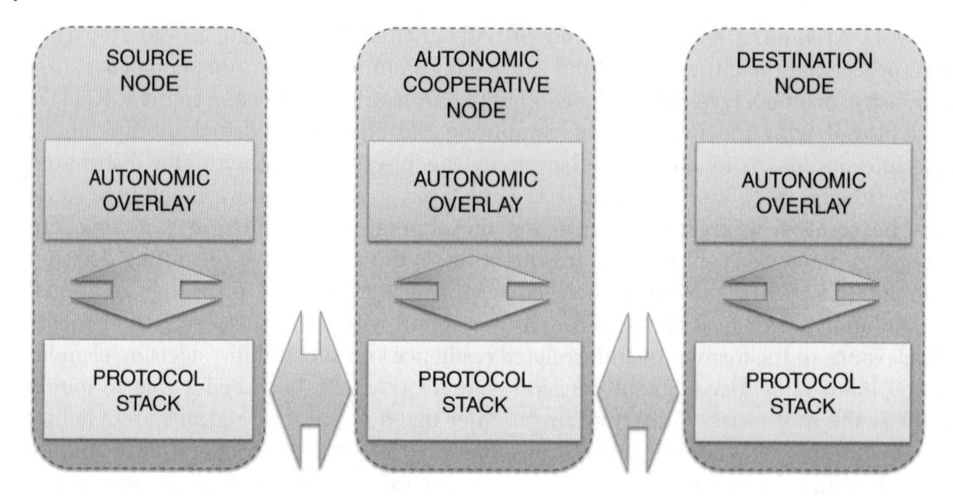

Figure 3.18 Interaction over protocol stack.

and encapsulating the packets coming from the network layer into link layer frames, as well as performing proper physical addressing, thereby making communication with the specified one-hop neighbour or neighbours possible (Wódczak, 2014). Finally, each link layer frame would be encapsulated into a physical layer frame, and, after having been modulated, physically transmitted over the wireless medium. Consequently, the dynamic composition of Autonomic Cooperative Behaviour using the cross-layer-orientated Autonomic Cooperative Networking Protocol would be intended to minimise the probability of occurrence of any undesirable propagation effects, thereby making the process of communication efficient and uninterrupted, as indicated by Wódczak (2011b). Since the protocol stack would need to be implemented by every Autonomic Cooperative Node, the orchestration shall take place at the level of the overall Autonomic Cooperative System Architectural Model, respectively.

3.4.2 Cooperative Transmission Decision Element

Having established the detailed rationale behind the Autonomic Cooperative Node, according to which its management takes place at all the levels of abstraction belonging to the Generic Autonomic Network Architecture, it becomes natural to provide an incrementally developing insight into the specific levels. Given the generally followed bottom-up approach, the CTDE, residing in the lowest protocol level, is going to be examined in the present chapter with special emphasis on its internal logic, in the first order, leaving the architectural integration aspects for later consideration, scheduled to follow closely behind. Such a pattern will be repeated in the upcoming chapters so that, eventually, a comprehensively composed and fully-fledged instantiation of the Autonomic Cooperative Networking Architectural Model may be presented. Going back to the CTDE, however, before its internal logic can be outlined it is important to account for the differences in the semantics of the word 'protocol', not only when perceived from the internally disjoint perspectives of the layers of the Open Systems Interconnection Reference Model, but also keeping in mind one of the levels of the Generic Autonomic Network Architecture. Next, the internal logic will be explored through the workings of the equivalent virtual multiple input multiple output radio channel of the physical layer in the context of the distributed space-time block coding of the link layer, yet without any function level orchestration. Finally, certain performance-related conclusions will be drawn on the basis of the obtained simulation evaluation results (Wódczak, 2012a).

Looking at the semantics, one can see that the bottom-up approach may, somewhat unintentionally, appear to create a conceptual mismatch, especially given the fact that the CTDE of the protocol level is apparently bound to interact directly with the physical layer. To account for such a situation it is necessary to understand that, even though, on the protocol side, the pertinent functionality of the distributed space-time block coding belongs much more to the link layer, the previously explained modulation-driven nature of the general version of space-time block coding obviously calls for substantial involvement of the physical layer, too. What is more, however, there is also the question of the above-mentioned internally disjoint perspectives of the layers of the Open Systems Interconnection Reference Model itself, stemming from the difference between the notion of link layer and network layer protocols. In other words, while the distributed space-time block coding undoubtedly falls into what is most often categorised as cooperative relaying protocols, a much more prudent naming approach would rather involve the notion of a 'scheme', leaving 'protocol' where it should more naturally belong.[21] As the above issues are considered deeply in the chapters to follow, it may transpire that the overall Autonomic Cooperative Networking Architectural Model will be strongly cross-layer driven, starting with ACSs, intended to bind both the physical layer and link layer, to arrive at the Autonomic Cooperative Networking Protocol, aiming to synergise both the link layer and network layer, yet to much higher an extent

The CTDE is assumed to be responsible for the orchestration of two-hop cooperative transmission on the path from the source node (SN), denoted as x, towards the destination node (DN), denoted as $n^{(2)}$, with the support of a set of Autonomic Cooperative Nodes, denoted as $ACS(x, n^{(2)})$ and located between them (Wódczak, 2014).[22] In particular, a set of channel coefficients, denoted as $C(n^{(2)})$, is maintained for all the radio links between the Autonomic Cooperative Nodes, conceptually belonging to the one-hop neighbourhood of x, and $n^{(2)}$, located in the two-hop neighbourhood of the same x (Wódczak, 2012a). The parameters of these radio links are continually monitored within the hierarchical autonomic control loop pertinent to the CTDE, so that they may be verified against a preset threshold value, denoted as β. Consequently, and taking into account any policies imposed, the CTDE is responsible for the decision whether the cooperative transmission to take place over the equivalent virtual multiple-input multiple-output radio channel should be assisted by either three or four intermediary Autonomic Cooperative Nodes, acting according to either the G_3 or G_4 code matrix. Other schemes could obviously also be applied, yet these two have been chosen to allow immediate comparison of the results, as each of them is characterised by the same code rate of $\frac{1}{2}$ (Tarokh et al., 1999a). Clearly, $ACS(x, n^{(2)})$ constituted in this way should be perceived as a group of Autonomic Cooperative Nodes expressing Autonomic

21 The complementary perspective of the role and place of a protocol will be further elaborated on and discussed in the next chapter.

22 While a two-hop case is investigated, the reader should note that, theoretically, a multi-hop setup should by no means be excluded. Given the context of the physical layer, however, it becomes conspicuous that, in practice, no sooner than at the network layer could one realistically think about such challenges. This is more or less the approach assumed by the author, where the entire concept scales in the upward direction.

Cooperative Behaviour through the instantiation of the distributed space-time block coding (Wódczak, 2014).

The logic of the CTDE may be summarised as adapted in Algorithm 3.1.[23] In particular, the G_3 mode becomes selected should the minimum value of the moduli contained in the $ACS(x, n^{(2)})$ set turn out lower than the threshold β (Wódczak, 2012a). As a result, one of the four allowable Autonomic Cooperative Nodes may no longer remain in the ACS, so that only the other three could still be involved in cooperative transmission. Otherwise, all four Autonomic Cooperative Nodes would be used, thereby making the distributed space-time block coding operate in G_4 mode, as outlined by Wódczak (2011a). In this way, the worst radio link would be autonomically discarded until its parameters, monitored within the hierarchical autonomic control loop, once again meet the selection criterion of the logic behind the CTDE. However, one should also keep in mind that the monitoring data pertinent to such radio links may equally well be overridden by certain policies imposed by a higher-level decision element, such as the cooperation management decision element to be yet analysed. Should this happen, this type of situation would be the most representative manifestation of the cross-layer nature of the entire Autonomic Cooperative Networking Architectural Model, as the above link layer could become notified by the network layer that, for example, for quality of service reasons, it would be better to move one of the Autonomic Cooperative Nodes to another ACS, of which there may be many, to increase the overall routing efficiency.

Algorithm 3.1 Logic of CTDE.

```
1: n = min(C(n^(2)))
2: if C(n^(2))[n] < β then
3:     ACS(x, n^(2)) ← ACS(x, n^(2))\{n}
4:     DSTBC ← G_3
5: else
6:     DSTBC ← G_4
7: end if
```

The evaluation results were achieved assuming a special configuration of an EVMIMO[24] flat fading Rayleigh channel with a single receiving antenna at the DN, where the channel coefficient for each of the links between a given Autonomic Cooperative Node and a DN is calculated according to the previously introduced simulation model (Zheng and Xiao, 2003). Moreover, the quadrature phase-shift keying (QPSK) modulation scheme was employed and the power emitted by each of the entitled Autonomic Cooperative Node was always normalised so that the overall transmitted power would amount to unity, while the received signal was perturbed by additive white Gaussian noise characterised by a zero mean and $N_0/2$ variance per dimension. The results for $\beta = 0.8$ in comparison with the reference curve for a regular G_4 code matrix are presented in Figure 3.19, where a gain of about 1 dB becomes achievable. Detailed results pertaining to the bit error rate improvement for a specific value of

23 For additional information the reader is also referred to the source concept, at that time still nonautonomic, in Wódczak (2005).

24 Such a configuration could also be referred to as an equivalent virtual multiple-input single-output (EVMISO) channel.

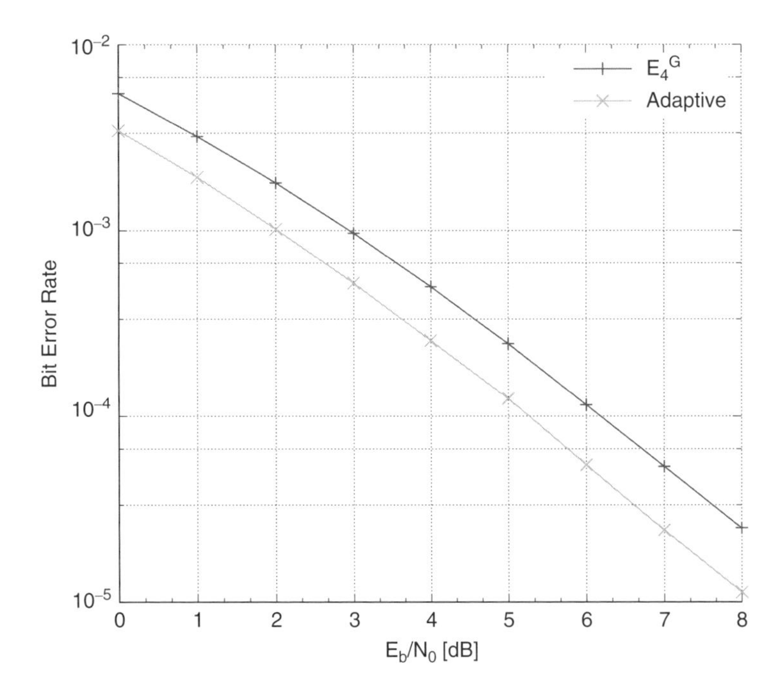

Figure 3.19 CTDE versus G_4 system for $\beta = 0.8$.

the E_b/N_0 ratio, given a certain threshold β, are presented in Figure 3.20.[25] In fact, the improvement resulting from the operation of the CTDE may be observed regardless of the E_b/N_0 ratio, while the optimum region seems to fall into the range delimited by $\beta = 0.5$ and $\beta = 0.8$. Finally, the percentage of G_3 encoder usage is presented in Figure 3.21, being about 2% for $\beta = 0.1$ and about 95% for $\beta = 1.2$. This means that in the former case autonomic cooperative transmission (ACT) works almost all the time in the G_4 mode, while in the latter case the three best radio links are mostly selected and the signal is processed in accordance with the G_3 code matrix.

3.4.3 Architectural Integration Aspects

Apparently, as it might have already transpired, due to certain semantic and functional differences between its building blocks, the conceived blueprint of the Autonomic Intelligence Evolved Cooperative Networking (AIECN) is, to a great extent, supposed to be based on cross-layer-driven approaches. In fact, the word 'based' appears to be of crucial importance here, as the cross-layered nature may clearly refer to the OSI-RM-rooted components only, while the overlaps with the workings of the Generic Autonomic Network Architecture definitely fall under a different category with a more cross-systemic connotation. For this reason, the analysis of the architectural integration aspects will need to encompass both these dimensions to provide the most comprehensive conceptual design, where all the potentially outstanding notions are clearly categorised. As the same pattern is going to be repeated throughout the

25 For additional results, covering the odd E_b/N_0 values, the reader is also referred to Figure A.6.

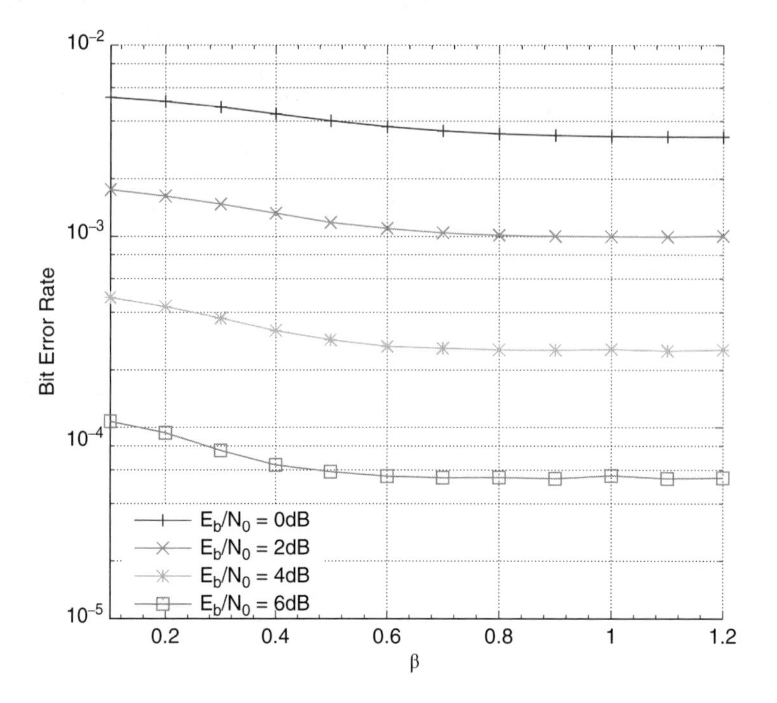

Figure 3.20 CTDE in relation to β and E_b/N_0.

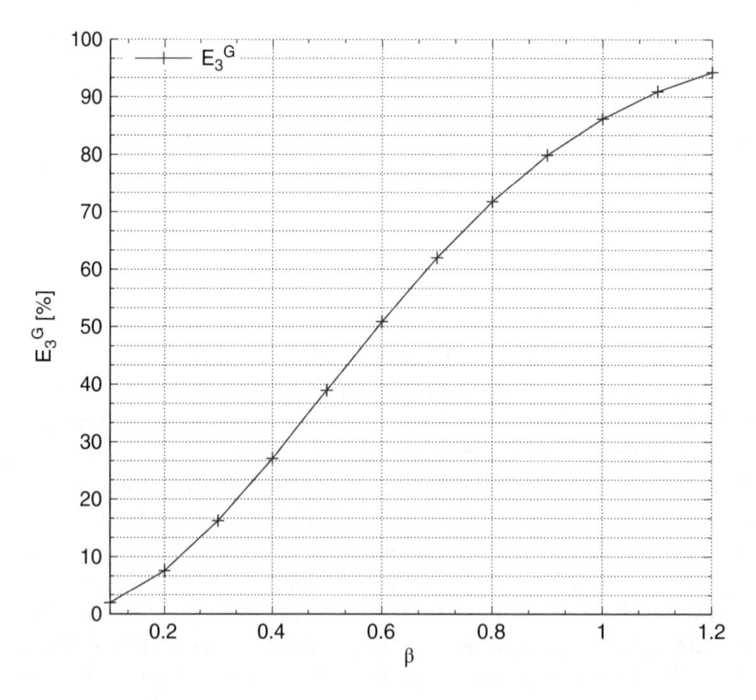

Figure 3.21 E_3^G usage percentage.

remaining chapters, at a given stage of advancement the considerations will usually be mostly limited to involving the routines of both the immediately higher layer of the Open Systems Interconnection Reference Model and the upper level of the Generic Autonomic Network Architecture. Yet, certain building blocks residing over there may equally well span even further, as is the case, for instance, for the Autonomic Cooperative Behaviour. Thus, the proper context for analysis must first be established, so that it is then possible to account for relevant architectural considerations, making it more straightforward to conclude with an outlook on the accommodation of the routines to come on top at later stages of analysis.

In order to anticipate certain presentation-related challenges expected to appear at this lowest part of the entire incremental design, to be further rolled out sequentially as the plot develops, one should note that the previously introduced split into Vertical Technological Pillars (VTPs) and Horizontal Architectural Extensions (HAEs) resulted in a certain conceptual perpendicularity or orthogonality between the layers of the Open Systems Interconnection Reference Model and the levels of the Generic Autonomic Network Architecture. In fact, such a mutually aligned or modified orientation is expected to prove to perform well in all those joint analyses, where both the layers and the levels will need to appear together under the umbrella of a comprehensive graphical representation. On the other hand, however, it would be rather awkward to maintain such an approach when only one of the dimensions is to be scrutinised. Such a limitation would not apply to the horizontally allocated levels, as this is how they are positioned within the description of the Generic Autonomic Network Architecture (ETSI-GS-AFI-002, 2013). Yet this could clearly interfere with the normally horizontal configuration of the layers, which apparently are never arranged otherwise, as presented, for example, by Tanenbaum and Wetherall (2011). For this reason, the modified representation is to be applied to cases where a two-dimensional drawing space is to be used for visualisation purposes. Otherwise, regardless of whether layers or levels are referred to, a horizontal layout is always followed so that the above justification may be fully exercised.

In fact, the approach explained above is immediately applied to the lowest-level discussion pertaining to the relation between routines of the physical layer and the link layer, as depicted in Figure 3.22. Given the overall context established for the CTDE, the

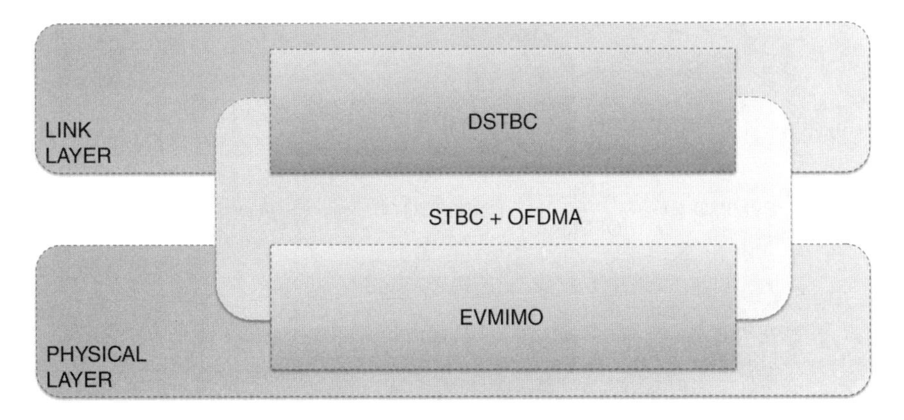

Figure 3.22 Dependencies between the physical layer and link layer routines.

relation in question most apparently boils down to the place and role of the distributed space-time block coding. This technology, to be examined in detail in the next chapter, when aspects more directly pertinent to the link layer are to be discussed, should be perceived as a kind of a protocol, or, following the previous discussion, rather a scheme, intended to orchestrate the factual cooperative processes ongoing at the physical layer. Not surprisingly, it may gradually transpire that this type of process should be instrumental in accounting for a certain duality of distributed space-time block coding, manifesting itself through the fact that it stems directly from space-time block coding. Following this line of thought, one may quickly stumble across what has also already been mentioned: that there is seemingly much more to space-time block coding in terms of its resemblance to a modulation than a coding technique. Based on this, it becomes more and more apparent that space-time block coding along with orthogonal frequency-division multiple access becomes a kind of a linking or glueing entity in this respect, overlapping with the distributed space-time block coding of the link layer and delving into the equivalent virtual multiple-input multiple-output radio channel of the physical layer. Specific interactions involving both the layers and levels are outlined in Figure 3.23.

Given the above, as well as referring to the overview of the Autonomic Cooperative Networking Architectural Model outlined in Figure 2.21, one may extract fragments thereof for the sake of elaborating on its upgraded workings even further. As such, the information related to the application of the most appropriate space-time block coding scheme, or, as it is yet to be introduced, the most proper operation mode of the so-called equivalent distributed space-time block encoder (EDSTBE), just to follow Wódczak (2014), is passed to the distributed space-time block coding entity embedded in the link layer and positioned in such a way that the cooperative transmission scheme may be executed properly across all the Autonomic Cooperative Nodes to be potentially involved. One should note that proper arrangement is key in this respect, since the

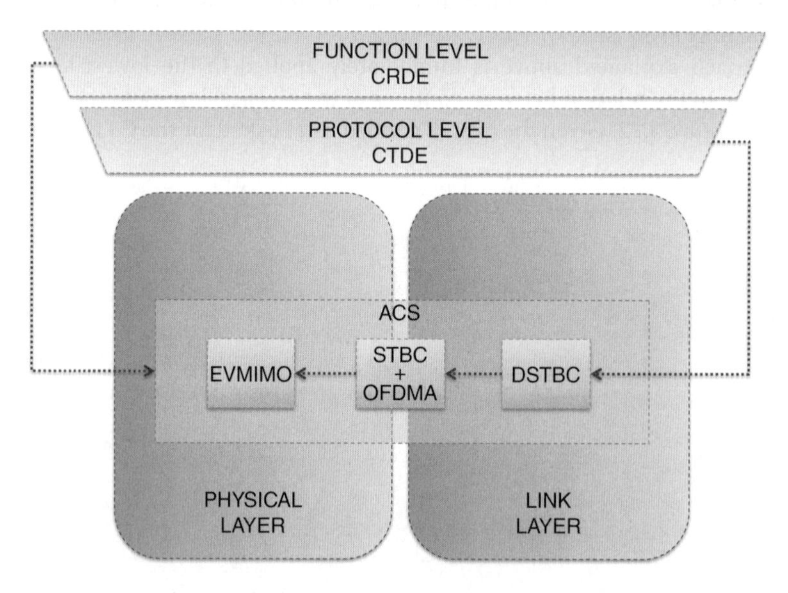

Figure 3.23 Architectural relations stemming from the physical layer and protocol level.

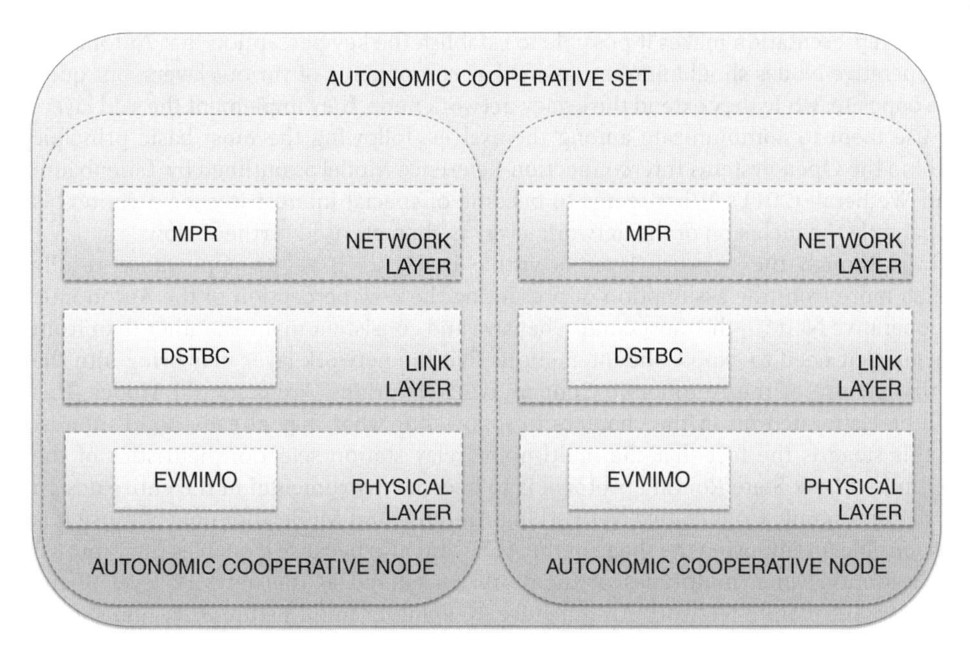

Figure 3.24 ACNs from the layer-driven orthogonal perspective.

described operation does not yet go beyond the link layer, where orchestration possibilities are not as extensive as at the network layer. In other words, the routing information available at the latter is undoubtedly driving the behavioural frames of the former, yet specific decisions related to the distributed space-time block coding remain at the layer to which they directly belong. Thus, the Autonomic Cooperative Nodes belonging to the same ACS need to apply the appropriate parts of their space-time block coding scheme by following its modulation-like pattern, denoted by the assigned column of its matrix, to perform cooperative transmission over the equivalent virtual multiple-input multiple-output radio channel under the assumption of tight synchronisation guaranteed by the underlying orthogonal frequency-division multiple access.[26]

In particular, considering an example ACS composed of two Autonomic Cooperative Nodes, as depicted in Figure 3.24, it is possible to obtain a better insight into the above-mentioned mutually orientated relations. This appears to be especially so should an attempt be made to perceive such an ACS from the perpendicular or orthogonal perspective, where the layers of the Open Systems Interconnection Reference Model rather than the levels of the Generic Autonomic Network Architecture are involved. Consequently, once again the horizontal approach may be followed in general, since only one dimension is to be presented. In order to be even more precise, one should avoid becoming misled by the notions of ACN or ACS, clearly referred to in Figure 3.24, since they do not belong to the Generic Autonomic Network Architecture, but are rather inherent in the Autonomic Cooperative Networking Architectural Model, which makes them agnostic to the question of graphical orientation. Nonetheless,

26 Keeping in mind the nowadays prevailing role of approaches driven by orthogonal frequency-division multiplexing, one may also refer to Zhang *et al.* (2016), where the most recent advancements in waveform design are examined and future developments anticipated.

such a representation makes it possible to establish the key perception that Autonomic Cooperative Nodes should not be classified as components of various layers, but quite the opposite: while they extend the legacy network node, they implement the said layers to use them to communicate among themselves, following the most basic principle behind the Open Systems Interconnection Reference Model as outlined by Tanenbaum and Wetherall (2011). What seems to become of special interest in such a context is apparently the inclusion of the network layer, as characterised further below.

Even though the network layer is visible in Figure 3.24, its appearance results much more from the assumption of presenting the new perception of the Autonomic Cooperative Node in the most comprehensive and consistent manner, rather than from any explicit need to expose the interactions that the network layer may bring into the global picture of the Autonomic Cooperative Networking Architectural Model. This will be elaborated on in the chapters to follow, but what may already be of interest at this stage is the fact that the multi-point relay station selection heuristics of the Optimised Link State Routing protocol is to become instrumental in the entire design of the Autonomic Cooperative Networking Architectural Model, not only because it is responsible for imposing the shape of the ACSs, but also because it enables their mutual orchestration when multitudes of them are considered at the network level of the Generic Autonomic Network Architecture. Keeping in mind Figure 3.23, one should also note that the roles of both the cooperative transmission decision element and cooperation management decision element are pivotal in this respect. While the former is responsible for a more direct interaction with the distributed space-time block coding, the latter rather attempts to orchestrate the ACS with its Autonomic Cooperative Nodes. All in all, the above architectural integration aspects will undoubtedly play a crucial role in the following chapters, paving the way for the rollout of various entities of the Autonomic Intelligence Evolved Cooperative Networking.

3.5 Conclusion

In this chapter the rollout of specific architectural considerations was commenced within the context set by the previous chapter. In particular, the presentation started with the foundations of the protocol level spatio-temporal processing, where the initial emphasis was laid on developments related to the multiple-input multiple-output channel to provide a good understanding of its workings. Then, its diversity-rooted origins were discussed, so that it became possible to clearly justify the role of and the necessity for the later deployment of spatio-temporal processing. Moreover, the question of radio channel virtualisation was addressed, where the singular-value decomposition theorem was explained in order to introduce the notion of an equivalent virtual multiple-input multiple-output radio channel to be deployable among Autonomic Cooperative Nodes. In this respect, the radio channel capacity was brought into the bigger picture of the opening analysis to account for its linear scaling with the number of generic transmitters or generic receivers. Finally, a specific external model for radio channel coefficient calculation, to be referred to throughout this book, was described, as well as the difference between coding gain and diversity gain was addressed. Given such a context, the focus advanced more towards space-time coding techniques, to account for their superiority over the above-mentioned diversity techniques, and to prepare the ground for their later elevation towards networked configurations, where the concept of distributed space-time block coding will prevail.

In particular, the most baseline approach to space-time coding was presented with a special emphasis on space-time block coding, where the question of its being perceived more as a modulation than a coding technique was touched upon. Then, the derivation of the decoding metrics for a selected set of space-time block coding matrices was outlined with the aim of clarifying certain inconsistencies the author came across in the source materials. Based on this, an extension towards space-time trellis coding was also presented, where additional coding gain became obtainable. Eventually, after all the aforementioned technological aspects were analysed, their mutual relation with the protocol level control logic was discussed for architectural integration purposes. To this end, the notion of an Autonomic Cooperative Node was introduced as one of the major building blocks of the proposed concept. On this occasion, not only was the relation between autonomics and cooperation discussed, but the internal structure of the Autonomic Cooperative Node was scrutinised. Then, the cooperative transmission decision element was brought into the global analysis picture as belonging to the protocol level, while being mostly responsible for the interaction with the routines of the physical layer. Given such a context, not only was the concept of a protocol discussed, but certain adaptive logic was presented whereby the relevant code matrices are switched on the basis of the radio channel parameters. Finally, all the pertinent architectural integration aspects were outlined in preparation for further extensions.

References

Alamouti S 1998 A simple transmit diversity technique for wireless communications. *IEEE Journal on Selected Areas in Communications* **16**(8), 1451–1458.

Alamouti S, Tarokh V, and Poon P 1998 Trellis-coded modulation and transmit diversity: Design criteria and performance evaluation. *IEEE International Conference on Universal Personal Communications, ICUPC*, pp. 703–707.

Bicocchi N and Zambonelli F 2007 Autonomic communication learns from nature. *IEEE Potentials* **26**(6), 42–46.

Bjornson E, Larsson E, and Marzetta T 2016 Massive MIMO: Ten myths and one critical question. *IEEE Communications Magazine* **54**(2), 114–123.

Carvalho E, Bjornson E, Sorensen J, Popovski P, and Larsson E 2017 Random access protocols for massive MIMO. *IEEE Communications Magazine* **55**(5), 216–222.

Chae CB, Forenza A, Heath R, McKay M, and Collings I 2010 Adaptive MIMO transmission techniques for broadband wireless communication systems. *IEEE Communications Magazine* **48**(5), 112–118.

Doppler K, Redana S, Wódczak M, Rost P, and Wichman R 2009 Dynamic resource assignment and cooperative relaying in cellular networks: Concept and performance assessment. *EURASIP Journal on Wireless Communications and Networking* **2009**(475281), 1–14.

ETSI-GS-AFI-002 2013 *Autonomic network engineering for the self-managing Future Internet (AFI); Generic Autonomic Network Architecture (An Architectural Reference*

Model for Autonomic Networking, Cognitive Networking and Self-Management). ETSI Group Specification.

Foschini GJ and Gans MJ 1998 On limits of wireless communications in fading environment when using multiple antennas. *Wireless Personal Communications* **6**, 311–335.

Goldsmith A 2005 *Wireless Communications.* Cambridge University Press.

Gong Y and Letaief K 2002 Concatenated space-time block coding with trellis coded modulation in fading channels. *IEEE Transactions on Wireless Communications* **1**(4), 580–590.

Jankiraman M 2004 *Space-Time Codes and MIMO Systems.* Artech House.

Jewel M and Rahman M 2009 Full rate general complex orthogonal space-time block code for 4-transmit antenna. *Fifth International Conference on Wireless Communications, Networking and Mobile Computing.*

Jung TJ, Chae CH, and Lim HS 2008 Full-rate full-diversity space-time block codes for any odd number of transmit antennas. *IET Communications* **2**(9), 1205–1212.

Larsson E and Stoica P 2003a Mean square error optimality of orthogonal space-time block codes. *Proc. IEEE International Conference on Communications, ICC* pp. 2272–2275.

Larsson EG and Stoica P 2003b *Space-Time Block Coding for Wireless Communications.* Cambridge University Press.

Li J, Wódczak M, Wu X, and Hsing T 2012 Vehicular networks and applications: Challenges, requirements and service opportunities. *International Conference on Computing, Networking and Communications (ICNC),* Maui, Hawaii, USA.

Liu KH and Lin P 2015 Toward self-sustainable cooperative relays: State of the art and the future. *IEEE Communications Magazine* **53**(6), 56–62.

Lozano A, Farrokhi FR, and Valenzuela RA 2001 Lifting the limits on high-speed wireless data access using antenna arrays. *IEEE Communications Magazine* **39**(9), 156–162.

Meriem TB, Chaparadza R, Radier B, Soulhi S, Lozano-Lopez JA, and Prakash A 2016 *GANA: Generic Autonomic Networking Architecture.* ETSI White Paper.

Molisch A and Win M 2004 MIMO systems with antenna selection. *IEEE Microwave Magazine* **5**(4), 46–56.

Murthya N and Gowrib S 2012 Full rate general complex orthogonal space-time block code for 8-transmit antenna. *International Workshop on Information and Electronics Engineering (IWIEE).*

Nakano T 2011 Biologically inspired network systems: A review and future prospects. *IEEE Transactions on Systems, Man, and Cybernetics Part C: Applications and Reviews (IEEE Transactions on Human-Machine Systems)* **41**(5), 630–643.

Phan H, Duong TQ, Zepernick HJ, and Tsiftsis TA 2013 Distributed orthogonal space-time block coding in wireless relay networks. *IET Communications* 7(16), 1825–1835.

Shannon CE 1948 A mathematical theory of communication. *Bell System Technical Journal* **27**, 379–423 and 623–656.

Su W and Xia XG 2003 Two generalized complex orthogonal space-time block codes of rates 7/11 and 3/5 for 5 and 6 transmit antennas. *IEEE Transactions on Information Theory* **49**(1), 313–316.

Tanenbaum A and Wetherall D 2011 *Computer Networks.* Prentice Hall.

Tarokh V, Jafarkhani H, and Calderbank AR 1999a Space-time block codes from orthogonal designs. *IEEE Transactions on Information Theory* **45**(5), 1456–1467.

Tarokh V, Jafarkhani H, and Calderbank AR 1999b Space-time block coding for wireless communications: Performance results. *IEEE Journal on Selected Areas in Communications* **17**(3), 451–460.

Tarokh V, Seshadri N, and Calderbank AR 1998 Space-time codes for high data rate wireless communication: Performance criterion and code construction. *IEEE Transactions on Information Theory* **44**(2), 744–765.

Tarokh V, Seshadri N, and Calderbank AR 1999c Space-time codes for high data rate wireless communication: Performance criteria in the presence of channel estimation errors, mobility, and multiple paths. *IEEE Transactions on Communications* **47**(2), 199–207.

Telatar IE 1999 Capacity of multi-antenna Gaussian channels. *European Transactions on Telecommunications* **10**(6), 585–595.

Viterbi AJ 1967 Error bounds for convolutional codes and an asymptotically optimum decoding algorithm. *IEEE Transactions on Information Theory* **13**(2), 260–269.

Vucetic B and Yuan J 2003 *Space-Time Coding*. Wiley.

Wesołowski K 2002 *Mobile Communications Systems*. Wiley.

Wódczak M 2005 On the adaptive approach to antenna selection and space-time coding in the context of relay-based mobile ad hoc networks. *XI National Symposium of Radio Science URSI*, Poznań, Poland, pp. 138–142.

Wódczak M 2007 Extended REACT: Routing information enhanced algorithm for cooperative transmission. *16th IST Mobile & Wireless Communications Summit 2007*, Budapest, Hungary.

Wódczak M 2010 Future autonomic cooperative networks. In *Mobile Networks and Management* (Lecture Notes of the Institute for Computer Sciences, Social Informatics and Telecommunications Engineering), ed. Pentikousis K, Aguero R, Garcia-Arranz M, and Papavassiliou S, Springer.

Wódczak M 2011a Autonomic cooperation in ad hoc environments. *Fifth International Workshop on Localised Algorithms and Protocols for Wireless Sensor Networks (LOCALGOS) in conjunction with IEEE International Conference on Distributed Computing in Sensor Systems (DCOSS)*, Barcelona, Spain.

Wódczak M 2011b Convergence aspects of autonomic cooperative networks. *International Journal of Information Technology and Web Engineering (IJITWE)* **6**(4), 51–62.

Wódczak M 2012a *Autonomic Cooperative Networking*. Springer.

Wódczak M 2012b Cooperative emergency communications. *Sixth International Conference on Next generation Mobile Applications, Services and Technologies (NGMAST)*, Paris, France.

Wódczak M 2014 *Autonomic Computing Enabled Cooperative Networked Design*. Springer.

Zhang X, Chen L, Qiu J, and Abdoli J 2016 On the waveform for 5G. *IEEE Communications Magazine* **54**(11), 74–80.

Zheng YR and Xiao C 2003 Simulation models with correct statistical properties for Rayleigh fading channels. *IEEE Transactions on Communications* **51**(6), 920–928.

4

Function Level Relaying Techniques

4.1 Introduction

Following the previous chapter, where the notion of protocol level spatio-temporal processing was addressed, mostly pertaining to the developments of the physical layer such as the multiple-input multiple-output radio channel, space-time coding techniques, as well as their protocol level control logic, the time has come to move forward towards function level relaying techniques. First, conventional relaying and cooperative relaying are characterised, while the latter is further investigated from the viewpoints of forwarding strategy and protocol nature. This way, amplify-and-forward, decode-and-forward, and decode-and-reencode, as well as fixed, adaptive, and feedback protocol types are described. Then the focus shifts towards the topic of supportive protocols and collaborative protocols, presented as two subcategories of cooperation understood in a wider sense. Based on such a dual perception of what would normally be referred to as a cooperative protocol, it becomes possible to identify that the former can be considered as a preparatory phase for the latter, making the attempt at further classification highly constructive. Last, but not least, the concept of virtual antenna arrays is outlined, under the assumption of applying its most versatile multi-tier version where, based on a generalised cooperative transmission scheme, a special operation mode is presented in the form of distributed space-time block coding, which will play a crucial role in the further developments to be discussed in this book.

Given these classifications, attention is directed towards a fixed deployment concept, where both the previously described conventional relaying and cooperative relaying techniques could be equally applicable, yet the plot is evolved keeping in mind the assumption that a mobile deployment concept will eventually become the subject of subsequent analyses. In particular, first of all, the grid-based Manhattan scenario is presented, where fixed relay nodes are located in such a manner that the pattern formed by their surrounding buildings, on the one hand, becomes critically important for the suppression of unwanted interference, while, on the other hand, makes it literally impossible to instantiate any VAA-aided cooperative relaying that would involve the said fixed relay nodes. Thus, an evaluation is carried out to show that despite these cooperative relaying related limitations certain link layer and network layer performance optimisation is still possible. To this end a specific adaptation strategy is proposed, where the framing structure of the former and the buffer memory of the latter are both optimised on the grounds of a process interaction simulation method, so that it is possible to observe improvement in terms of the attainable packet throughput

Autonomic Intelligence Evolved Cooperative Networking, First Edition. Michał Wódczak.

at the network layer. Finally, a cooperation-enabled indoor scenario in the form of a relay-enhanced cell is analysed in a similar manner, yet with the major emphasis being placed on the link layer aspects. Even though fixed relay nodes are also used here, the scenario will be applied later in the book to address mobile deployment concept analyses.

Moving towards the end of this chapter, the narrative encompasses the function level overlay logic, where the pattern initially introduced in the previous chapter is strictly followed for consistency reasons. To this end, initially, for background setting purposes, yet, in fact, related to the entire book, the roots of Autonomic Cooperative Behaviour are outlined, so that it becomes possible not only to define its role, but also understand its complexity. In particular, its enablers in the physical layer, link layer, and network layer are identified, while major emphasis is laid on the definition of the concept of the equivalent distributed space-time block encoder. Next, the rationale behind the cooperative re-routing decision element is presented, with the utmost attention paid to the fact that, unlike in the author's two previous books, this time it is to reside at the function level and not at the node level. The logic behind cooperative re-routing is introduced, which allows the system to outperform the legacy fast re-routing if proper information is fed from both the fault management decision element and resilience and survivability decision element. Last, but not least, the architectural integration aspects are discussed in an incremental manner, complementing what was written in the previous chapter. This involves the depiction of the general dependencies between or among the routines of the above-mentioned three layers, as well as a more detailed insight into the architectural relations stemming from the link layer and the function level, along with the introduction of an extended version of the Autonomic Cooperative Node.

4.2 Conventional and Cooperative Relaying

4.2.1 Classification of Relaying Protocols

Approaching the classification of relaying techniques from the top-most perspective, one may distinguish between conventional relaying (CNR) and cooperative relaying (COR). As such, the concept of conventional relaying, also known as single-path relaying (SPR),[1] not only forms the basis for more advanced cooperative relaying schemes, but can also be found, for example, in the multi-point relay (MPR) station selection heuristics of the Optimised Link State Routing (OLSR) protocol, which will become of special interest for the overall idea of Autonomic Intelligence Evolved Cooperative Networking. Given the above-mentioned alternative naming pattern, it should be clear that while conventional relaying exploits a singular transmission path only, cooperative relaying uses at least two such paths,[2] either consecutively or in parallel. Apart from

1 One should note that, since both the are mutually independent they shall be used separately, as any collocation thereof would imply a rather tautological notion where the two could be perceived as somewhat mixed together. Thus, from now on, predominantly for consistency reasons, the naming pattern of conventional relaying will be followed because of its prevailing adoption in the literature, yet the interchangeability option will be kept in mind when it comes to the explanation of its workings.

2 As will be outlined in more detail later this chapter, the notion of a path may encompass one, two, or even more consecutive hops between the source node and the destination node, established over the intermediate relay nodes.

its routing-related applications, the concept of conventional relaying appears to be predominantly known from the developments stemming from the so-called Manhattan scenario, to be yet discussed in more detail, where its advantages and disadvantages become rather transparently exposed. The approach of cooperative relaying appears to go way beyond what conventional relaying may offer, leading to the ultimate concept of distributed space-time block coding (DSTBC) enabled virtual antenna arrays. Before these are analysed, however, both cooperative relaying schemes will be scrutinised from the technological viewpoint, so that it is possible to classify them on the grounds of both forwarding strategy and their protocol nature. Only once this has been done will it become possible to advance the discussion even further.

Moving into the details, irrespective aforementioned dual naming pattern, it is no accident that the method of conventional relaying, as depicted in Figure 4.1, is also referred to as a layer-3 decode-and-forward (L3DF) scheme (Zimmermann et al., 2005). This complementary naming option appears to carry a lot of useful characteristic information about the approach itself, since not only does it imply a certain operation mode from the perspective of the Open Systems Interconnection Reference Model, but implicitly it also attempts to provide guidelines on possible limitations related to the non-cooperative nature of the same, as it is discussed further below in terms of the gains it thereby may or may not offer. In particular, the conventional relaying scheme under investigation will be perceived as comprising two disjoint, yet consecutive, transmission phases. During the first phase, the source node (SN) sends the signal it is attempting to convey to its intermediate relay node (RN). The RN would then become bound to fully decode the received signal for regeneration needs, so that it could be encoded again and conveyed to the destination node (DN). Although the presented approach makes it feasible either to reduce the transmitted power or extend the transmission range,

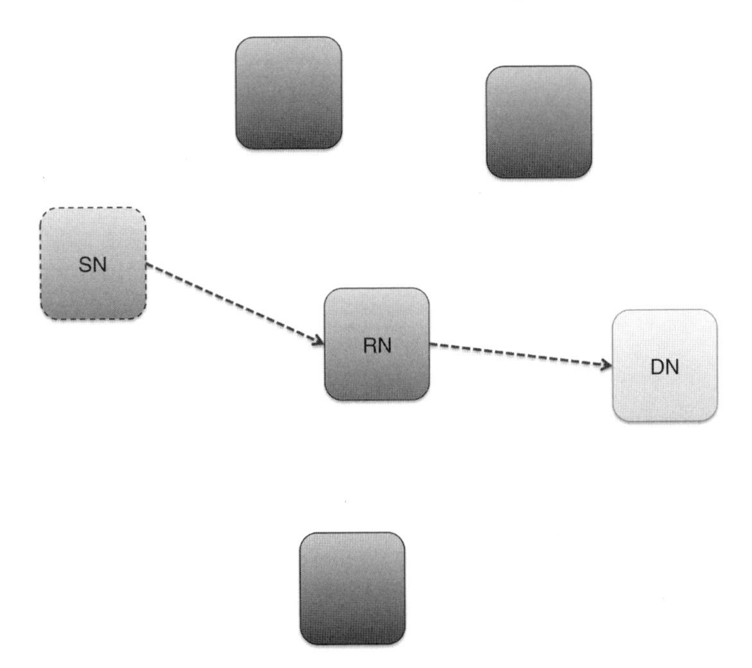

Figure 4.1 Conventional relaying.

it cannot offer any diversity gain due to its inherently noncooperative nature. This is because, as its technical name of L3DF indirectly suggests, the scheme in question does not take advantage of possible enhancements coming from the link layer, thereby remaining limited solely to network layer relaying (Herhold et al., 2005).

For this reason there arose a natural necessity to make up for the above-mentioned deficiency, given the fact that one could take advantage of the otherwise destructive characteristics of the wireless channel, especially visible when singular and nonparallel radio transmission paths are employed, manifesting itself through the introduction of the effect of fading. Consequently, the most baseline method of cooperative relaying is based on such an involvement of an RN[3] to assist the process of transmission between the SN and DN, similarly carried out in two consecutive phases as outlined in Figure 4.2. During the first phase, both the RN and DN receive the transmitted signal; during the second one, the RN may additionally resend its copy towards the DN so that the transmission reliability can possibly be improved thanks to the aforementioned diversity gain. In general, taking into account the work of Laneman et al. (2004), as well as adapting the later classification provided by Zimmermann et al. (2003), and, additionally, extended by Zimmermann et al. (2005) and Herhold et al. (2005), one could categorise the cooperative relaying protocols either with regard to their forwarding strategy or their protocol nature.[4] While the former category would encompass the amplify-and-forward (AF), decode-and-forward (DF), and decode-and-reencode (DR)

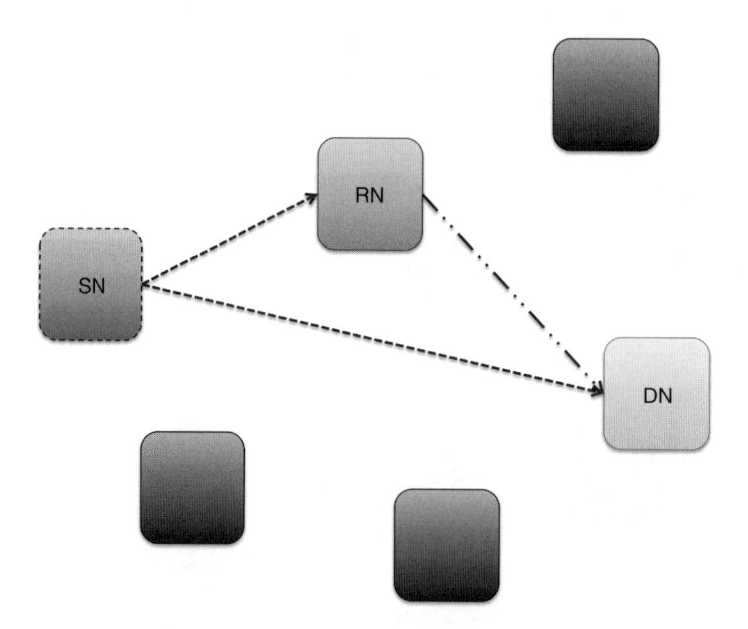

Figure 4.2 Cooperative relaying. Adapted from Zimmermann et al. (2005).

3 One may also involve more than one RN, as we shall see in the case of VAAs.

4 Given the fact that multiple network nodes may be involved in cooperative relaying, there is clearly no denying that it comes at a certain price from the energetic perspective. For a broader discussion of self-sustainability thereof, the reader is referred to Liu and Lin (2015), for instance.

protocols,[5] the latter would include fixed, adaptive, and feedback protocols. In order to provide a better insight into each of these classifications, their pertinent protocols are characterised in additional detail according to category in Tables 4.1 and 4.2.[6]

Examining the forwarding strategy first of all, one immediately comes across the most baseline among the related protocols, the said amplify-and-forward: before a retransmission may take place, a given RN, or a group of RNs in general, must exercise the functionality of a legacy analogue repeater (Pabst et al., 2004). Unfortunately, even though the received signal would become amplified, the noise would obviously also be strengthened, making this approach of less substantial interest. At the same time, the other two approaches in the form of decode-and-forward and decode-and-reencode, would seem

Table 4.1 Forwarding strategy.

Protocol	Description
Amplify-and-forward	Before a retransmission may take place, the device of a relay node, or a group of relay nodes in general, is used to execute the operation of analogue repeating, which is not only known to amplify the received signal, but also to enhance the noise level in the relay node transmission path.
Decode-and-forward	Before a retransmission may take place, the device of a relay node, or a group of relay nodes in general, is used to execute the operation of fully decoding, regenerating, and re-encoding the received signal, which may possibly result in propagation of decoding errors, and, therefore, translate into wrong decisions at the destination node.
Decode-and-reencode	Before a retransmission may take place, the device of a relay node, or a group of relay nodes in general, is used to execute the operation of fully decoding, regenerating, and constructing a new codeword, different from the source one, which may possibly also translate into propagation of errors, yet enables parallel channel coding.

Table 4.2 Protocol nature.

Protocol	Description
Fixed	The device of a relay node, or a group of relay nodes in general, performs the operation of the forwarding of the received signal, possibly after additional processing.
Adaptive	The device of a relay node, or a group of relay nodes in general, performs the operation of autonomously deciding whether to forward the received signal or not.
Feedback	The device of a relay node, or a group of relay nodes in general, performs the operation of assisting the transmission solely when an explicit request from the destination has been received.

5 One should note that, according to Herhold et al. (2005), the AF protocols are also referred to as non-regenerative, while the DF and DR protocols are classified as regenerative.
6 One could also consider the role and place of the seemingly complementary perspective of a user-centric approach to cooperative relaying, as advocated by Jamal and Mendes (2014).

to fulfil their purpose in a much more comprehensive manner. In particular, DF assumes that an RN, or a group of RNs in general, should attempt to fully decode, regenerate, and reencode the received signal before any retransmission may even take place. Quite interestingly, DR goes even further by not only assuming that an RN, or a group of RNs in general, would perform the stages of fully decoding and regenerating the received signal, but, additionally, facilitate the construction of a new codeword, different from the source one. Even though this approach makes it possible to enable advanced channel coding, one should still keep in mind the fact that the propagation of decoding errors is still possible, so one might expect some wrong decisions to be made as a result at the destination (Herhold et al., 2005).

Now looking at the protocol nature, one may begin with the fixed approach, where an RN, or a group of RNs in general, would always, possibly after having performed some relevant processing, forward the received signal. However, given the fact that such behaviour might not always be advisable in a networked system, some additional automation, so necessary for the proper operation of a higher-layer orchestration protocol, is introduced by both the adaptive and feedback schemes. In particular, while the former assumes that an RN, or a group of RNs in general, can autonomously[7] decide whether or not to forward the received signal, the latter requires that the RN, or group of RNs in general, could assist the SN in the process of transmission but only if it should receive an explicit request from the DN, as indicated by Herhold et al. (2005). As may be becoming clear, such adaptive and feedback schemes could possibly efficiently integrate into the overall concept of decision elements, the key components of the Generic Autonomic Network Architecture (Wódczak et al., 2011). For the sake of clarity, one should note, however, that such integration requires that the orchestration to be performed by a relevant decision element would need to take place within a hierarchical autonomic control loop running at the protocol level of the abstraction introduced by the Generic Autonomic Network Architecture, where either the adaptive or the feedback scheme would be overseen as a managed entity, following the analysis of Wódczak (2014).

4.2.2 Collaborative and Supportive Protocols

The most natural advancement of what has been described thus far calls for an attempt to be made at extending the classification of relaying techniques from the viewpoint of the actual meaning of cooperation.[8] To this end, before account can be given of the transition from conventional relaying or single-path relaying to cooperative relaying, it appears necessary to examine the broader context of the mutual relation between the notions of cooperation and collaboration, as well as discuss all the related topics the two might raise. In fact, not only does it seem that collaboration could be perceived as a different form of cooperation, but, additionally, one could also distinguish between two major forms of the latter by naming them as the collaborative and supportive ones. At the same time, it is necessary to keep in mind that such nomenclature could easily become vague or inconsistent should the following definitions not be taken into account. Given such a context, one could notice immediately that, as far as the resulting

7 It by all means appears that an autonomic operation could also apply in this respect.
8 In this respect, one could find it also interesting to scrutinise the role and place of device-to-device (D2D) communications, and network coding (NC) in particular. As this is beyond the scope of this book, the reader is referred, for example, to Pahlevani et al. (2014) for a broader context.

collocation 'supportive cooperation' could in general be considered as being more or less correct from the linguistic perspective, the mere juxtaposition of 'collaborative cooperation' brings the risk of being classified as unacceptably tautologous. Solely for this reason, the analysis begins with a clearer subdivision of cooperation into supportive protocols and collaborative protocols, so that any language-based inconsistencies may be avoided from the very outset. Then, the notion of a protocol is revisited in more detail, as opposed to the notion of a scheme, in order to pave the way for the eventual insight into the resulting question of cross-layering.

In fact, attempting to scrutinise the mutual relation between the semantic fields of the terms cooperation and collaboration, one may easily discern that the two remain in tight relation, as depicted in Figure 4.3. While cooperation is usually defined as a 'joint operation or action', collaboration appears to highlight more of the 'act of working with' other entities, thereby being also prone to be understood along with its negative connotation kept in mind, just to follow the entries of Sinclair (1997). Given such a context, it clearly transpires that a really very thin borderline exists between these notions, which appears to make cooperation slightly more generic and collaboration slightly more specific; in other words, more or less reflecting what would be needed for the proposed classification. Assuming a reverse order, the class of collaborative protocols could encompass all the concepts stemming from cooperative transmission or cooperative relaying rooted in the link layer, where joint processing of the signals broadcast by a given SN, with the intention of having them delivered to the appropriate DN, would be performed by a group of cooperating RNs, or rather Autonomic Cooperative Nodes (ACNs), in accordance with a specific space-time block coding (STBC) matrix. Assuming possibly the most general perspective, such a joint action would mean performing the same activity, or a consistent fragment thereof, through distributed processing of the same data, at the same time, using a shared resource, as projected in Figure 4.4. In other words, unity of action takes precedence as the most important factor, making all the Autonomic Cooperative Nodes work temporarily as a singular and unified entity.

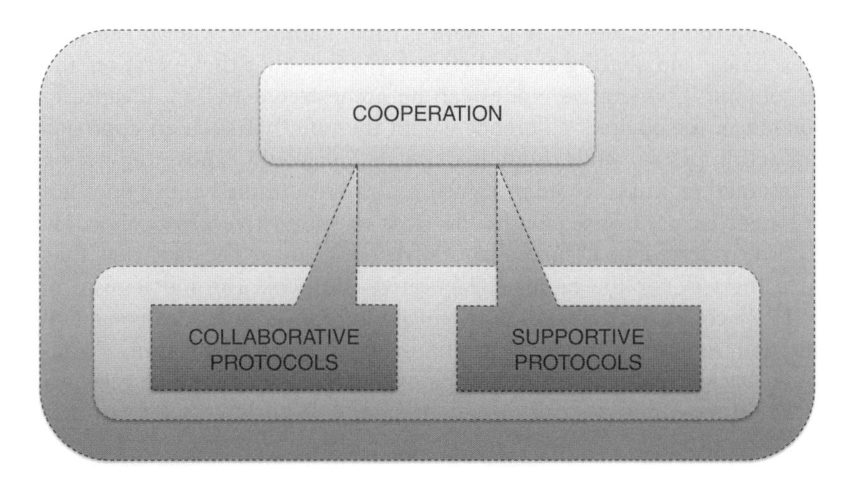

Figure 4.3 Classification of cooperation methods.

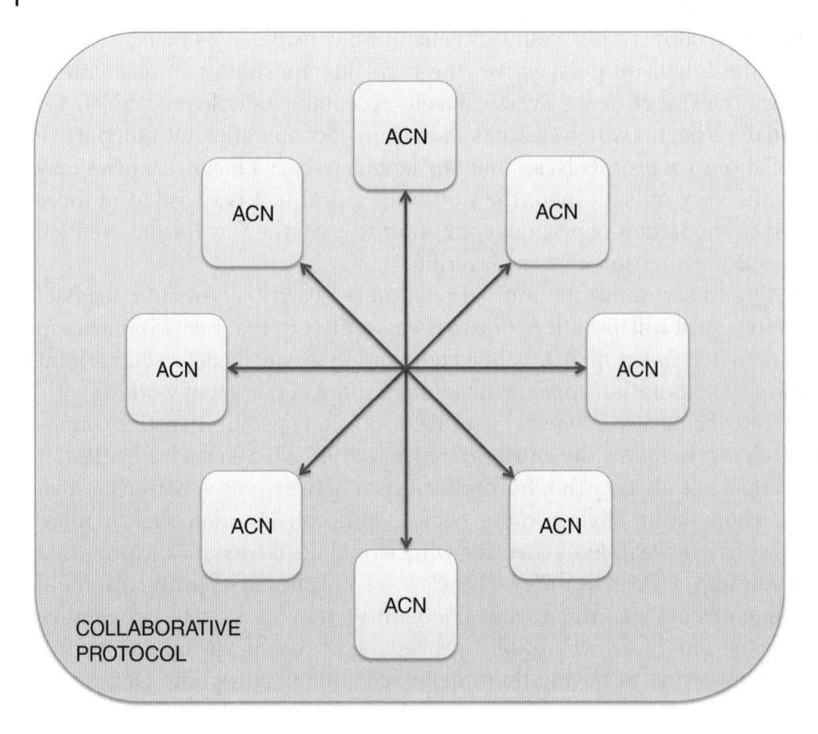

Figure 4.4 Projection of collaborative protocol.

This way, the second class of cooperation, already signalled under the name of supportive protocols, calls for a different, yet somewhat dual, perception, since it is understood as all the activities preceding the above-mentioned collaborative phase. In fact, one could also say that the routines of the network layer would play the key role in this respect, as they may normally feed and equip network nodes with all the cooperation-related data before the collaborative phase begins. As such, even though the described approach may bring some implicit reminiscence of the notion of cooperative routing, the focus is more on highlighting a widespread area of information exchange and sharing at the network layer, so that the operation and behaviour of Autonomic Cooperative Nodes can be orchestrated well in advance of the collaboration stage. An obvious difference would then be that such an approach requires no joint action per se, while, quite the contrary, it would expose the notion of introductory interaction, as projected in Figure 4.5, where a unified entity does not yet exist or no longer exists. Consequently, the class of supportive protocols could then be equally well referred to as precollaborative ones, given the fact that they may only prepare the way for the target collaborative protocol. The mechanism of fast re-routing (FRR) could form an example, since alternative paths are sought in advance to potentially quickly reroute the traffic over a backup connection (Wódczak, 2012b). Such backup paths may, in fact, be exploited for cooperative relaying, which, in turn, would most directly call for a consistent collaborative action to take place.

Given such a context, by no means should the link layer and network layer be considered separately, especially in the analysed case, where one may easily go even

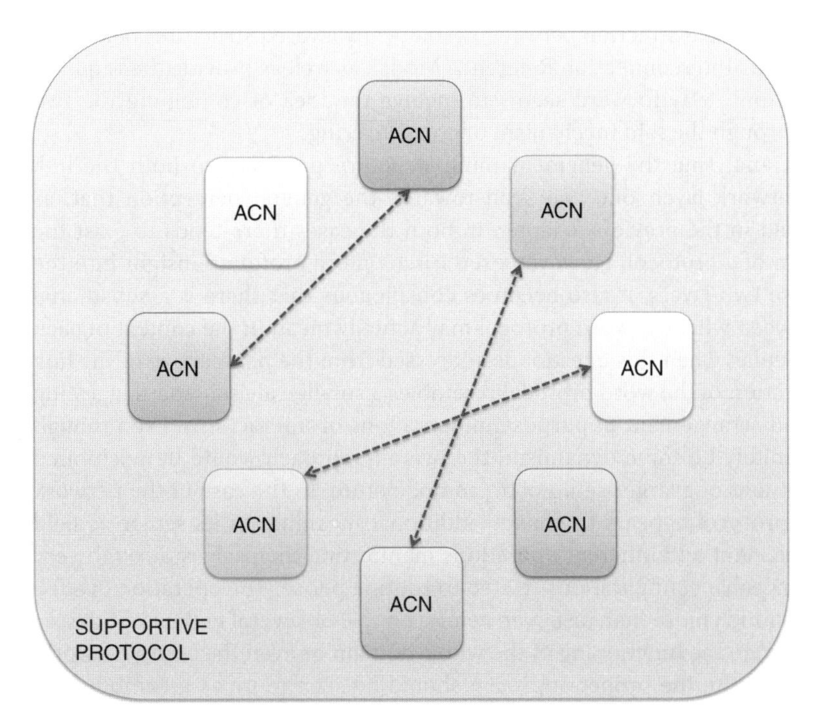

Figure 4.5 Projection of supportive protocol.

beyond what is referred to as a cross-layered approach.[9] In particular, keeping in mind the context settled previously with the introduction of collaborative protocols, presumably residing at the link layer, as well as taking into account the need for their orchestration with the aid of certain network layer routines, and the multi-point relay station selection heuristics in particular, it becomes of prime importance not only to understand the role of and the need for the said cross-layered approach, but also to contemplate the applicability of the related overlay developments calling for the introduction of external decision-making processes (Wódczak, 2011a). In fact, the Optimised Link State Routing protocol[10] already provides a range of functionalities that would readily facilitate the integration of the relevant link layer mechanisms, which by themselves are not really able or designed to accommodate the orchestration of concurrent occurrences of cooperative transmission on a broader scale. Most obviously, it would certainly be possible to implement all the necessary extensions within the link layer alone, but this way the operation thereof could then become somewhat overloaded and unbalanced, if not entirely warped (Wódczak, 2014). For this reason,

9 The reader is directly referred to the cross-systemic aspect induced by the proclaimed perpendicularity or orthogonality of the Open Systems Interconnection Reference Model and the Generic Autonomic Network Architecture, as explained in the previous chapter.

10 In fact, its nature of a Link-State Routing Protocol (LSRP) enables such operation. Before the pertinent mechanisms of the network layer are introduced in the next chapter, the reader might wish to become familiar with the broader context of the somewhat related Open Shortest Path First protocol, as outlined by Retvari et al. (2011).

in order to maintain the distinction between hermetically layered structures belonging to the Open Systems Interconnection Reference Model, as well as provide the required functionality, the only way forward seems to involve the idea of combining the two adjacent layers through the said mechanism of cross-layering.

Going further, analysing the general naming structure pertinent to both the link layer and the network layer, one may lean towards the general perception that, as already mentioned in the previous chapter, in both the cases there tends to exist the clear connotation of a protocol. However, exercising a more profound insight into the workings of these two layers, it also becomes conspicuous that there is a substantial discrepancy between what the word protocol may actually mean in the context of each of them. In particular, when this question is addressed from the perspective of the link layer, the connotation of the word protocol resembles a small-scale scheme, suggesting an uncomplicated arrangement of purposeful operations or interactions. Even though there would definitely be some dynamism, the overall approach would be positioned more as a constituent of a larger setup of an ad hoc nature. In the case of the network layer, the word protocol appears to convey additional meaning, as its semantic field covers all the pertinent and inherent operations manifesting themselves through very dynamically changeable configurations related to routing per se. The operation of such a protocol covers much more than just overseeing a couple or several network nodes, as it seeks to orchestrate the functioning of the whole domain or even the entire network. This is why, once again, the proper approach should be seen more as a combination of the two layers, emphasising the synergy between them in terms of the ability to use and combine certain specific network layer mechanisms, such as the multi-point relay station selection heuristics, to orchestrate the operation of the link layer (Wódczak, 2014).

All in all, as discussed above, the rationale behind cooperation apparently becomes rather many-faceted when scrutinised thoroughly, and therefore understandable in many different ways. Making an attempt to identify the semantic field of the term cooperation, one may quickly come to the conclusion that it is not only intangible, but also hardly quantifiable or definable. However, when perceived from various angles, the reason for the variety of forms by which it appears to be characterised appears to unveil as being more and more justifiable. In particular, as the commonly understood cooperation between or among Autonomic Cooperative Nodes may obviously follow different schemes or patterns, the notions of cooperative transmission or cooperative relaying pretend to become reduced to being, at most, its mere enablers. Given such an assumption, it becomes possible to make an attempt at generalising the idea of cooperation to such an extent that one might even come to the conclusion that it could represent any information exchange between Autonomic Cooperative Nodes. Regardless of the resulting classification into collaborative protocols and supportive protocols, the question of whether any sufficient justification exists to use the word 'protocol' in such a context becomes the next major topic of interest. In particular, as much as the notion of a protocol may be used at a very high level of discussion, a clear distinction would need to be made in comparison with a scheme, especially when the more specific context of a link layer or network layer is concerned, unless a clearly cross-layered approach to system design is considered.

4.2.3 Virtual Antenna Arrays

In general, either of the equivalent approaches usually referred to as cooperative relaying or cooperative transmission[11] that may be applied to improve radio transmission reliability in wireless networks, appears to be additionally referred to by a number of different naming schemes, for example, according to the original classification provided by Zimmermann et al. (2005): cooperation diversity, cooperative diversity, virtual antenna arrays, or coded cooperation. In fact, the application of the spatio-temporal processing (STP) techniques discussed in the previous chapter to such cooperative transmission schemes seems highly advantageous because in most cases it shall hold true that due to significant separation between RNs, or their elevated incarnation in the form of Autonomic Cooperative Nodes, the resulting equivalent virtual multiple-input multiple-output (EVMIMO) channel formed among them would usually be considered as entirely uncorrelated (Wódczak, 2014). Consequently, the theoretical increase in radio channel capacity envisaged for the case of the classic multiple-input multiple-output (MIMO) systems would be perfectly applicable to the case of cooperative transmission, should one assume that tight synchronisation can be guaranteed (Wódczak, 2012b). In order to introduce virtual antenna arrays, a generalised model of a cooperative transmission scheme will first be outlined. Then, the DSTBC-driven approach will be incorporated as one of the available options. Finally, the rationale behind virtual antenna arrays will be presented with emphasis on a generalised multi-hop approach fully agnostic of the employed cooperative transmission technology.

In fact, one may attempt to define the said generalised cooperative transmission scheme in the way proposed by Herhold et al. (2004b) and depicted in Figure 4.6, where the following notation is used:

$$p = \{M, m_{\mathrm{Tx}}, m_{\mathrm{Rel}}, m_{\mathrm{Rx}}\}. \tag{4.1}$$

Here, p denotes a set of parameters, including the number of RNs, denoted by M, and the numbers of antennae deployed at each of the SN, RN or RNs, and DN, indicated by m_{Tx}, m_{Rel}, and m_{Rx}, respectively. For instance, a system exploiting a single-input single-output (SISO) radio channel only and featuring a single RN would be defined as $p = \{1, 1, 1, 1\}$, whereas a system employing an SN equipped with two transmit antennae, i.e. employing a multiple-input single-output (MISO) radio channel, would be described as $p = \{1, 2, 1, 1\}$. In order to establish the background for the further analysis it is necessary to note that the most generic approach to cooperative transmission, known under the name of virtual antenna arrays, has a certain relation to its fairly dual concept of distributed space-time block coding. As such, the idea of virtual antenna arrays was introduced by Dohler et al. (2002b) with the assumption that the intermediary RNs, located between the SN and the DN, will become grouped into tiers able to employ different cooperative transmission protocols, possibly even at the same time. Thus, in a particular situation, as in the one-tier system of interest in this book, a given virtual antenna array could employ the desired logic stemming from distributed space-time block coding, as defined by Laneman and Wornell (2003).

11 Although they both seem to be generic and technology agnostic, and both terms may be used interchangeably, the author will give preference to the latter for consistency reasons (Wódczak, 2006), at least at this stage of the analysis.

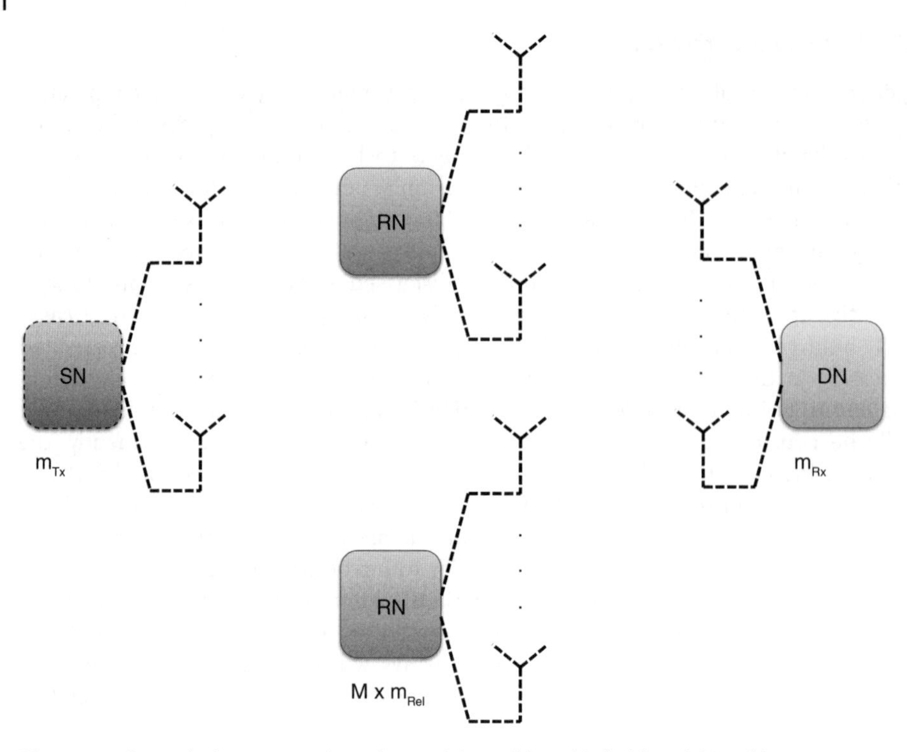

Figure 4.6 General relay cooperation scheme. Adapted from Herhold et al. (2004b).

In fact, though the concept of distributed space-time block coding appears to have been originally introduced by Laneman and Wornell (2003), where the techniques of repetition-based cooperative diversity (RBCD) and space-time-coded cooperative diversity (STCCD) were analysed and compared, one needs to immediately mention the works of Dohler et al. (2002a, 2003b), where the reference approach of space-time block coding (STBC) was evaluated directly with the use of virtual antenna arrays. In particular, the system model depicted in Figure 4.7 was analysed, where the process of transmission between the SN and the DN comprises two distinct stages. During the first stage, the SN broadcasts its signal so that it may be received not only by the DN but also by the potential RN or RNs. This signal is then processed by the intermediate RN or RNs, and during the second stage is resent directly towards the DN. Most importantly, such a retransmission may be performed in either the aforementioned repetition-based or a space-time-coded manner, yet, as shown by Laneman and Wornell (2003), even though both approaches are capable of achieving full diversity in the spatial domain, STCCD tends to outperform RBCD since it is more effectively exploitable for higher spectral efficiencies. Besides, there are also different retransmission schemes for the second stage, for example simple adaptive decode-and-forward (SAdDF) or complex adaptive decode-and-forward (CAdDF), as detailed by Herhold et al. (2005).

The difference between the above-mentioned schemes chiefly consists in the operation to be undertaken by each of them should the RN or RNs not be in a position to accurately convey the received signal. In the case of SAdDF, as explained by Herhold et al. (2004a), it is assumed that should the RN decline to participate in the retransmission phase it would become bound to remain silent, while in the case of CAdDF, as advocated by Laneman et al. (2004), the SN would retransmit the signal instead. In fact, as

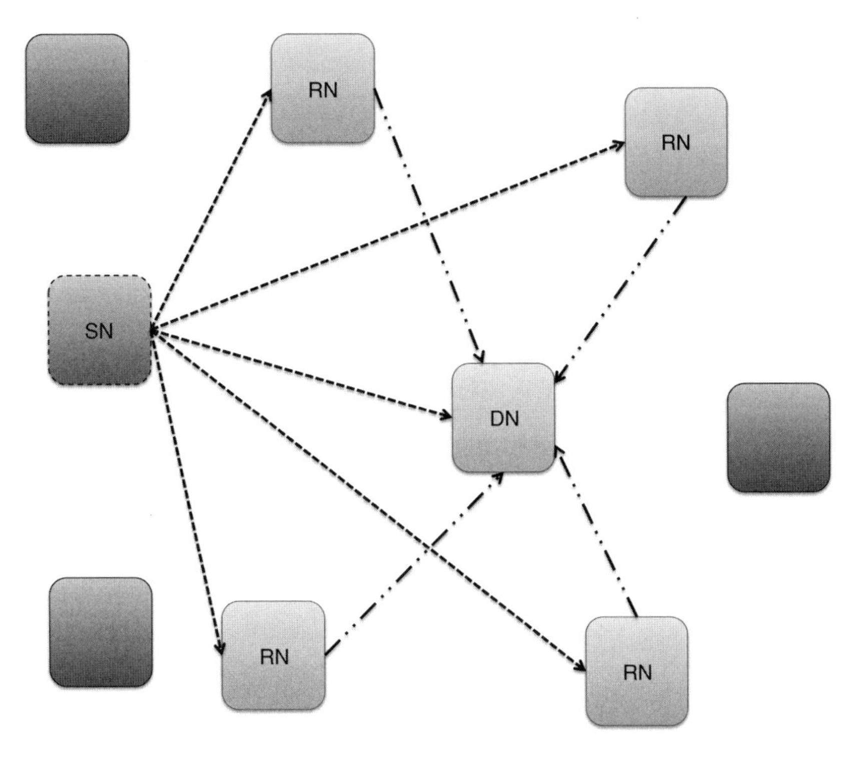

Figure 4.7 Distributed space-time block coding system. Adapted from Laneman and Wornell (2003).

it was already mentioned, both the ideas of distributed space-time block coding and virtual antenna arrays complement each other: both Laneman et al. (2004) and Laneman and Wornell (2003) clearly mention that the terminals share their antennae and other resources to create a virtual array through distributed transmission and signal processing. As already indicated, however, one should note that virtual antenna arrays may also employ other spatio-temporal processing techniques, for example space-time trellis coding (STTC) or layered space-time coding (LSTC). What is more, the ability of the concept of virtual antenna arrays to employ multiple tiers of RNs immediately brings additional scalability and the possibility for the resulting system to naturally support the desired capability of multi-hop relaying. In particular, originally a special case of a two-tier distributed MIMO system was analysed by Dohler et al. (2003a) and, following, it was additionally extended to multi-tier transmission by Dohler et al. (2004), as will be discussed in more detail below.

In fact, looking at the setup outlined in Figure 4.8, one notices that the existence of a number of tiers of RNs may make it impossible to establish a direct connection between the SN and the DN. Consequently, depending on the specifics of a given deployment scenario, one may by no means exclude a seemingly not so infrequent situation where end-to-end connectivity could solely be provided by multiple tiers of RNs acting as virtual antenna arrays (Dohler and Li, 2010).[12] Going further, one could equally well consider a one-tier scenario with such a spread between the SN and the VAA, as well as between the same and the DN, that exploitation of the direct link would also remain at

12 Such a multi-tiered configuration may well be referred to as a generalised virtual antenna array (GVAA) (Wódczak, 2014).

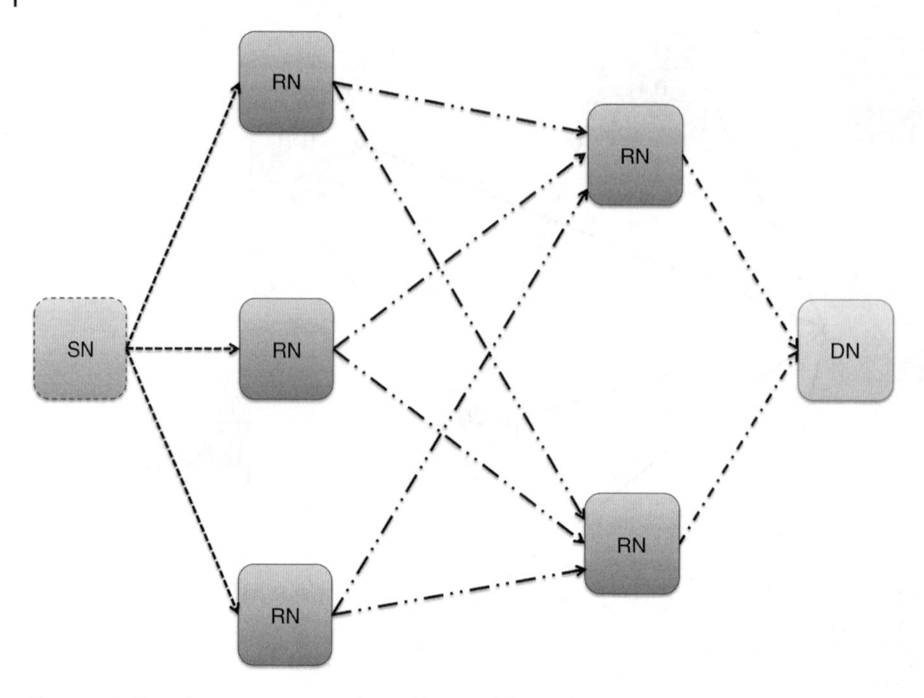

Figure 4.8 Virtual antenna arrays. Adapted from Dohler et al. (2004).

least impractical. One way or another, a fusion between the concept of a virtual antenna arrays and the idea of distributed space-time block coding is critically important for the conceptual advancement related to the internal structure and characteristics of the Autonomic Cooperative Networking Architectural Model. Putting these technologies together leads to the notion of Autonomic Cooperative Behaviour being quite a specific development, encompassing certain routines of the physical layer, link layer, and network layer on the side of the Open Systems Interconnection Reference Model, and also remaining in interaction with all the levels of abstraction on the side of the Generic Autonomic Network Architecture. The ground for such a concept will be laid with the upcoming introduction of the equivalent distributed space-time block encoder.

4.3 Fixed Relay Deployment Concepts

4.3.1 Grid-Based Manhattan Scenario

Given the above context, the instantiation of conventional relaying is presented with the aid of a fixed deployment concept, known as the Manhattan scenario, as depicted in Figure 4.9.[13] The fixed nature thereof results from the fact that the scenario comprises non-moving infrastructure network nodes in the form of a base station (BS) situated at the centre with four fixed relay nodes (FRNs) separated by an evenly distributed grid of buildings, hence the name, as outlined by Esseling et al. (2005). At the same time, there

13 For additional background, the reader may also find it useful to refer to Pabst et al. (2005).

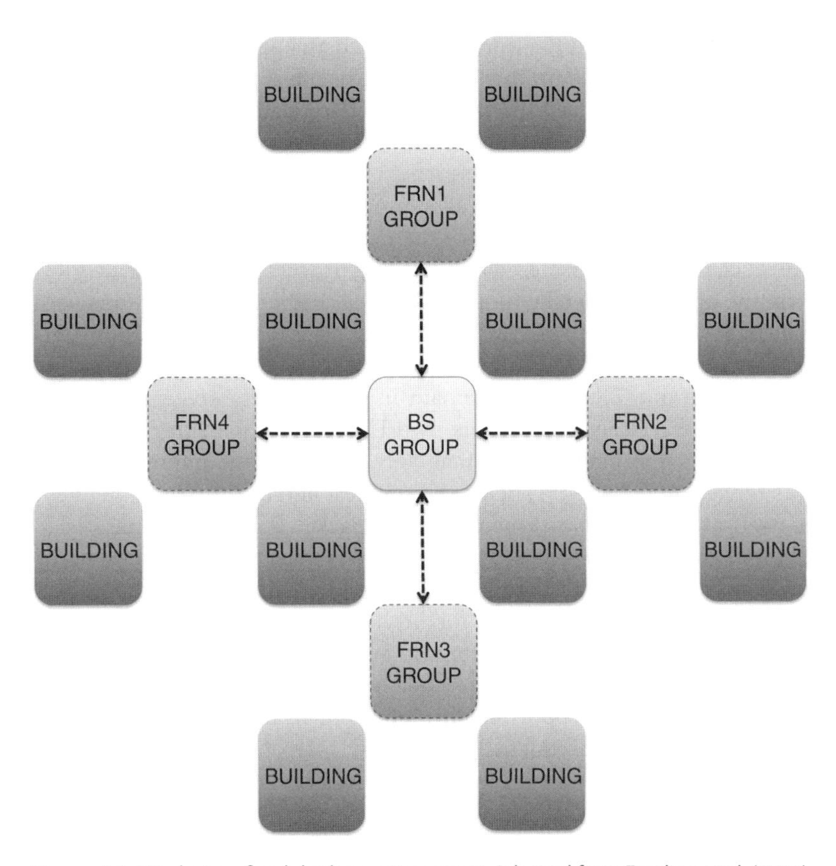

Figure 4.9 Manhattan fixed deployment concept. Adapted from Esseling et al. (2005).

are also a number of user terminals (UTs)[14] around each of these fixed network nodes, together with them forming a BS group and FRN groups, respectively, as highlighted in Figure 4.9. While different approaches to frame structuring are possible (Esseling et al., 2004), the one presented in Figure 4.10 will be applied below. The major advantage of such a multi-frame boils down to the fact that its being subdivided into three phases guarantees that spatially independent pairs of FRNs may work in parallel, as proposed by Schultz et al. (2003). Looking into the details, one may see that during the first phase, all four FRNs are served by the BS, one by one, in consecutive order, and in the second phase the two spatially independent couples of FRNs serve their respective

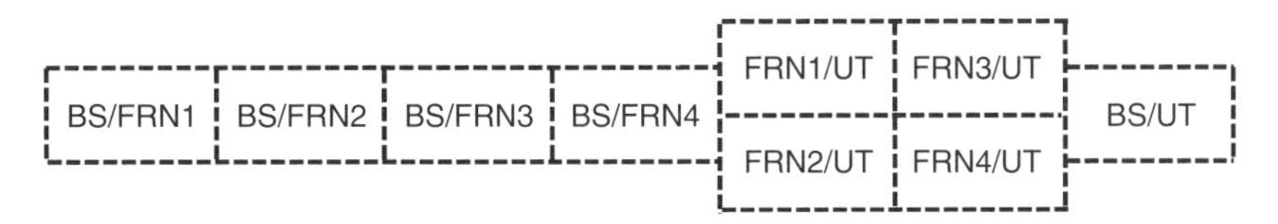

Figure 4.10 Multi-frame structure.

14 At this stage of the analysis, UTs will be perceived as being of the recipient nature only. However, one may consider their employment as RNs, just to reflect the most recent trends in this area (Nishiyama et al., 2014).

UTs. Finally, the BS serves its UTs in the third phase. Such an approach may clearly be further optimised by taking into account the traffic patterns offered by distinct network nodes at the network layer, as well as exploiting knowledge related to buffer memory utilisation (Głąbowski and Wódczak, 2006).

The related analysis of the above setup is performed by means of computer-assisted simulation (CAS) investigations using the so-called process interaction (PI) method, as defined by Tyszer (1999). Consequently, each of the network nodes, including the BS, the FRNs, and the UTs, becomes considered a computing process (CPR). Such a CPR is then scheduled on the agenda at specific time instances according to interactions with other CPRs (or, in other words, network nodes) and also as a result of certain events occurring in the simulated system, as outlined in Figure 4.11. Following the framing structure strictly, there is one BS that can remain in five different stages, including interaction with each of the FRNs and with its own UTs. An FRN, in turn, is always scheduled by the BS and can remain in one stage only, related to the interaction with the UTs within its range. Finally, the UT, being the simplest and the most independent CPR, schedules itself at the moments when the next packet should be generated.[15] Traffic-wise, the total number of packets sent by each UT is depicted for buffer lengths of 30 and 60 in Figures 4.12 and 4.13, respectively. As is clearly visible, the packet generation intensity is on average the same for both the nonadaptive and adaptive strategy yet to be detailed. In both cases, all the UTs belonging to the first and second FRN send many more packets compared to those belonging to the third or fourth FRN, or to the BS. Thanks to such a disproportion the behaviour of the adaptive approach will be much more visible in the subsequent analyses.

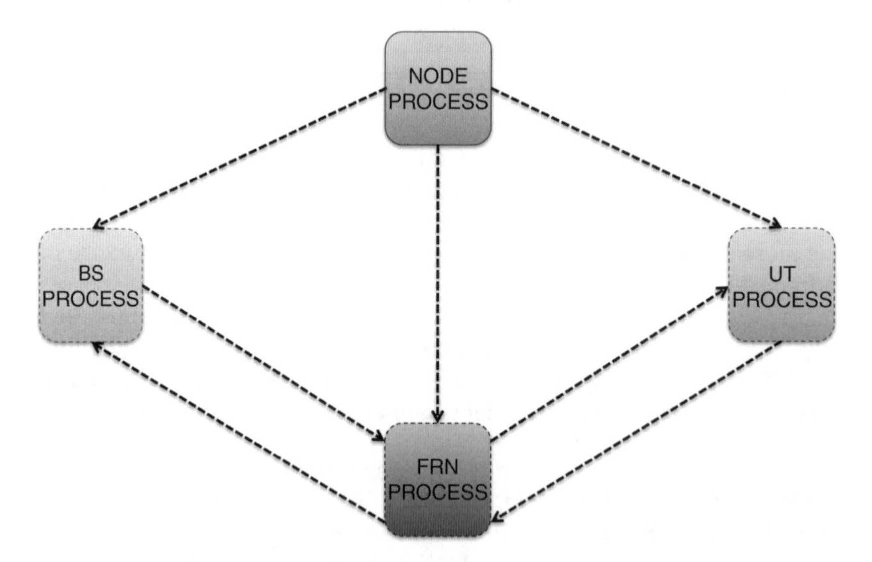

Figure 4.11 Process-based interactions.

15 One should note that since all the listed CPRs inherit from the same base class, the scheduler is not even aware of the classes of the CPRs executed at specific time instances. Such knowledge is not necessary because all the CPRs are involved in interaction without the aid of the scheduler, which is only responsible for the proper execution and processor time assignment.

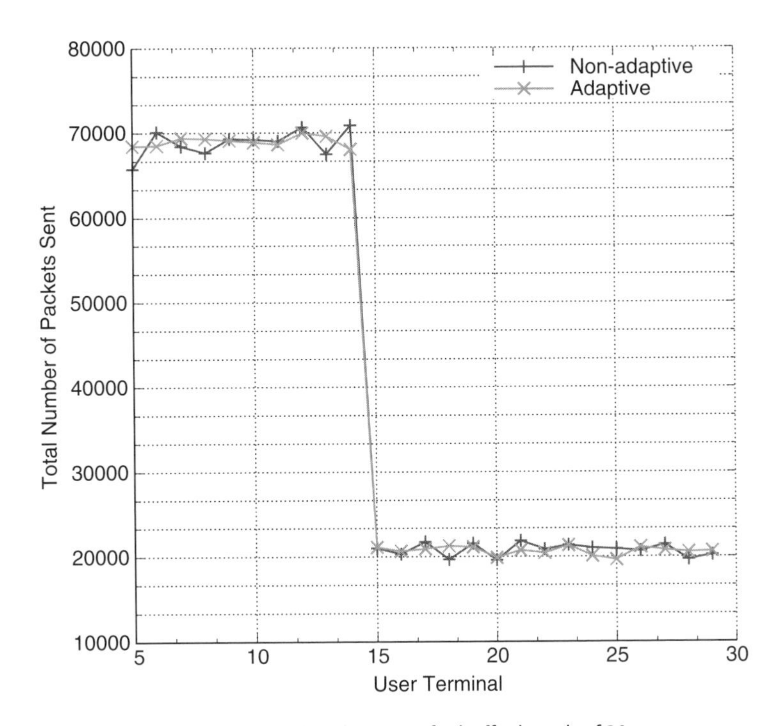

Figure 4.12 Total number of packets sent for buffer length of 30.

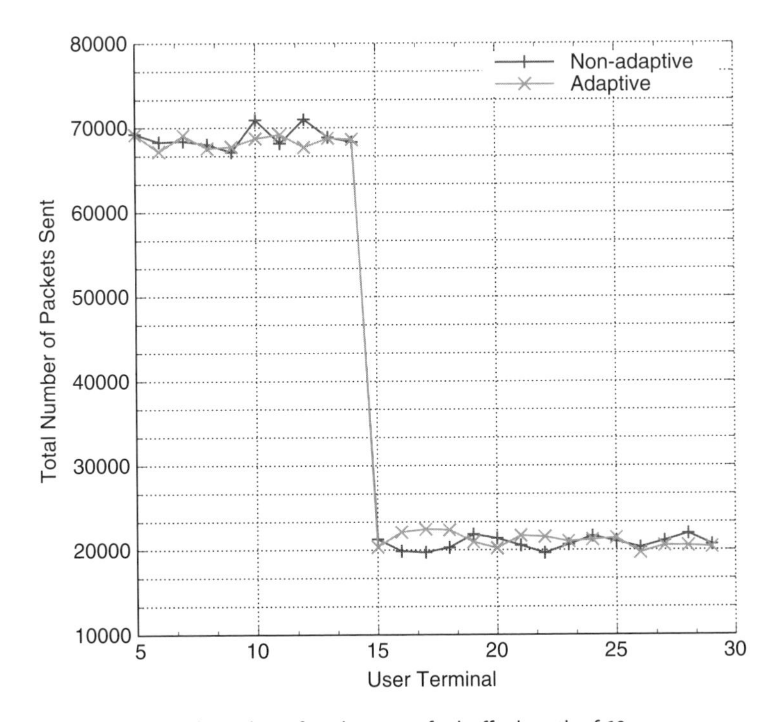

Figure 4.13 Total number of packets sent for buffer length of 60.

In such a system, where one may gain chiefly from network layer optimisation, different levels of quality of service (QoS) may be supported, depending on the requirements of the specific UTs and their applications. However, to enable QoS provision from a wider perspective, there must be a resource management mechanism active at the BS capable of allocating service times accordingly. In particular, both the BS and the FRNs are equipped with buffer memory where they store packets.[16] Such memory is limited, and to minimise the number of packets lost the system would need to be sensitive to any changes in the traffic offered by distinct network nodes. This problem can be handled effectively with the aid of a scheduler; while different scheduling algorithms are available, weighted round-robin (WRR) appears to be the most flexible solution, since users are allowed to utilise the bandwidth with respect to their weights, as outlined, for instance, by Raha et al. (1996). The presented approach to throughput maximisation for conventional relaying, as outlined in Algorithm 4.1, is based on such a scheduler which exploits additional feedback regarding the buffer load in the previous cycle $t-1$, denoted by $L_{j,i}^{t-1}$, where the total length of the multi-frame remains unchanged. However, the lengths of specific frames, corresponding to the duration of the service time, and referred to as slots, denoted by j, are calculated adaptively, as outlined by Głąbowski and Wódczak (2006). The calculation is based on the buffer utilisation percentage in the previous cycle, denoted by $CL_{j,i}^{t-1}$, where C represents the reciprocal of the overall load of the system in question.

Algorithm 4.1 Throughput maximisation for conventional relaying.

1: **for** $i = 1$ to 4 **do**
2: $\quad L_{1,i}^{t} = \text{load}(\text{FRN}_i, t)$
3: $\quad L_{2,i}^{t} = \text{load}(\text{UTs} \in \text{FRN}_i, t)$
4: **end for**
5: $L_{3,0}^{t} = \text{load}(\text{UTs} \in \text{BS}, t)$
6: $C = (\sum_{i,j} L_{j,i}^{t-1})^{-1}$
7: **switch** (j):
8: $\quad$ **case** 1:
9: $\quad\quad$ **for** $i = 1$ to 4 **do**
10: $\quad\quad\quad \text{slot}_t^{\text{BS}\to\text{FRN}_i} \sim CL_{j,i}^{t-1}$
11: $\quad\quad$ **end for**
12: $\quad$ **case** 2:
13: $\quad\quad \text{slot}_t^{\text{FRN}_1\to\text{UTs}} = \text{slot}_t^{\text{FRN}_2\to\text{UTs}} \sim C(L_{j,1}^{t-1} + L_{j,2}^{t-1})$
14: $\quad\quad \text{slot}_t^{\text{FRN}_3\to\text{UTs}} = \text{slot}_t^{\text{FRN}_4\to\text{UTs}} \sim C(L_{j,3}^{t-1} + L_{j,4}^{t-1})$
15: $\quad$ **case** 3:
16: $\quad\quad \text{slot}_t^{\text{BS}\to\text{UTs}} \sim CL_{j,0}^{t-1}$
17: **end switch**

4.3.2 Noncooperative Approach Limitations

According to Algorithm 4.1, the length of the time slot $\text{slot}_t^{\text{BS}\to\text{FRN}_i}$, assigned by the BS to itself during the first phase to serve FRN_i, is proportional to the value of $CL_{1,i}^{t-1}$,

16 Being radio access points (RAPs), the BS or an FRN are denoted by i, which is assigned the value of 0 for the former, while the latter may be assigned a number of 1 through 4.

where $L^t_{1,i}$ is defined as the buffer utilisation of FRN_i in the cycle t and is denoted by load(FRN_i, t). During the second phase, when the two spatially independent FRNs serve their respective UTs, the slot length for each such couple is calculated similarly in regard to the average buffer utilisation of these UTs in the previous cycle. Finally, during the third phase, when the BS serves its UTs, the slot length similarly becomes proportional to the buffer utilisation for these UTs in the previous cycle. As to system parameters, the assumptions made during the simulation were that five UTs per RAP may be active, and each UT may set up a session during which it sends packets to another UT with a steady intensity. In particular, the UTs with addresses 26–30 belong to the BS of address 0, while those with addresses 5–10, 11–15, 16–20, and 21–25 are assigned to the FRNs of addresses 1, 2, 3, and 4, respectively. The simulation termination condition shall be met when the last packet has reached its destination. The major reason for the inclusion of such an evaluation is to show that despite the fact that one may expect gains from the application of adaptive framing at the link layer and buffer memory management at the network layer, any additional improvement that could potentially result from the previously introduced diversity gain, even if it were achievable, is not seamlessly applicable.[17]

In the light of the above, the values for a system with a buffer memory length of 30 packets are outlined for both the nonadaptive case, i.e. with a fixed slot length, and the adaptive case, i.e. with a dynamically adjusted slot length. The respective results are depicted in Figures 4.14 and 4.15, where the number of packets lost and the packet loss ratio curves are presented, proving that adaptive conventional relaying is much more stable and the overall throughput is significantly maximised. A similar analysis is performed for both nonadaptive and adaptive systems with double the buffer memory length, 60 packets; this time, far better performance can be observed, as depicted in Figures 4.16 and 4.17. Most importantly, however, comparing Figures 4.15 and 4.17 one can see that the adaptive system with half the amount of buffer memory may perform no worse than the nonadaptive one. Going further, the average delay is computed between the time a packet is generated by a UT and the time this packet leaves the FRN. The results are presented in Tables 4.3 and 4.4, where one may notice substantial gains. Last, but not least, yet another parameter evaluated during the simulations is the difference between the average time slot length for both systems and the sizes of the buffer memory. The results, as presented in Figures 4.18 and 4.19, show that the maximum average variation in the time slot length is less than 8%, which is really not much in the context of the achieved throughput improvement and delay time reduction. As was mentioned concerning the link layer, however, the entire improvement is limited to the framing adaptation process only.

4.3.3 Cooperation-Enabled Indoor Scenario

Moving forward from the Manhattan scenario, where the instantiation of virtual antenna arrays was impossible due to the physical separation between FRNs, a much more appealing cooperation-enabled indoor scenario will now be investigated, also

17 For the sake of clarity, one could, theoretically, think of the establishment of the most baseline cooperative relaying scheme, as initially outlined in Figure 4.2, yet the layout of the Manhattan scenario would not allow for cooperative transmission in the VAA mode being of major interest in this book.

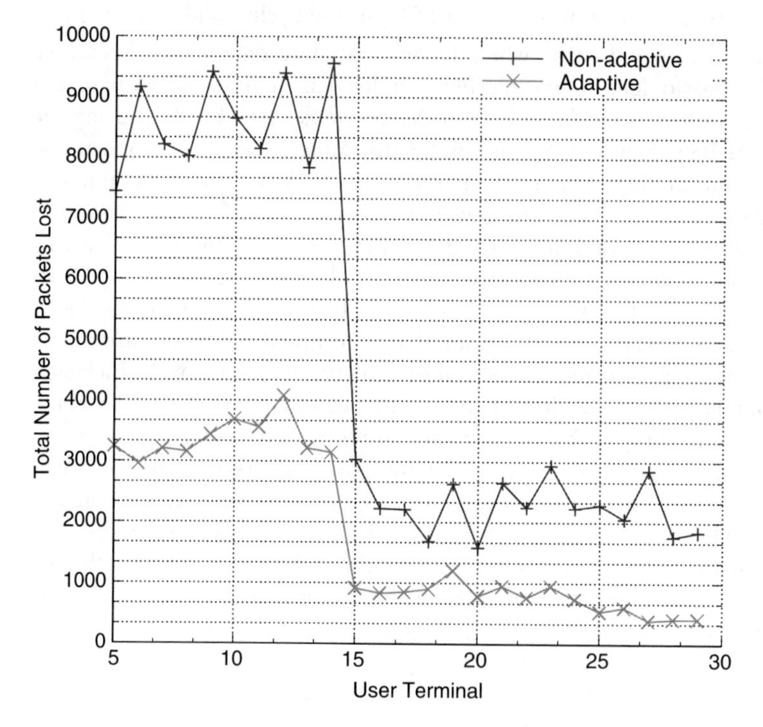

Figure 4.14 Total number of packets lost for buffer length of 30.

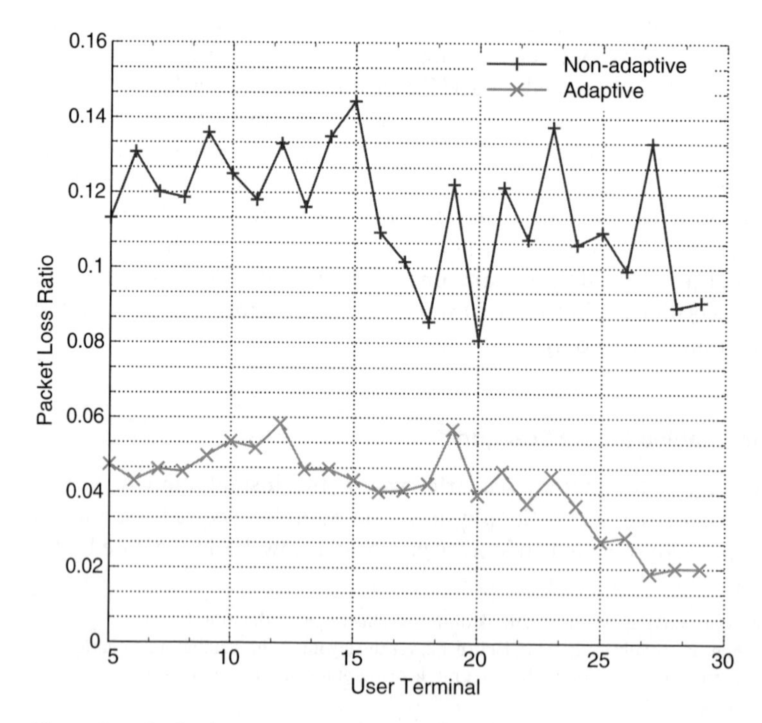

Figure 4.15 Packet loss ratio for buffer length of 30.

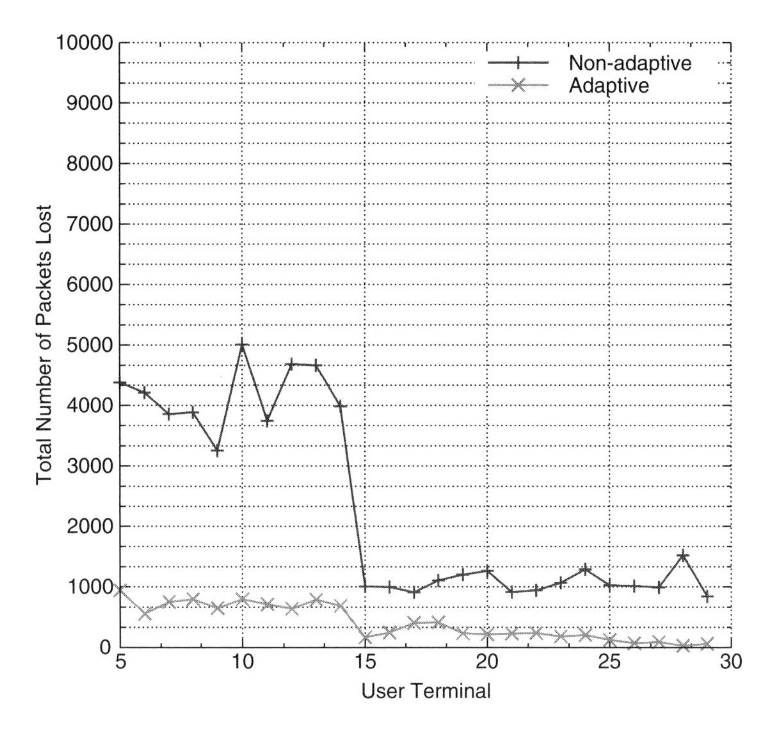

Figure 4.16 Total number of packets lost for buffer length of 60.

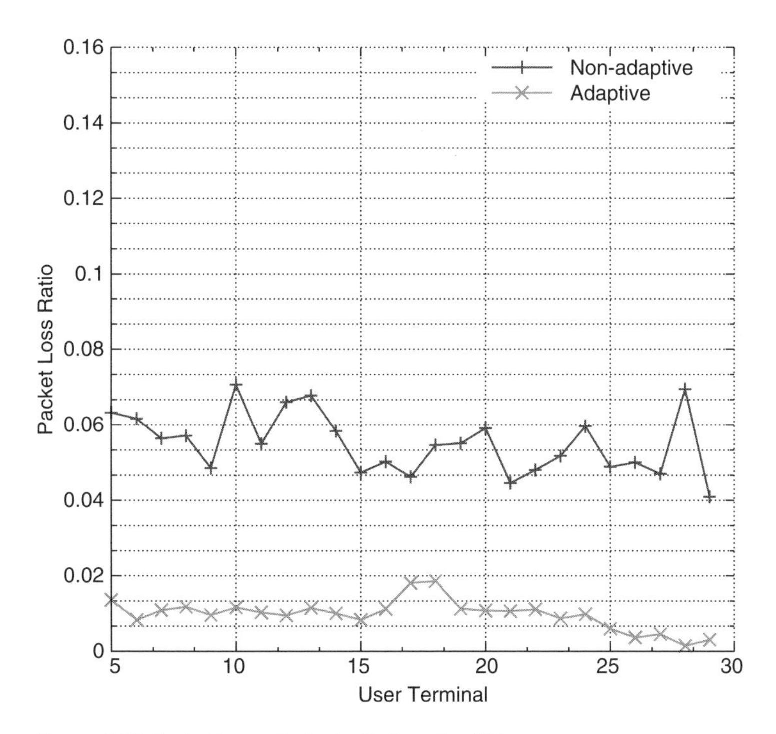

Figure 4.17 Packet loss ratio for buffer length of 60.

Table 4.3 Average delay for buffer length of 30.

FRN	Nonadaptive [unit]	Adaptive [unit]	Gain [%]
1	98.8	57.3	42
2	103.4	60.7	41
3	93.6	55.4	41
4	95.9	57.1	41

Table 4.4 Average delay for buffer length of 60.

FRN	Nonadaptive [unit]	Adaptive [unit]	Gain [%]
1	163.5	67.8	59
2	176.1	70.4	60
3	150.0	71.5	52
4	153.4	70.5	54

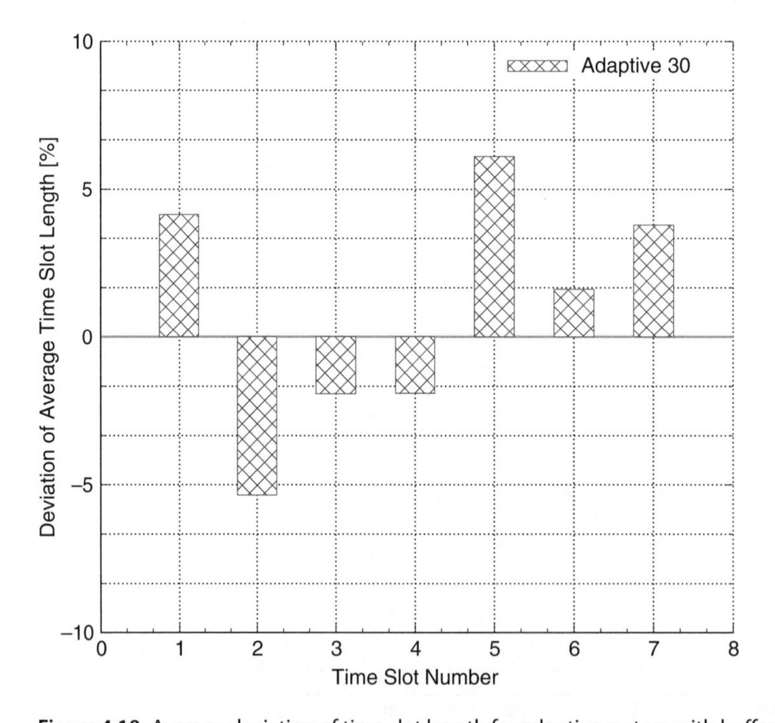

Figure 4.18 Average deviation of time slot length for adaptive system with buffer size of 30.

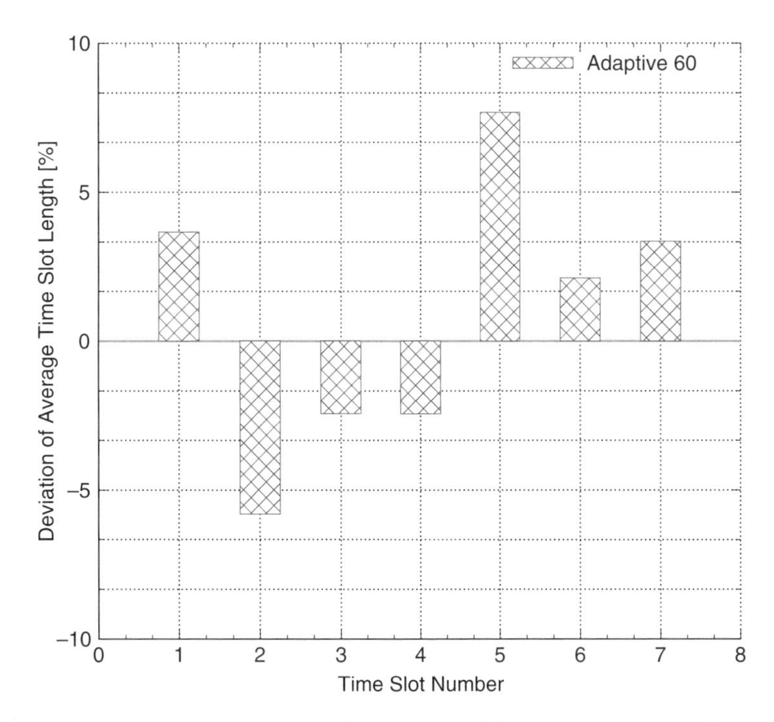

Figure 4.19 Average deviation of time slot length for adaptive system with buffer size of 60.

known as a relay-enhanced cell (REC), as outlined by Dottling et al. (2009).[18] This scenario is characterised by a considerable user density, high shadowing and heavy signal attenuation due to the existence of obstacles represented by numerous walls (Dottling et al., 2009). Thanks to such an isolated nature thereof, it is possible to easily observe advantages such as low interference, especially when compared to outdoor cases. As shown in Figure 4.20, it consists of one floor of a height of 3 m in a building, where two corridors characterised by the dimensions of 5 m × 100 m and 40 rooms characterised by the dimensions of 10 m × 10 m are situated. In the case of the reference configuration, there are four FRNs positioned in the corridors, while the transmission is coordinated by the BS mounted in the very centre. Such a base deployment assumes that the FRNs are placed in the middle of the corridors, 25 m and 75 m away from the left or right sides of the storey.[19] First, a more detailed insight into system parameters will be provided, so that it is possible to transition to the topic of radio resource partitioning, where the structure of the so-called super-frame will be presented. Based on this, a computer-assisted simulation analysis will be carried out to show whether the setup in question may benefit from cooperative transmission enabled with VAA technology employing distributed space-time block coding.

18 This scenario, coming from the European Union Sixth Framework Programme Integrated Projects Wireless World Initiative New Radio I (WINNER I) and Wireless World Initiative New Radio II (WINNER II), will be referred to again towards the end of the book, since it allows to present the potential of an autonomic system to be able to instantiate cooperative transmission between FRNs belonging to a bigger mesh of radio access points, as outlined by Wódczak (2010).

19 In the light of the more advanced deployments to be discussed later, the reader may find it useful to additionally refer, for example, to Wódczak (2012a).

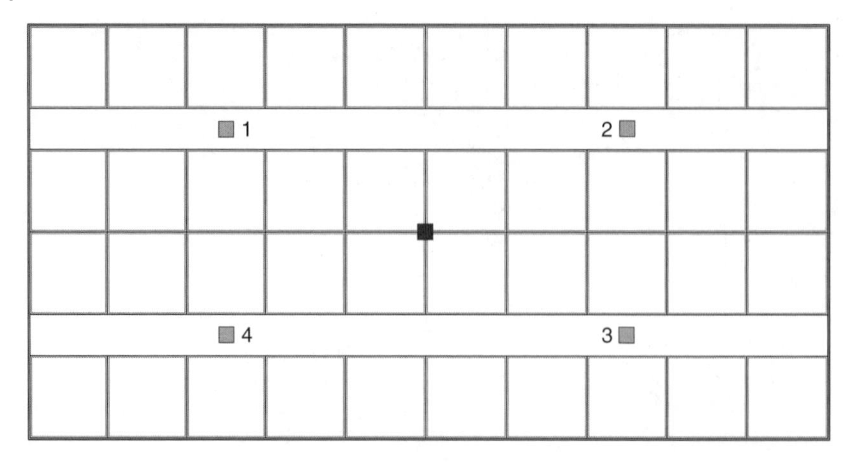

Figure 4.20 Baseline relay deployment indoor scenario. Adapted from Dottling et al. (2009).

In general, although the analysis will be performed according to the parameters for the REC provided by Dottling et al. (2009), certain additional assumptions will be made. Most importantly, given the fact that the scenario is symmetric, the simulation area can be limited to a set of 10 rooms located in the bottom right part of Figure 4.20, adjacent to the corridor where the FRN is positioned. Such an approach is followed to maintain the clarity of the visualisation of the results; however, at a later stage of the book the entire area will be considered, too. What is more, a fixed modulation and coding scheme is (MCS) is employed consisting of quadrature phase-shift keying (QPSK) modulation and a (4, 5, 7) convolutional code, and an additive white Gaussian noise (AWGN) channel is assumed along with the A1 radio propagation model for the radio links between FRNs and UTs, and also between the BS and FRNs (Dottling et al., 2009). Depending on the presence of walls, either the line-of-sight (LOS) or non-line-of-sight (NLOS) version of the propagation model is applied. The path loss $L(d)$ for the LOS model is defined as

$$L(d) = 18.7 \log_{10}(d) + 46.8 + \sigma \quad [\text{dB}], \tag{4.2}$$

where d denotes the distance in meters between the SN and the DN,[20] while σ represents the standard deviation of shadow fading and is equal to 3 dB. For the NLOS propagation model, $L(d)$ is described as

$$L(d) = 20.0 \log_{10}(d) + 46.4 + 5n_{\text{w}} + \sigma \quad [\text{dB}], \tag{4.3}$$

where n_{w} denotes the number of walls between SN and DN, while σ is equal to 6 dB. This means that all the walls are assumed to be of the same light type.

Looking at the more detailed parameters, as outlined in Table 4.5, orthogonal frequency-division multiple access (OFDMA) is employed in time division duplex (TDD) mode at a carrier frequency of 5.0 GHz and with a channel bandwidth of 100 MHz. The transmission power for the BS, FRNs, and UTs is 21 dBm, whereas the antenna gains are 14 dBi, 7 dBi, and 0 dBi, respectively. The noise figure at the receiver is 7 dB, while the noise power spectral density amounts to -174 dBm Hz^{-1}. For the

20 One should note that the nomenclature of source node and destination node pertains to a single hop only. Should a two-hop cooperative transmission scheme be investigated, the DN of the first hop may, in fact, play the role of the RN, being also the SN of the second hop.

Table 4.5 System parameters.

Parameter	Value	Comments
Carrier frequency	5.0 GHz	TDD mode
Channel bandwidth	100 MHz	OFDMA
Spatial processing	DSTBC	FRN–FRN cooperation
BS antenna count	1	Omnidirectional
FRN antenna count	1	Omnidirectional
UT antenna count	1	Omnidirectional
BS transmission power	21 dBm	Antenna gain of 14 dBi
FRN transmission power	21 dBm	Antenna gain of 7 dBi
UT transmission power	21 dBm	Antenna gain of 0 dBi
Channel modelling	AWGN channel	A1 NLOS model room–room and room–corridor for BS–FRN and FRN–UT
Link adaptation	Fixed CMS	QPSK and (4, 5, 7) convolutional code
Mobility support	Yes	UTs
Resource scheduling	Fixed	1 chunk per UT (8 subcarriers and 15 OFDM symbols)
RAP selection	Signal power	At destination
Traffic model	CBR	Constant bit rate

baseline deployment presented in Figure 4.20, each UT is assigned one chunk of radio resources spread over 8 subcarriers and 15 orthogonal frequency-division multiplexing symbols. Moreover, the average interference power level per subcarrier is −125 dBm. The structure of the super-frame for this base deployment is defined according to Doppler et al. (2007a), and radio resources are partitioned in both the temporal and spectral domains, as outlined in Figure 4.21; the spatial dimension is exploited, too. In particular, following the preamble, a very similar pattern is repeated twice during one super-frame, following Doppler et al. (2007b). First, the resources are assigned to the BS and then to different combinations of FRNs, so that three of the latter may be active simultaneously, with two thereof operating cooperatively in spatio-temporal processing mode, as explained by Wódczak (2008). The simulation results presented

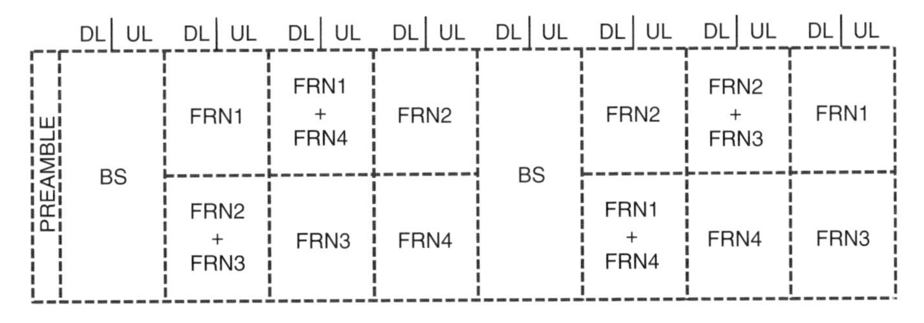

Figure 4.21 Super-frame structure.

below apply to the following three cases: direct transmission, conventional relaying, and cooperative relaying.

In the cooperative relaying mode, pairs of selected FRNs form VAAs in order to perform the operation of distributed space-time block coding using the G_2 code matrix in Equation 3.17. For each of the three cases, the attainable QoS is evaluated from the perspective of the relative user throughput, which is calculated as the ratio between the number of bits transmitted successfully and the total number of bits sent. All the results are obtained assuming that a given FRN always takes part in cooperative transmission, even if decoding has been unsuccessful, in order to purposely augment any degradation in performance resulting from wrong positioning thereof, translating into its inapplicability. The results obtained show that it is possible to make up for the performance degradation visible in Figure 4.22(b) with the aid of conventional relaying, as shown in Figure 4.22(c). Unfortunately, the gain provided by cooperative relaying, as outlined

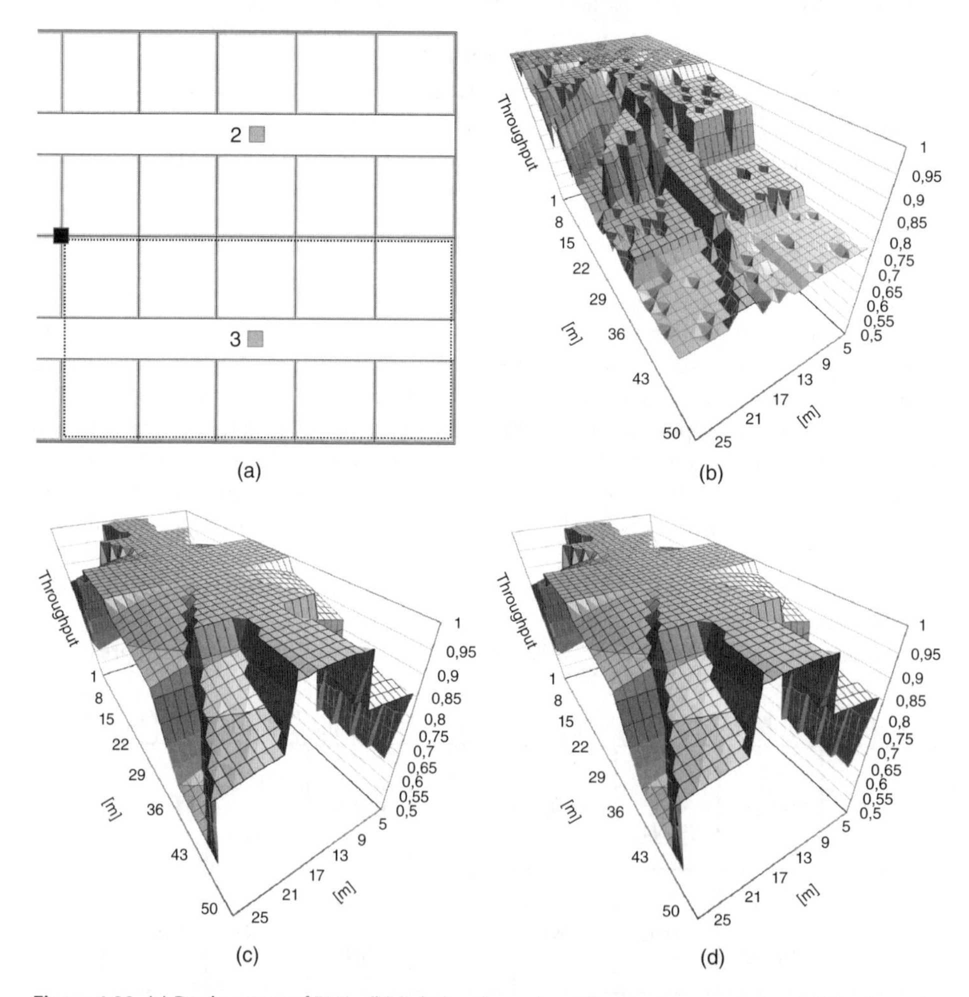

Figure 4.22 (a) Deployment of FRNs. (b) Relative throughput for direct transmission. (c) Single-path relaying via FRN3. (d) FRN2–FRN3 cooperation.

in Figure 4.22(d), seems almost negligible when compared to the conventional relaying case. The reason is that the signal coming from the remote FRN to the destination UT is usually too heavily attenuated by the higher number of walls. According to the aforementioned A1 NLOS propagation model, as defined by Dottling et al. (2009), each of the walls between rooms attenuates the transmitted signal by 5 dB. As a result, the power levels of the signals received by the destination UT from the cooperating FRNs may differ, even by up to about 15 dB. This is why the notion of Autonomic Cooperative Behaviour is to be introduced, thanks to which the system shall be able to self-manage in this respect.

4.4 Function Level Overlay Logic

4.4.1 Roots of Autonomic Cooperative Behaviour

In general, the concept of Autonomic Cooperative Behaviour (ACB), initially defined by Wódczak (2014),[21] will constitute the foundation of the final five-tier Autonomic Cooperative Networking Architectural Model, as previously depicted from the hierarchical perspective in Figure 2.19. In particular, the Autonomic Cooperative Behaviour is constituted by the two bottom tiers thereof, including Autonomic Cooperative Nodes and autonomic cooperative sets (ACSs), while, itself forming tier three, it also becomes the major component of the two top-most ones, in the form of the Autonomic Cooperative Networking Protocol and the overall umbrella of the Autonomic Cooperative Networking Architectural Model. In fact, looking at Autonomic Cooperative Behaviour from this viewpoint, one may conclude that not only is it multifaceted, but may also be hardly definable. Attempting to account for its role and meaning, it is necessary to distinguish between collaborative and supportive protocols, as introduced at the beginning of this chapter. Given such a distinction, one may better understand that the notion of behaviour becomes a kind of link between the physical interpretation of the underlying STBC-driven cooperative transmission and the complementary role of the orchestration performed by the multi-point relay station selection heuristics of the Optimised Link State Routing protocol. To provide the most comprehensive account of Autonomic Cooperative Behaviour, first of all, the technological background will be further explored. Based on this, the equivalent distributed space-time block encoder (EDSTBE) will be defined and the reference curves governing such Autonomic Cooperative Behaviour will be plotted.

In the light of what was proposed by Wódczak (2014), looking bottom-up from the physical layer, one of the key enablers for the aforementioned Autonomic Cooperative Behaviour, as defined in this chapter, appears to be the previously introduced equivalent virtual multiple-input multiple-output radio channel. This is because it allows for efficient instantiation of distributed space-time block coding through its properly tailored application among the Autonomic Cooperative Nodes involved. Going further, one may discern, however, that there is much more to Autonomic Cooperative Behaviour, as its ultimate composition involves the incorporation of ACSs stemming directly from the concept of VAAs. What is more, the process of creation of ACSs is assumed to be

21 For additional context, the reader may also refer to Wódczak et al. (2013a).

integrated into and orchestrated by the multi-point relay station selection heuristics of the Optimised Link State Routing protocol. Most of all, the above-mentioned components clearly stem solely from the Open Systems Interconnection Reference Model, leaving additional room for the Generic Autonomic Network Architecture entities responsible for the autonomic dimension. In essence, the interaction, either in a more direct or an indirect form, is expected to take place with all the appropriate levels of abstraction through the related orchestration to be performed over hierarchical autonomic control loops by the cooperative transmission decision element (CTDE), cooperative re-routing decision element (CRDE), cooperation management decision element (CMDE), and cooperation orchestration decision element (CODE) (Wódczak, 2014).

What is more, as much as the said equivalent virtual multiple-input multiple-output radio channel may be called the enabler of Autonomic Cooperative Behaviour, its real roots stem much more from the concept of the equivalent distributed space-time block encoder, as introduced in Definition 4.1 – originally outlined by Wódczak (2006), and reinforced in Wódczak (2012b).[22]

Definition 4.1 ***(Equivalent distributed space-time block encoder)*** A perfectly synchronised set of distributed Autonomic Cooperative Nodes connected to the SN with error-free links,[23] and able to cooperatively encode received signals according to a given STBC matrix X_Y, conceptually forms and is defined as an equivalent distributed space-time block encoder E_Y^X, where $X_Y = \{G_2, G_3, G_4, H_3, H_4\}$.[24] In this context, the Autonomic Cooperative Nodes may also be referred to as a generic transmitters; the DN, acting as a single receiving antenna, may similarly be referred to as a generic receiver.[25]

Pursuant to the above definition, three different equivalent distributed space-time block encoders are evaluated, in the form of E_2^G, E_3^G, E_4^G, E_3^H, and E_4^H, where the single generic receiver is equipped with a multi-element array of size up to 8. To this end, a MIMO[26] flat fading Rayleigh channel is employed, characterised by the lack of correlation between any of the wireless radio links. The results are presented in Figures 4.23, 4.24, 4.25, 4.26, and 4.27, respectively, where the previously defined diversity gain is visible.[27]

---

22 Following Wódczak (2014), the relative throughput results for the Autonomic Cooperative Behaviour based on the respective modes of operation of the EDSTBE, assuming both AWGN and Rayleigh channels, are presented in Figures A.7–A.16.

23 Analysing the existence of error-free links at the first hop from the perspective of cooperative transmission, such an assumption may seem a bit unrealistic. Yet, as will be shown in the next chapter, an approximation of this type may make a lot of sense when a realistic scenario is investigated, where RAPs powered from the electrical grid may offer a much higher transmission power than battery-driven UTs, thereby providing a much better signal-to-interference-plus-noise ratio (SINR), as described by Wódczak (2012b).

24 One should note that other STBC matrices stemming from complex orthogonal designs (CODs) are by no means excluded.

25 The notions of generic transmitter and generic receiver are introduced to align with the nomenclature outlined in the previous chapter.

26 In the case of a single receiving antenna such a channel would obviously reduce to MISO.

27 For the sake of comparison, the results for an AWGN channel are outlined in Figures A.17, A.18, A.19, A.20, and A.21, where no diversity gain is observable.

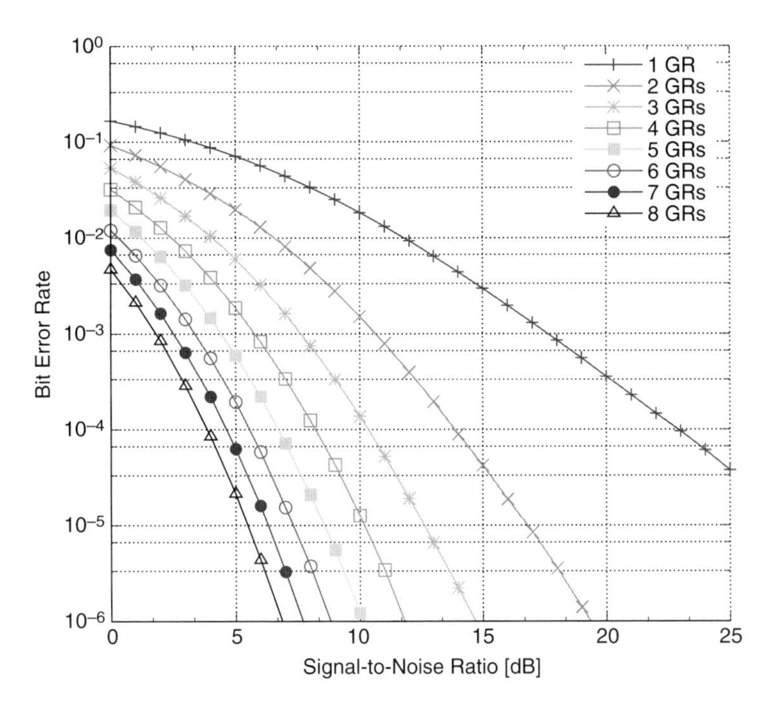

Figure 4.23 E_2^G equivalent distributed space-time block encoder in a Rayleigh channel.

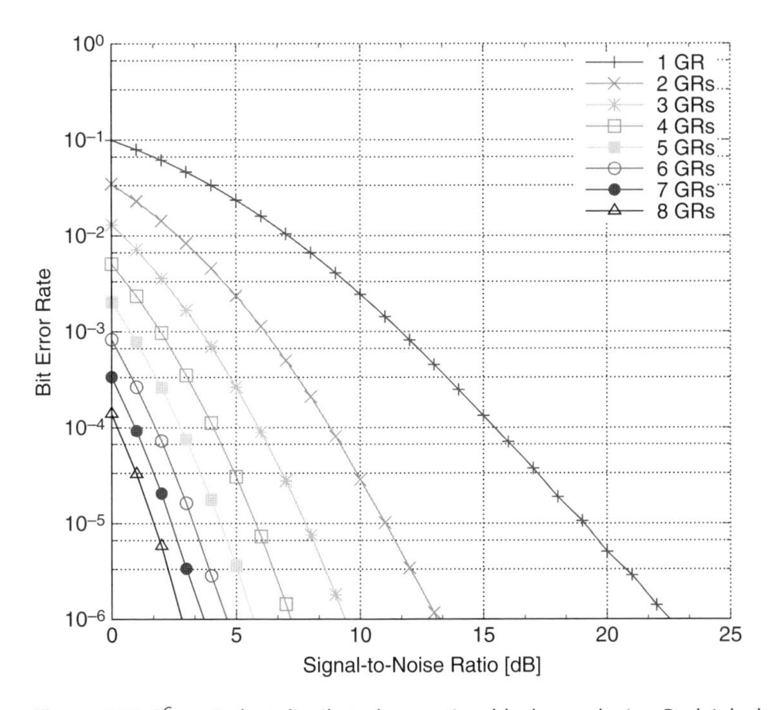

Figure 4.24 E_3^G equivalent distributed space-time block encoder in a Rayleigh channel.

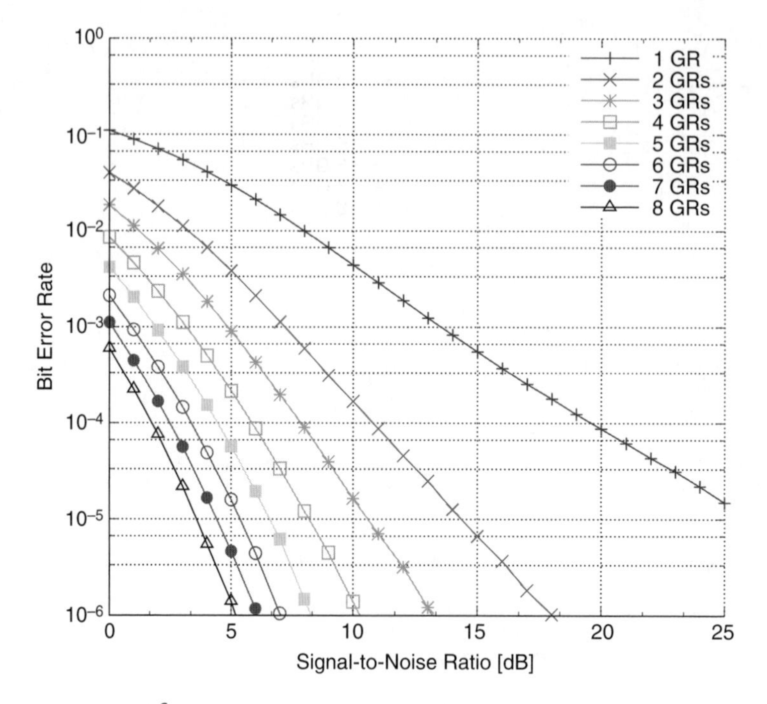

Figure 4.25 E_4^G equivalent distributed space-time block encoder in a Rayleigh channel.

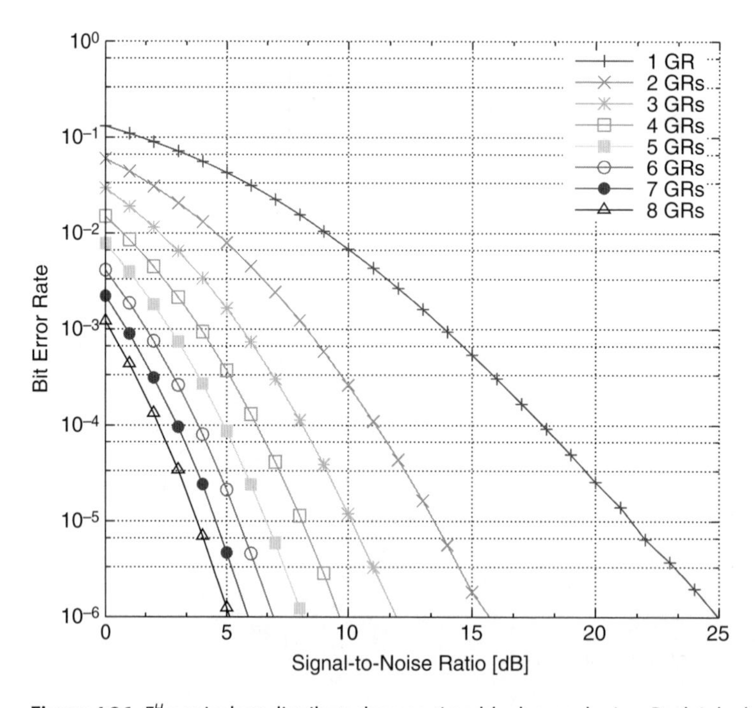

Figure 4.26 E_3^H equivalent distributed space-time block encoder in a Rayleigh channel.

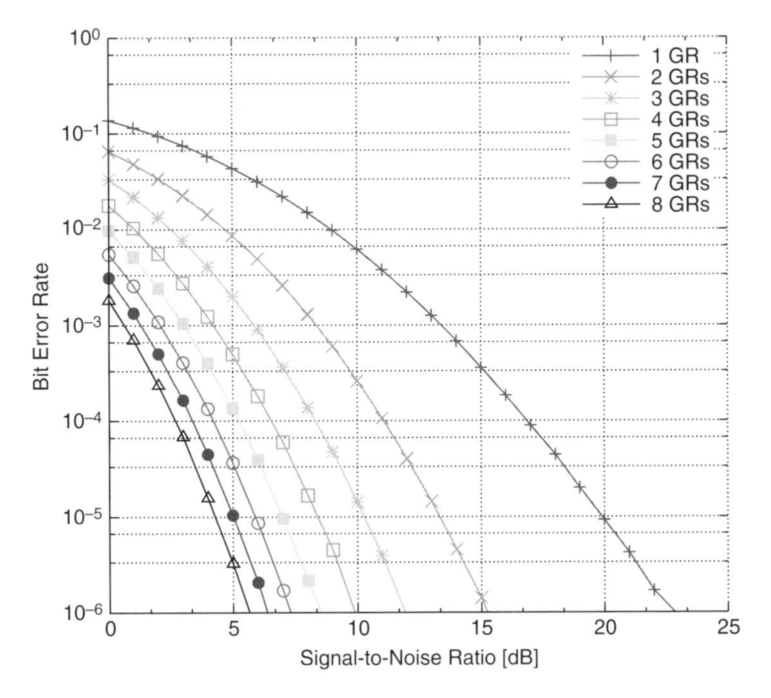

Figure 4.27 E_4^H equivalent distributed space-time block encoder in a Rayleigh channel.

4.4.2 Cooperative Re-Routing Decision Element

Moving upwards to encompass a yet wider perspective one comes across the CRDE, of special importance should the necessity arise for the reconfiguration of ongoing cooperative or noncooperative transmissions. Such a reconfiguration could be performed on the basis of additionally acquired information, potentially coming either from the resilience and survivability decision element or the fault management decision element, as indicated by Wódczak (2011b).[28] In particular, keeping in mind that, according to ETSI-GS-AFI-002 (2013), both those decision elements are located at the node level, one should also think about the broader context of the somewhat pertinent network layer, especially because of the CMDE and its related routing mechanisms. In fact, the operation of cooperative re-routing stems directly from the concept of FRR, which is usually considered to reside somewhere between the link layer and the network layer. Based on this, especially keeping in mind the evolved Autonomic Cooperative Networking Architectural Model, and quite contrary to what was assumed by Wódczak (2014), the CRDE becomes shifted to the function level. Given the above, first of all the investigated setup is described in more detail, so that it is then possible to introduce the logic behind the CRDE. Finally, evaluation results are outlined to serve as justification of its applicability (Wódczak, 2012b, 2014).

28 Resilience, as defined by Chaparadza et al. (2013), is perceived as being directly related to system dependability understood as outlined by Wódczak (2013).

Looking at the scenario depicted in Figure 4.28, in a normal situation the packet stream between nodes A and C would be routed over node B.[29] Should there be a failure of any of link 1 or 2, or node B itself, the system would need to react properly. In the legacy case orchestrated with the FRR approach, one of the readily available alternative paths formed from links (3, 4, 5), (3, 6), or (7, 8) would be used instead almost immediately. This process could obviously be enhanced with the proper application of Autonomic Cooperative Behaviour among the nodes involved, as outlined in Figure 4.29 (Wódczak, 2012b). In particular, following Algorithm 4.2, adapted[30] with the notation introduced within the description of the CTDE, the links between the source node, x, and its one-hop neighbour, n, playing the role of an Autonomic Cooperative Node, as well as between the latter and the destination node, $n^{(2)}$, would need to be checked for bit error (BER) rate compliance. Depending on whether one or both offers the requested transmission quality, further steps directed towards a cooperative action could be undertaken.[31] The evaluation was carried out using an AWGN radio channel with QPSK modulation. The curves obtained, as outlined in Figures 4.30, 4.31, 4.32, and 4.33, illustrate[32] that for a given relative throughput threshold of 99%, 96%, 93%, and

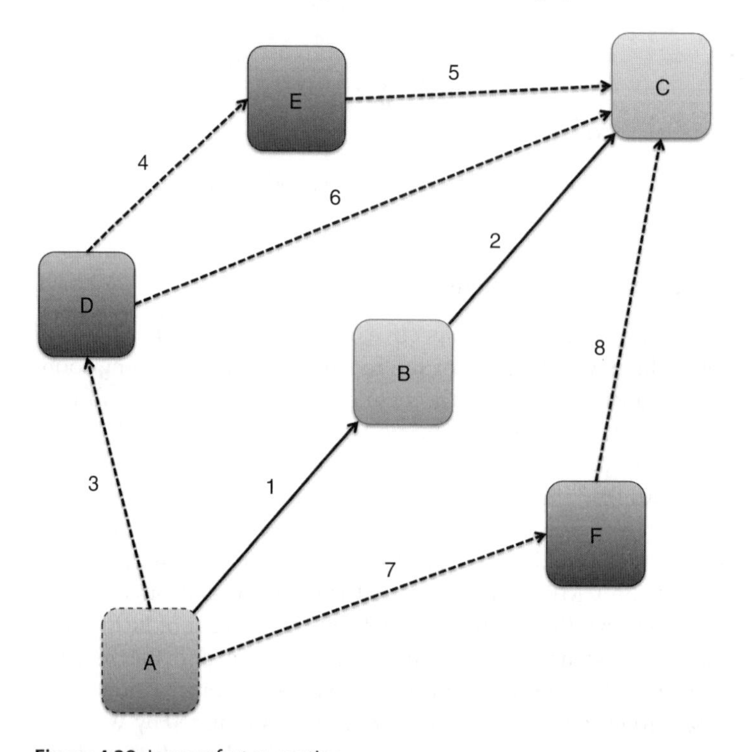

Figure 4.28 Legacy fast re-routing.

29 'Node B' shall by no means be perceived as pertaining to what is otherwise referred to by the 3rd Generation Partnership Project (3GPP).
30 For the previous version, the reader is referred to Wódczak (2014).
31 For a broader background, related to the design and evaluation of techniques enabling resilience and survivability of a routing network node, the reader is referred to Wódczak et al. (2013b).
32 For additional results the reader may also refer to Figures A.22, A.23, A.24, and A.25, where a different, more bit error rate orientated perspective, is assumed.

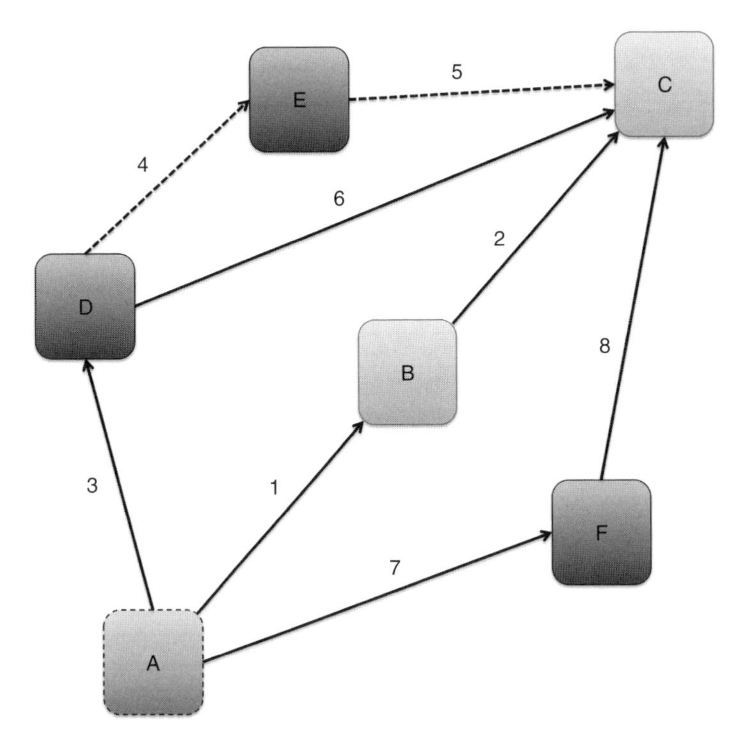

Figure 4.29 Cooperative re-routing.

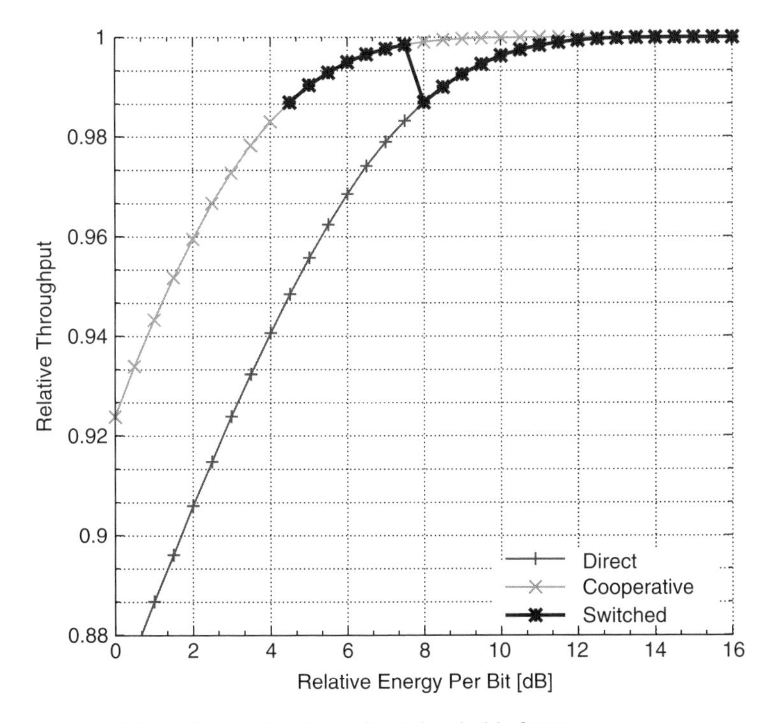

Figure 4.30 CRDE at relative throughput threshold of 0.99.

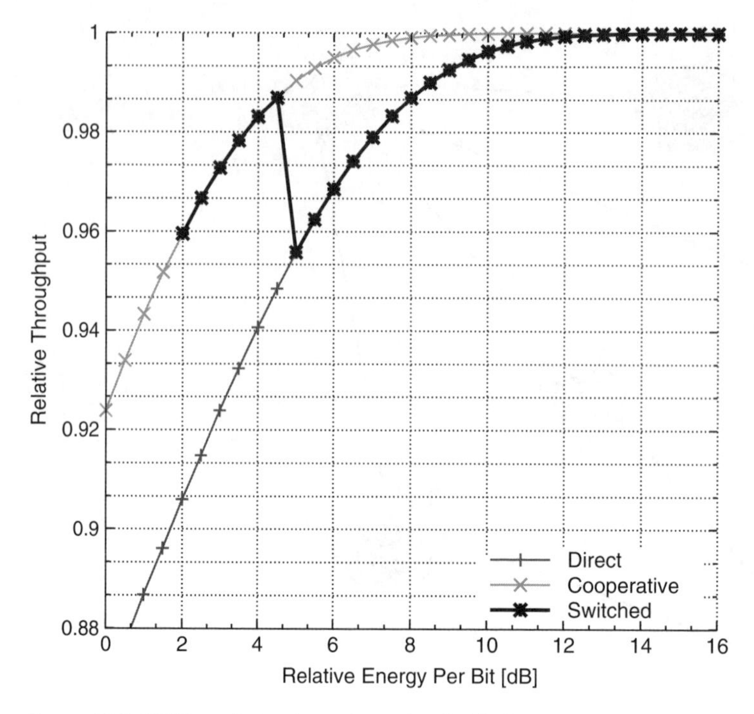

Figure 4.31 CRDE at relative throughput threshold of 0.96.

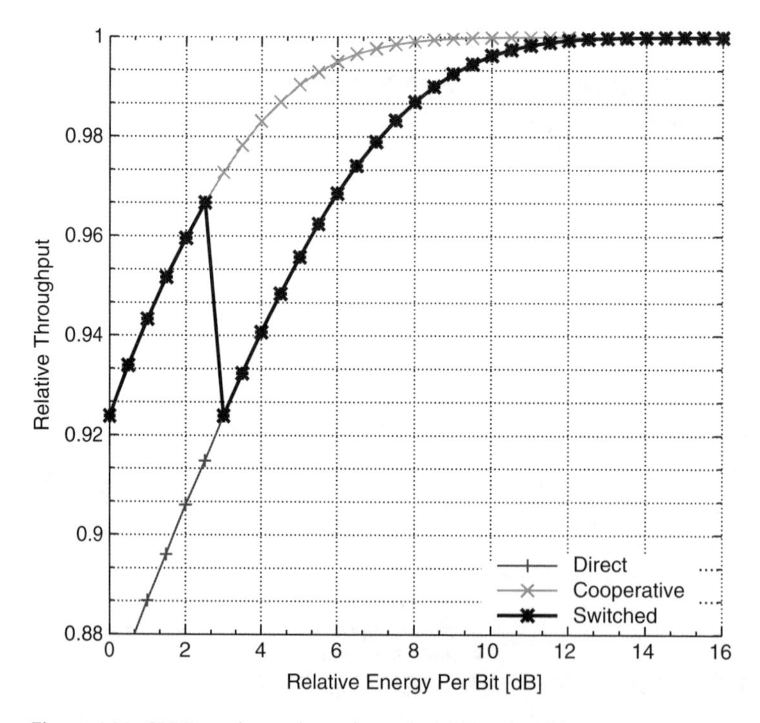

Figure 4.32 CRDE at relative throughput threshold of 0.93.

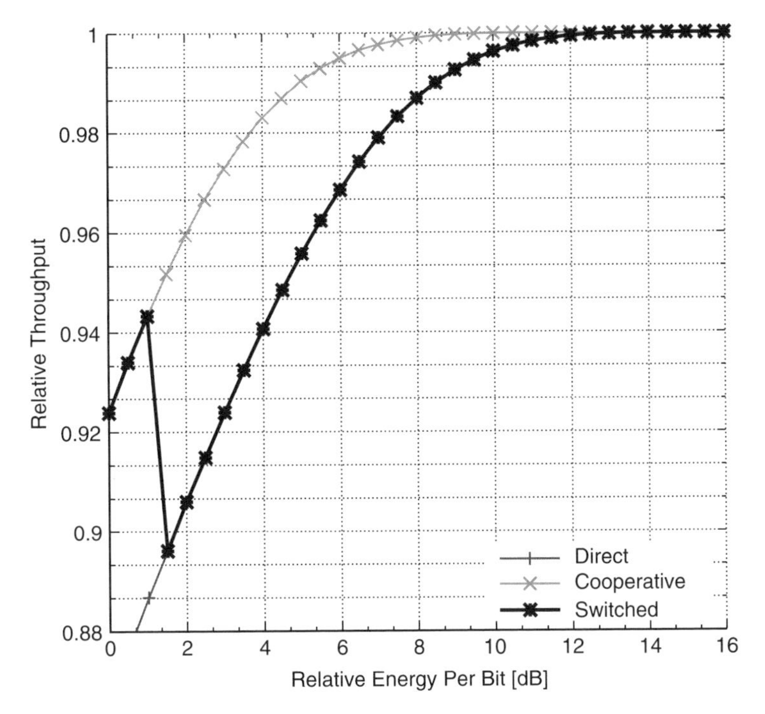

Figure 4.33 CRDE at relative throughput threshold of 0.90.

90%, the CRDE may switch from noncooperative transmission operation to cooperative transmission mode, and thereby instantiate Autonomic Cooperative Behaviour among the Autonomic Cooperative Nodes (Wódczak, 2014).[33]

Algorithm 4.2 Logic of CRDE.

1: **if** ($\mathbf{BER}(x, n) < \theta$ **or** $\mathbf{BER}(n, n^2) < \theta$) **then**
2:   **if** $ACS(x, n^{(2)}) \neq \emptyset$) **then**
3:     $\mathbf{ACB}(ACS(x, n^{(2)}))$
4:   **else**
5:     $\mathbf{FRR}(x, n^{(2)})$
6:   **end if**
7: **end if**

4.4.3 Architectural Integration Aspects

Shifting the focus to the aspects of architectural integration once again, as well as expanding what was previously discussed in the context of the CTDE, this time the scope of analysis becomes even broader, given the background role of the Autonomic Cooperative Behaviour. Not being straightforwardly definable, the notion of the

33 In other words, this way the related SINR gain may be observed, allowing to avoid the necessity of choosing another path, as would normally be the case for the legacy FRR. Quite the contrary – in the setup discussed, it would be possible to maintain the connection over a diversified set of cooperative paths, despite some problems with a single link.

Autonomic Cooperative Behaviour appears to span all the layers of interest belonging to the Open Systems Interconnection Reference Model. For this reason, not only is the link layer inspected along with its adjacent network layer, positioned right above, but additionally the properly upgraded entities of the physical layer, located directly below, are taken into account. What is more, looking into the perpendicular or orthogonal dimension of the relevant orchestration entities of the Generic Autonomic Network Architecture, one may similarly discern that, as much as the focal function level needs to interact with the protocol level, it also requires certain views of the node level to be fed back for resiliency-related optimisation or imposed for proper and durable operation. One should also keep in mind that there is still the third dimension to the entire setup stemming from the fact that there are the ACSs physically instantiated for the sake of grouping Autonomic Cooperative Nodes. In the following, first the dependencies among the routines of the physical layer, link layer, and network layer will be discussed. Then, the necessity for interaction with the node level will be outlined, so that it is possible to prepare the ground for a more detailed incorporation and integration of the node level in the next chapter.

As depicted in Figure 4.34, the relations between or among the routines of all of the physical layer, link layer, and network layer will be examined, keeping in mind the background in the form of Autonomic Cooperative Behaviour. As much as the Autonomic Cooperative Behaviour could be considered as being elevated more towards the orchestration of these layers, it also encompasses the vital entities thereof, consequently making any strict separation of no interest in the analysed case. What is more, similarly

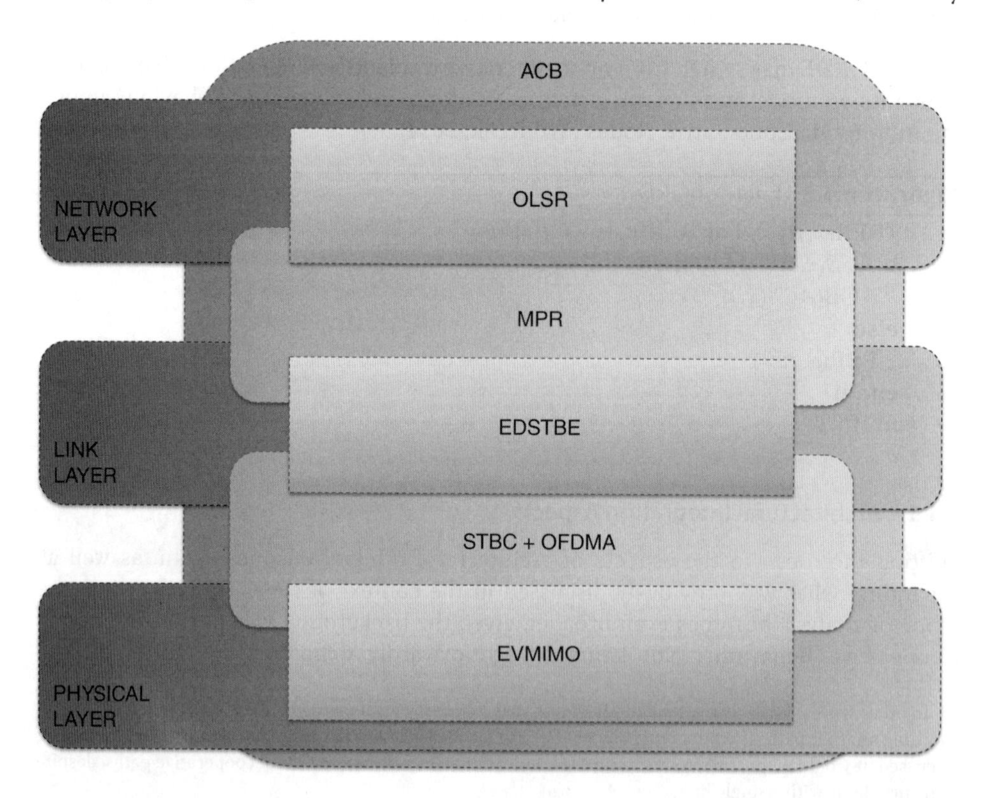

Figure 4.34 Dependencies among the routines of the physical layer, link layer, and network layer.

to the considerations related previously to the CTDE, this time also the horizontal orientation is exercised to enhance the clarity of presentation. Looking at the central point in the form of the link layer, one may discern that, following the reference Figure 3.22, the notion of distributed space-time block coding is now upgraded to encompass the most recently defined equivalent distributed space-time block encoder, while the linkage between the link layer and the physical layer remains unaltered and based on the concatenation of both STBC and OFDMA. Changing the direction towards the network layer, it becomes clear that the major role and responsibility over there is assigned to the Optimised Link State Routing protocol, with its inherent multi-point relay station selection heuristics, similarly providing integration with the equivalent distributed space-time block encoder of the link layer. Given the fact that the blueprints representing dependencies should be understood as being of a higher-level and context-setting nature only, additional insight into the architectural workings is provided below.

In fact, as depicted in Figure 4.35, it is the complexity of the architectural relations driven by the tandem of the link layer and the function level that results in a substantial incorporation of, in a sense external, components to be described in the following chapter, where the related workings of both the network layer and the node level are to be addressed. Such a situation appears to be enforced by the fact that, already at this stage, it becomes necessary to highlight certain outer, compared to the present advancement of the description, triggers in the still partial incarnation of the Autonomic Intelligence Evolved Cooperative Networking. Consequently, it seems much more conspicuous that even though the CRDE of the function level is supposed to directly coordinate the creation of ACSs, whenever it becomes necessary to commence the routine of cooperative re-routing, there are also other dependencies, for example the requirement to know which Autonomic Cooperative Nodes should be assigned to a given ACS. In the system configuration described here, such information may only

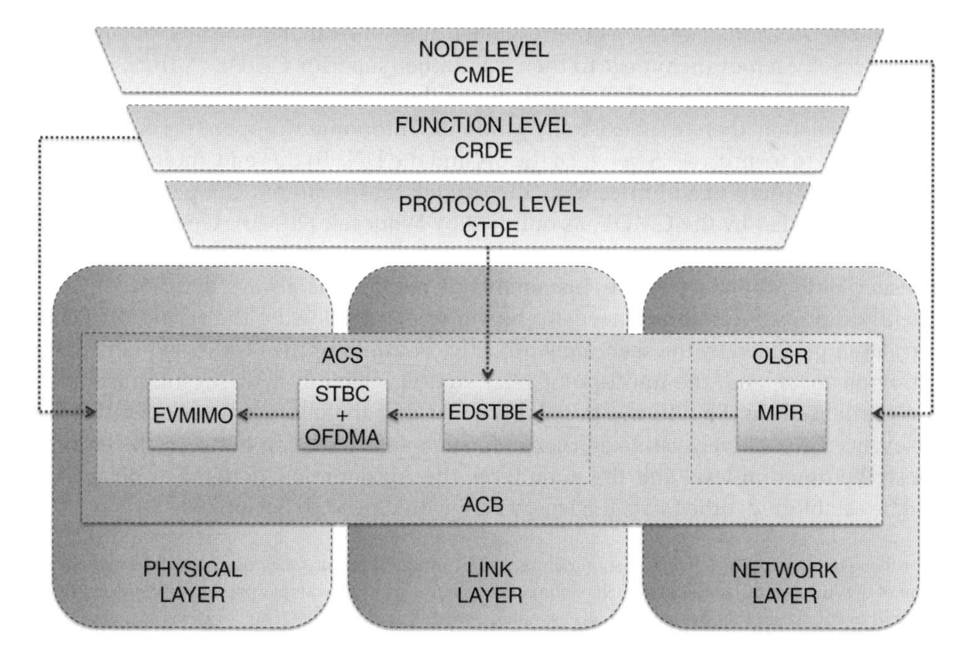

Figure 4.35 Architectural relations stemming from the link layer and function level.

be provided by the multi-point relay station selection heuristics of the Optimised Link State Routing protocol, which is later to be encapsulated in another entity, referred to as the Autonomic Cooperative Networking Protocol (ACNP). As such, the Autonomic Cooperative Networking Protocol will be overseen by the CMDE of the node level, while the Autonomic Cooperative Behaviour is left untouched, though it will play a key role in terms of subsystem atomisation when the network level is introduced.

On the operational side, not only will the decision elements of the Autonomic Cooperative Networking Architectural Model be capable of inferring that certain failures may be imminent, based on the root causes identified by the fault management decision element (FMDE), but also react immediately for the purposes of orchestrating any remediation action by employing its complementary resilience and survivability decision element (RSDE). The role of the inherent CMDE may then be indispensable in such a context, as it might impose the instantiation of cooperative transmission with the aid of the CRDE in the cases where a noncooperative transmission would normally remain the only option. In other words, should a failure be about to occur, the CRDE could attempt to avoid the default approach of the legacy FRR, consisting in switching to alternative paths, through the instantiation of Autonomic Cooperative Behaviour among appropriate Autonomic Cooperative Nodes well in advance. Thus, the CMDE would be used along with its Autonomic Cooperative Networking Protocol to route packets over multiple cooperative paths. Such a multi-path operation should not be confused with the Equal Cost Multipath Protocol (ECMP), aimed at addressing load balancing, as outlined by Hopps (2004) and yet to be discussed, whereas the Autonomic Cooperative Networking Protocol would gain from the increased robustness of the DSTBC-enabled cooperative transmission over the equivalent virtual multiple-input multiple-output radio channel, as indicated by Wódczak (2014).

In order to provide a wider context for the above, an appropriately extended version of the Autonomic Cooperative Node is outlined in Figure 4.36, where as well the fault management decision element, as the resilience and survivability one are visualised at the node level, each of them next to their still rather superior CMDE.[34] Thus, looking from the viewpoint of the CRDE, belonging to the node level, it becomes necessary to identify whether there may exist any group of Autonomic Cooperative Nodes that would be able to form an ACS between the SN and the DN. To this end, interaction with the node level would need to occur, as the candidate Autonomic Cooperative Nodes should be provided by the CMDE, as outlined by Wódczak (2012b). One should note, however, that such a process could by no means be reactive like the rationale behind the legacy FRR, where a virtually instantaneous response is always demanded. Consequently, a proactive solution needs to be put in place, allowing the readily available information provided by the seemingly proactive routines of the Optimised Link State Routing protocol, and, the multi-point relay station selection heuristics in particular, to be capitalised on. Should the possibility of having the autonomic cooperative set of relevance be positively validated thanks to such an interaction between the entities of both the function level and the node level, the Autonomic Cooperative Behaviour could be established; otherwise, the legacy FRR would need to be applied.

34 For the sake of clarity, however, one should note that for practical reasons the depiction of the extended version of the Autonomic Cooperative Behaviour contains a single entity, whereas previously it was always duplicated to emphasise that the network level should be treated as a more detached or elevated component thereof.

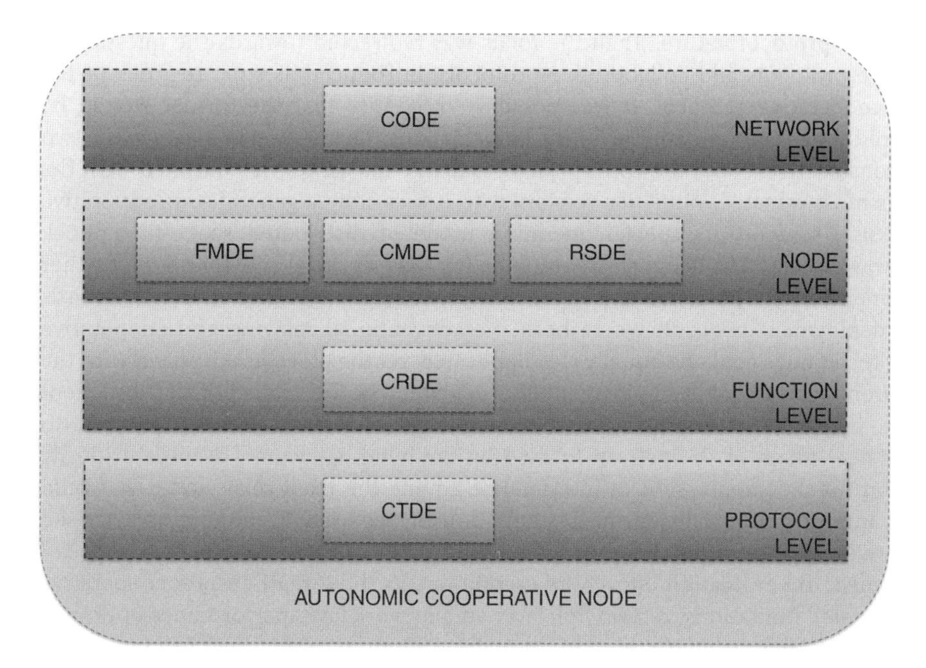

Figure 4.36 Extended version of an Autonomic Cooperative Node.

Last, but not least, and to some extent related to summarising the architectural advancements of this chapter, as well as smoothing the transition to the next one, there appears to arise the question of the incrementally growing role and scope of the consecutive levels of the Generic Autonomic Network Architecture. In fact, looking into this topic, one may find the prevailing perspective, where the viewpoint becomes more and more shifted in the direction of decision elements, substantially helpful in terms of structuring what has been presented. This is mostly because of the introduction of the Autonomic Cooperative Behaviour spanning all of the involved layers of the Open Systems Interconnection Reference Model, as this way a certain gradation becomes conspicuous making it necessary to perceive the related role of the decision elements residing in the consecutive levels as a whole range of possible scopes. Given such a context, the CTDE inherent in the protocol level could be classified as being of a rather more narrow scope, restricted to the responsibility for informing the CMDE inherent in the node level, and thereby classified as being of a much broader scope, about the necessity of involving additional intermediary Autonomic Cooperative Nodes to form ACSs. Between the two there would reside the presently discussed CRDE of the function level, acting as a kind of a modifier imposing cooperative re-routing whenever necessary, while on the very top there would appear the CODE of the network level.

4.5 Conclusion

In this chapter, first the topic of conventional relaying and cooperative relaying was addressed from the classification perspective, where the two approaches were characterised and the latter was further investigated in respect of the forwarding strategy and protocol nature. Then the focus was redirected towards the question of supportive protocols and collaborative protocols, introduced as subcategories of the generic cooperative protocol. It was possible to identify that the former should be considered as a preparatory phase for the latter, making the interaction between the two highly constructive. Next, the concept of virtual antenna arrays was outlined with the aid of its most versatile multi-tier version, where, assuming a generalised cooperative transmission scheme, its special operation mode of distributed space-time block coding was discussed as being clearly intended to play a crucial role for all the further developments in the book. Given such a context, attention was then directed towards a fixed deployment concept, where both the conventional relaying and cooperative relaying techniques could be equally well applicable, yet the plot was advanced with the assumption that the mobile deployment concept would be the subject of all subsequent analyses. In particular, the grid-based Manhattan scenario was initially outlined to stress that as much as the pattern formed by the buildings would become critically important for the suppression of interference among the fixed relay nodes, it would make it impossible to call for the instantiation of any virtual antenna array based cooperative relaying among them.

In essence, the evaluation effort was carried out to show that, despite cooperative relaying related limitations, certain link layer and network layer performance optimisation would still be possible. To this end a specific adaptation strategy was proposed with regard to framing structure and buffer memory, so that, using the process interaction simulation method, it became possible to observe improved packet throughput at the network layer. Similarly, a cooperation-enabled relay-enhanced cell type indoor scenario was analysed where the major emphasis was put on aspects related to the link layer, bearing in mind its applicability to any later mobile deployment concept considerations. Eventually, the focus was shifted towards the function level overlay logic, where, first of all, the roots of Autonomic Cooperative Behaviour were outlined to account for its role and complexity, including its enablers, and the equivalent distributed space-time block encoder in particular. Then, the rationale behind the cooperative re-routing decision element was presented, including its transition from the node level to the function level, and the logic behind cooperative re-routing involving the role of the fault management decision element, as well as resilience and survivability decision element. Finally, the architectural integration aspects were discussed to account for the general dependencies between the routines of all three layers of interest, as well as to provide a more detailed insight into the architectural relations driven by the tandem of the link layer and the function level, complemented by the introduction of an extended version of the Autonomic Cooperative Node.

References

Chaparadza R, Wódczak M, Meriem TB, De Lutiis P, Tcholtchev N, and Ciavaglia L 2013 Standardization of resilience and survivability, and autonomic fault-management, in evolving and future networks: An ongoing initiative recently launched in ETSI. *Ninth International Conference on the Design of Reliable Communication Networks (DRCN)*, Budapest, Hungary.

Dohler M and Li Y 2010 *Cooperative Communications: Hardware, Channel & PHY*. Wiley.

Dohler M, Gkelias A, and Aghvami H 2003a 2-hop distributed MIMO communication system. *IEE Electronics Letters* **39**(18) 1350–1351.

Dohler M, Gkelias A, and Aghvami H 2004 A resource allocation strategy for distributed MIMO multi-hop communication systems. *IEEE Communications Letters* **8**(2), 99–101.

Dohler M, Lefranc E, and Aghvami H 2002a Space-time block codes for virtual antenna arrays. *13th IEEE Proceedings on Personal, Indoor and Mobile Radio Communications, PIMRC 2002*, pp. 414–417.

Dohler M, Lefranc E, and Aghvami H 2002b Virtual antenna arrays for future wireless mobile communication systems. *International Conference on Telecommunications, ICT*.

Dohler M, Said F, and Aghvami H 2003b Higher-order space-time block codes for virtual antenna arrays. *10th International Conference on Telecommunications, ICT*, pp. 198–203.

Doppler K, Osseiran A, Wódczak M, and Rost P 2007a On the integration of cooperative relaying into the WINNER system concept. *16th IST Mobile & Wireless Communications Summit 2007*, Budapest, Hungary.

Doppler K, Redana S, Wódczak M, Rost P, and Wichman R 2007b Dynamic resource assignment and cooperative relaying in cellular networks: Concept and performance assessment. *EURASIP Journal on Wireless Communications and Networking* **2009**(475281), 1–14.

Dottling M, Irmer R, Kalliojarvi K, and Rouquette-Leveil S 2009 System model, test scenarios, and performance evaluation. In *Radio Technologies and Concepts for IMT-Advanced*, eds. Dottling M, Mohr W, and Osseiran A. Wiley.

Esseling N, Pabst R, and Walke B 2005 Delay and throughput analysis of a fixed relay concept for next generation wireless systems. *11th European Wireless Conference*, pp. 273–279.

Esseling N, Walke B, and Pabst R 2004 Performance evaluation of a fixed relay concept for next generation wireless systems. *15th IEEE International Symposium on Personal, Indoor and Mobile Radio Communications, PIMRC*.

ETSI-GS-AFI-002 2013 *Autonomic network engineering for the self-managing Future Internet (AFI); Generic Autonomic Network Architecture (An Architectural Reference Model for Autonomic Networking, Cognitive Networking and Self-Management)*. ETSI Group Specification.

Głąbowski M and Wódczak M 2006 On throughput maximization oriented approach to buffer memory management in context of the relay-based Manhattan-type deployment concept. *IST Mobile Summit*.

Herhold P, Zimmermann E, and Fettweis G 2004a A simple cooperative extension to wireless relaying. *2004 International Zurich Seminar on Communications*, Zurich, Switzerland.

Herhold P, Zimmermann E, and Fettweis G 2004b On the performance of cooperative amplify-and-forward relay networks. *ITG Conference on Source and Channel Coding (SCC)*, Erlangen, Germany.

Herhold P, Zimmermann E, and Fettweis G 2005 Cooperative multi-hop transmission in wireless networks. *Computer Networks Journal* **49**(3), 299–324.

Hopps C 2004 *Analysis of an Equal-Cost Multi-Path Algorithm*. RFC 2992.

Jamal T and Mendes P 2014 Cooperative relaying in user-centric networking under interference conditions. *IEEE Communications Magazine* **52**(12), 18–24.

Laneman JN and Wornell GW 2003 Distributed space-time-coded protocols for exploiting cooperative diversity in wireless networks. *IEEE Transactions on Information Theory* **49**(10), 2415–2425.

Laneman JN, Tse D, and Wornell G 2004 Cooperative diversity in wireless networks: Efficient protocols and outage behavior. *IEEE Transactions on Information Theory* **50**(12), 3062–3080.

Liu KH and Lin P 2015 Toward self-sustainable cooperative relays: State of the art and the future. *IEEE Communications Magazine* **53**(6), 56–62.

Nishiyama H, Ito M, and Kato N 2014 Relay-by-smartphone: Realizing multihop device-to-device communications. *IEEE Communications Magazine* **52**(4), 56–65.

Pabst R, Esseling N, and Walke B 2005 Fixed relays for next generation wireless systems: System concept and performance evaluation. *Journal of Communications and Networks, Special Issue 'Towards the Next Generation Mobile Communications'* 7(2), 104–114.

Pabst R, Walke B, Schultz DC, Herhold P, Yanikomeroglu H, Mukherjee S, Viswanathan H, Lott M, Zirwas W, Dohler M, Aghvami H, Falconer D, and Fettweis G 2004 Relay-based deployment concepts for wireless and mobile broadband radio. *IEEE Communications Magazine* **42**(9), 80–89.

Pahlevani P, Hundeboll M, Pedersen M, Lucani D, Charaf H, Fitzek F, Bagheri H, and Katz M 2014 Novel concepts for device-to-device communication using network coding. *IEEE Communications Magazine* **52**(4), 32–39.

Raha A, Malcolm N, and Zhao W 1996 Hard real time communications with weighted round robin service in ATM local area networks. *1st International Conference on Engineering of Complex Computer Systems*, pp. 96–104.

Retvari G, Nemeth F, Prakash A, Chaparadza R, Hokelek I, Fecko M, Wódczak M, and Vidalenc B 2011 A guideline for realizing the vision of autonomic networking: Implementing self-adaptive routing on top of OSPF. In *Formal and Practical Aspects of Autonomic Computing and Networking: Specification, Development, and Verification*, ed. Cong-Vinh P. IGI Global.

Schultz DC, Walke B, Pabst R, and Irnich T 2003 Fixed and planned relay based radio network deployment concepts. *10th Wireless World Research Forum*.

Sinclair J 1997 *Collins COBUILD English Dictionary*. HarperCollins Publishers.

Tyszer J 1999 *Object-Oriented Computer Simulation of Discrete-Event Systems*. Kluwer Academic Publishers.

Wódczak M 2013 Dependability aspects of autonomic cooperative computing systems. In *Advances in Intelligent Systems and Computing*, ed. Zamojski W, Mazurkiewicz J, Sugier J, Walkowiak T, and Kacprzyk J, Springer.

Wódczak M 2006 *On Routing information Enhanced Algorithm for space-time coded Cooperative Transmission in wireless mobile networks*. PhD thesis, Faculty of Electrical Engineering, Institute of Electronics and Telecommunications, Poznań University of Technology, Poland.

Wódczak M 2008 Cooperative relaying in an indoor environment. *ICT Mobile Summit*, Stockholm, Sweden.

Wódczak M 2010 Future autonomic cooperative networks. *Second International ICST Conference on Mobile Networks and Management*, Santander, Spain.

Wódczak M 2011a Aspects of cross-layer design in autonomic cooperative networking. *IEEE Third International Workshop on Cross Layer Design*, Rennes, France.

Wódczak M 2011b Resilience aspects of autonomic cooperative communications in context of cloud networking. *IEEE First Symposium on Network Cloud Computing and Applications*, Toulouse, France.
Wódczak M 2012a Autonomic cooperative communications for emergency networks. *Fourth Computer Science and Electronic Engineering Conference (CEEC)*, University of Essex, Colchester, United Kingdom.
Wódczak M 2012b *Autonomic Cooperative Networking*. Springer.
Wódczak M 2014 *Autonomic Computing Enabled Cooperative Networked Design*. Springer.
Wódczak M, Meriem TB, Radier B, Chaparadza R, Quinn K, Kielthy J, Lee B, Ciavaglia L, Tsagkaris K, Szott S, Zafeiropoulos A, Liakopoulos A, Kousaridas A, and Duault M 2011 Standardizing a reference model and autonomic network architectures for the self-managing future internet. *IEEE Network* **25**(6), 50–56.
Wódczak M, Szott S, and Chaparadza R 2013a Autonomic Cooperative Behaviour in ETSI AFI scenario for autonomicity enabled ad-hoc and mesh network architecture. *Fifth IEEE MENS at GLOBECOM 2013*, 9–13 December, Atlanta, Georgia, USA.
Wódczak M, Tcholtchev N, Vidalenc B, and Li Y 2013b Design and evaluation of techniques for resilience and survivability of the routing node. *International Journal of Adaptive, Resilient and Autonomic Systems (IJARAS)* **4**(4), 36–63.
Zimmermann E, Herhold P, and Fettweis G 2003 On the performance of cooperative diversity protocols in practical wireless systems. *58th VTC*, Orlando.
Zimmermann E, Herhold P, and Fettweis G 2005 On the performance of cooperative relaying in wireless networks. *European Transactions on Telecommunications* **16**(1), 5–16.

5

Node Level Routing Mechanisms

5.1 Introduction

Following the previous chapter, where the notion of function level relaying techniques was analysed, starting with the developments related to both conventional relaying and cooperative relaying, advanced with the rationale behind the related fixed deployment concepts, to eventually prepare the ground for the introduction of approaches compliant with the mobile deployment concept, orchestrated by the Autonomic Cooperative Behaviour, the time has come to progress further with the node level routing mechanisms. The description is opened with the pertinent workings of the Optimised Link State Routing protocol, where special emphasis is laid on the experimentation-related version thereof. First, the functional and structural characteristics are analysed in more detail, translating generally into the field of applicability and the assumed messaging structure. Not only is the proactivity of the Optimised Link State Routing protocol underlined as being highly relevant to the mobile ad hoc network scenarios that are predominantly of interest, but its inherent multi-point relay station selection heuristics is presented, incorporating certain small alignments in light of its pivotal role in the concept of the Autonomic Intelligence Evolved Cooperative Networking. Additionally, the information storage repositories are scrutinised to provide the necessary context for further developments, and to introduce new elements such as the virtual antenna array selector set and its related virtual antenna array selector tuples, so necessary to advance the plot to encompass routing information enhanced cooperative transmission.

Then follow developments stemming from the routing information enhanced algorithm for cooperative transmission, originally devised by the author as a method for exploiting the additional data collected at the network layer by the Optimised Link State Routing protocol, and its accordingly modified version in particular, for the sake of not only enabling, but also orchestrating the said cooperative transmission at the link layer. For this reason the justification for the introduction of the routing information enhanced algorithm for cooperative transmission is provided, with special emphasis on its algorithmic description, which incorporates certain elements and nomenclature of the Optimised Link State Routing protocol, mostly due to the direct usage of the outcome of the multi-point relay station selection heuristics. The reliance of the routing information enhanced algorithm for cooperative transmission on the multi-point relay station selection heuristics is key in this respect, as these multi-point relays are to form the virtual antenna arrays intended to instantiate the distributed space-time block coding over the virtual multiple-input multiple-output radio channel.

Autonomic Intelligence Evolved Cooperative Networking, First Edition. Michał Wódczak.

Then, as a more accurate concept in the form of the extended routing information enhanced algorithm for cooperative transmission is outlined, which will lead directly to the Autonomic Cooperative Networking Protocol, the evolved messaging structure is presented with appropriate commentary. Last, but not least, both the critical issues of address auto-configuration and duplicate address detection are analysed on the basis of certain externally proposed solutions.

The focus then shifts towards the ultimate function level overlay logic intended to allow the fusion of all the technological advancements presented thus far under the umbrella of the entities inherent in the Autonomic Cooperative Networking Architectural Model. In particular, the description starts with the aforementioned Autonomic Cooperative Networking Protocol: the reasoning for the prior introduction of its source extended routing information enhanced algorithm for cooperative transmission is provided, with all the updates necessary to ensure proper transition and integration, as well as the justification of the role and importance of the evolved messaging structure in the process of Autonomic Cooperative Node preselection is given, along with the rationale behind the specifically designed routing table. In the light of the above, an extended algorithmic description of the logic governing the cooperation management decision element is presented with reference to the previous analysis of the original routing information enhanced algorithm for cooperative transmission, so that it is not only possible to evaluate the advantages thereof with the aid of simulation analysis, but also to perform further investigations into the overhead aspects of the above-mentioned evolved messaging structure. As usual, the chapter is concluded with the architectural integration aspects, where, initially, the roots of the Autonomic Cooperative Networking Protocol are discussed, complemented with an overview of certain pertinent conceptual transitions, so that more complex dependencies among the routines and their architectural relations can be scrutinised.

5.2 Optimised Link State Routing Protocol

5.2.1 Functional and Structural Characteristics

In this book the emphasis is laid on the original version of the Optimised Link State Routing (OLSR) protocol, 'left in place for further experimentation'[1] (Clausen et al., 2014). In particular, it is the multi-point relay (MPR) station selection heuristics, aimed at reducing the control overhead (understood as the number of control messages broadcast to disseminate network topology information), that is to be of special interest for cooperative transmission integration. In general, this idea consists in the transmission of the so-called topology control (TC) messages with the aid of exclusive sets of carefully selected neighbour nodes, which not only belong to the one-hop

1 While the OLSR protocol version 2, as defined by Clausen et al. (2014), retains and leverages all the advantages of the original OLSR protocol, as proposed by Clausen and Jacquet (2003), the former is especially characterised by a much more modular design, where certain parts of the previously monolithic structure are extracted as standalone modules to be usable by other protocols. While such an approach is more than desirable from the commercial perspective, the experimental flavour of the original is much more appealing to the considerations and developments analysed here, mostly because additional complexity is avoided.

neighbourhood of a given network node, but are able to jointly cover the entire strict two-hop neighbourhood of the same, as outlined by Qayyum et al. (2002). Such one-hop network nodes are recognised with Hello messages, which are received by each of them, but never directly retransmitted any further. These messages are generated on the basis of the information stored in the local link set, neighbour set, and multi-point relay set, as defined by Clausen and Jacquet (2003). The broadcasting of Hello messages takes place in a periodic manner by every active network node over all the interfaces thereof for the sake of link sensing, which is indispensable for detecting the existence of a radio link.[2] Given this context, once the functional routing classes have been presented, the structural aspects of both the Optimised Link State Routing protocol packet and its inherent Hello message are described so that, following encapsulation, they may be conveyed with the aid of the User Datagram Protocol (UDP).

As proposed originally by Clausen and Jacquet (2003) and then expanded by Clausen et al. (2014), the Optimised Link State Routing protocol is targeted at mobile ad hoc networks.[3] Since such environments are typically characterised by very dynamic fluctuations in terms of a continually changing network topology, the protocol should be tailored accordingly so that, keeping the control overhead at a reasonable level, it is still possible to follow any changes and provide accurate routing information, as summarised by Wódczak (2012a). Most generally, one may distinguish between three major routing classes[4] attributable to mobile ad hoc network environments,[5] as outlined by Abolhasan et al. (2005). First of all, there is the proactive class, where each network node performs topology recognition on a regular basis, so that routing tables (RTBs) are always up to date. Unfortunately, unless optimised, such an approach could be costly in terms of the above-mentioned control overhead – frequently referred to as protocol overhead. Then one may distinguish the reactive class where topology recognition is performed once the routing tables need to be updated, which reduces the overhead, but unfortunately increases the routing path selection delay. Last, but not least, there is the hybrid class, being a combination of the advantages of the previous two methods, taking into account the activity of mobile nodes (MNs) in specific regions of the network. As long as the topology changes can be classified as insignificant the reactive mode should most naturally prevail, while, should the opposite hold true, the proactive one would apply.

Moving forward, the Hello messages allow each MN to discover its entire one-hop neighbourhood and to identify its overall two-hop neighbourhood, while the information gathered becomes directly exploitable by the multi-point relay station selection heuristics. In order to provide sufficient context for further introduction of the relevant extensions to the protocol in question, the packet format will be described along with the format of the Hello message it normally encapsulates, as originally introduced by Clausen and Jacquet (2003). In fact, one should note that the aforementioned TC messages are encapsulated in similar protocol packets, but as their analysis appears to remain beyond the scope of this description, the reader is referred directly to the

---

2 A radio link may exist in both directions, merely in one, or maybe even none (Adjih et al., 2003).

3 For both information theory related and novelty driven aspects thereto, the reader is referred to Andrews et al. (2008), as well as Conti and Giordano (2014), respectively.

4 For extended context, the reader may also refer to Sholander et al. (2002).

5 In essence, the OLSR protocol may well serve military-grade purposes, as indicated by Plesse et al. (2004). For this reason, depending on mission stage and specifics of the scenario, it may well appear similarly suited for emergency communications (Wódczak, 2012b).

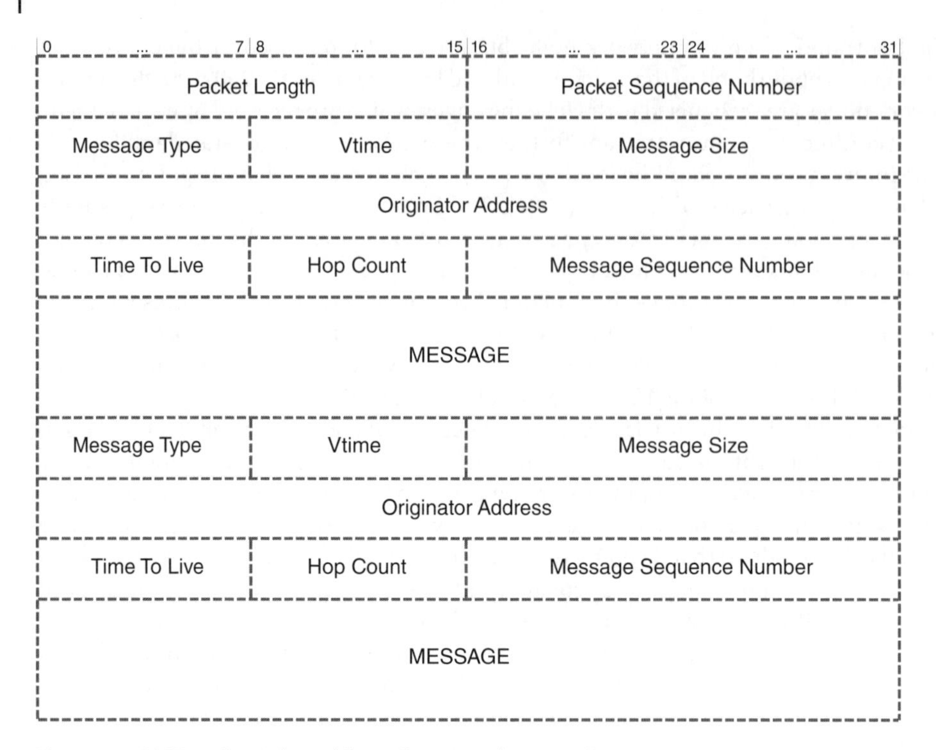

Figure 5.1 OLSR packet. Adapted from Clausen and Jacquet (2003).

specification by Clausen and Jacquet (2003). Scrutinising the said packet format, as depicted in Figure 5.1, one should discern immediately that its structure always opens with the Packet Length field of 16 bits which specifies the packet length in bytes. It is followed by the Packet Sequence Number field of 16 bits which is incremented each time a new packet is transmitted. Then, distinct messages are included, preceded by a header composed of a number of relevant fields. First of all, there is the Message Type of 8 bits which not only indicates the type of the carried message, but also accommodates sufficient amount of space to make any future attempts at defining new message types possible. Second is the Vtime field of 8 bits, also known as Validity time, which defines for how long the received data shall be considered valid in the absence of further updates, and is represented by mantissa a and exponent b values.

These values are contained in the four most significant bits (MSBs) and the four least significant bits (LSBs) of the Vtime field, respectively, and the resulting Validity time is calculated according to the following formula:

$$V_t = C\left(1 + \frac{a}{16}\right)2^b, \tag{5.1}$$

where C is a constant scaling factor which, according to Clausen and Jacquet (2003), is assumed to be equal to

$$C = \frac{1}{16} = 0.0625 \text{ s.} \tag{5.2}$$

Next is the Message Size field of 16 bits containing the size of the message in bytes, as counted from the beginning of the Message Size field to the beginning of the next

Message Size field, or the end of the entire Optimised Link State Routing protocol packet if there are no more messages. What follows is the Originator Address field of 32 bits containing the main address of the network node which originally issued this message. According to Clausen and Jacquet (2003), it is crucial to note that this address does not correspond to the Source Address of the Internet Protocol (IP) header, which changes each time to the address of the intermediate retransmitting interface. Then, there is the Time To Live (TTL) field of 8 bits, indicating the maximum number of hops over which a given message may be retransmitted. It is decremented by 1 before a retransmission, and a message must not be retransmitted should its TTL become equal to 0 or 1. Moving forward, there are the Hop Count field of 8 bits which contains the number of hops a given packet has traversed so far and the Message Sequence Number field of 16 bits which contains a unique identification number to ensure that a specific message is transmitted only once.

The final component in the form of the MESSAGE field has a variable size and carries the relevant contents, such as the Hello message. According to Figure 5.2, Hello messages also comprise a number of important fields (Clausen and Jacquet, 2003). First, there is the Reserved field of 16 bits that must be set to 0000000000000000.[6] This is followed by the Htime field of 8 bits, also known as the Holding time, which is used to specify the Hello message emission interval over a given interface. Such an interval is similarly represented in the form of a mantissa a, the four MSBs, and an exponent b, the four LSBs. Consequently, the emission interval can be calculated according to the following formula:

$$H_t = C\left(1 + \frac{a}{16}\right)2^b, \tag{5.3}$$

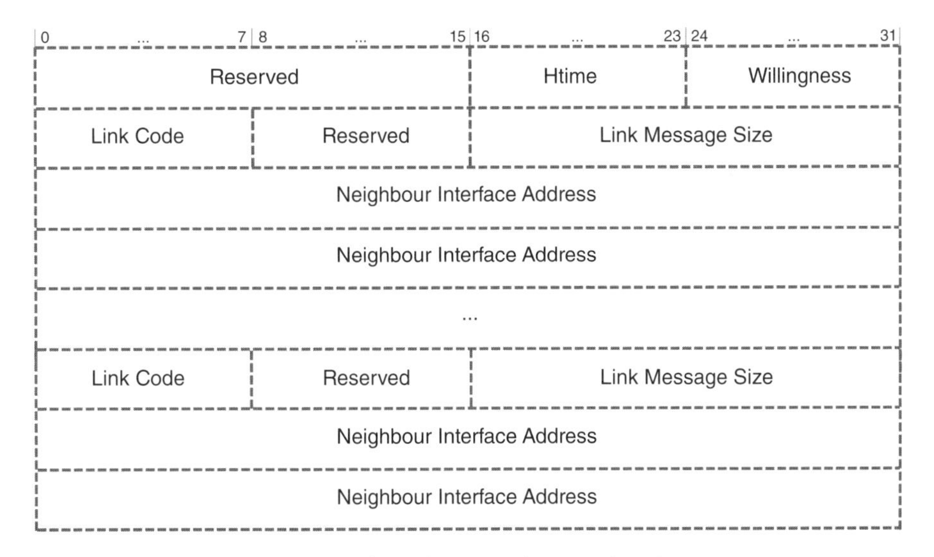

Figure 5.2 Hello message. Adapted from Clausen and Jacquet (2003).

6 In the case of the specification by Clausen and Jacquet (2003) only 13 zero values are given instead of 16; the author sees no point in leaving three of them unset (see also Wódczak, 2014a).

where C is the constant scaling factor defined in Equation 5.2. Although the predefined Hello message emission interval is 2 s, it can range from 62.6 ms up to almost 2.28 hr. Next, there is the Willingness field of 8 bits specifying whether a given network node is willing to carry and forward traffic or not. There are the following levels of willingness: WILL_NEVER (0), WILL_LOW (1), WILL_DEFAULT (3), WILL_HIGH (6), and WILL_ALWAYS (7). One should note that for a Willingness of 0 or 7 a given network node must never or must always be selected as a multi-point relay, respectively. Then comes the Link Code of 8 bits defining the type of the link between an interface of a given network node and the listed interfaces of its neighbours, as well as the Neighbour Type. Last is the Neighbour Interface Address of 16 bits, denoting the interface address of a given neighbour node.

Given the above, it transpires that, due to a unified packet format, not only is the Optimised Link State Routing protocol easily extensible, but its packets normally undergo further encapsulation with the use of the User Datagram Protocol (UDP), as defined by Postel (1980), before transmission over the network. For reference reasons, the format of a UDP datagram is outlined in Figure 5.3. Its structure begins with the Source Port field of 16 bits which is optional, yet when exploited indicates the port of the sending process in case such information is useful. If it is not specified a zero value is used; otherwise, this is the port to which a reply shall be sent should any other information be missing. Then comes the Destination Port field of 16 bits indicating the port of the destination process to which the datagram is to be delivered. Next comes the Length field of 16 bits which specifies the length of the datagram in octets including both the header and the data sections. Consequently, the minimum value of this field is 8. Then there is the Checksum field of 16 bits containing the checksum of the so-called pseudo-header, which is composed of the IP header, the UDP header, and the DATA field, padded with zero octets at the end, as described by Postel (1980). Last, but not least, is the DATA field of variable size which contains the data octets in the form of, for instance, the Hello messages described above, encapsulated into packets of the Optimised Link State Routing protocol, which was assigned the exclusive port number of 698 by the Internet Assigned Number Authority (IANA).

5.2.2 Multi-Point Relay Station Selection Heuristics

One of the main advantages of the Optimised Link State Routing protocol is related to its inherent ability to utilise a set of preselected network nodes for the purposes of control data dissemination. Such network nodes, known as multi-point relays, are

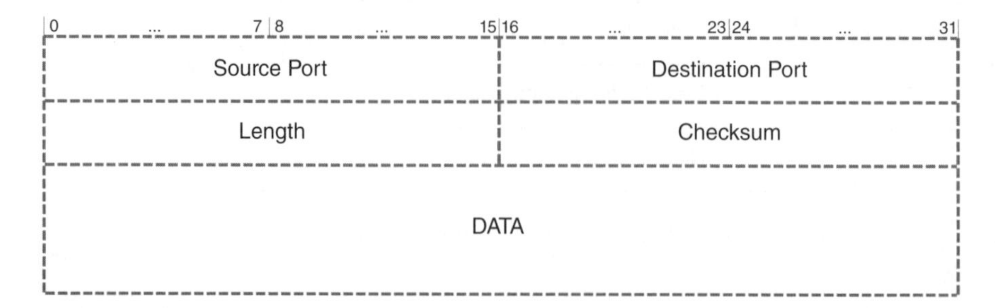

Figure 5.3 UDP datagram. Adapted from Postel (1980).

identified by a given source node on the basis of an analysis of its entire symmetric one-hop and two-hop neighbourhoods. In fact, even though all the neighbour nodes located within range of such a source node are always entitled to receive and process the control messages it may broadcast, given the obvious nature of the wireless transmission medium, they are not allowed to retransmit those messages any further unless they have been included in the MPR set. In other words, this approach aims to minimise the amount of redundantly retransmitted control data, so that the overall protocol overhead may be reduced, if not optimised. In order to perform the relevant multi-point relay station selection heuristics, a given source node, denoted as x, needs to acquire all the pertinent information related to its one-hop and two-hop neighbourhoods. Since the aforementioned Hello messages are periodically emitted by each one-hop neighbour node n, and they also include the statuses of the corresponding links, the source node may identify both its one-hop and two-hop symmetric neighbourhoods. Given the above, an optimised algorithmic description of the multi-point relay station selection heuristics is provided below, and the relevant neighbourhood types are classified for reference reasons. Finally, the distinction between routing and flooding multi-point relays is outlined, as defined by Clausen et al. (2014).

Before the multi-point relay station selection heuristics,[7] as depicted in Figure 5.4, is described, it is necessary to introduce the notation to be used. $N(x)$ and $N^{(2)}(x)$ refer to the sets of one-hop and two-hop neighbour nodes of a given source node, respectively. Similarly, $MPR(x)$ represents the multi-point relays of that source node,

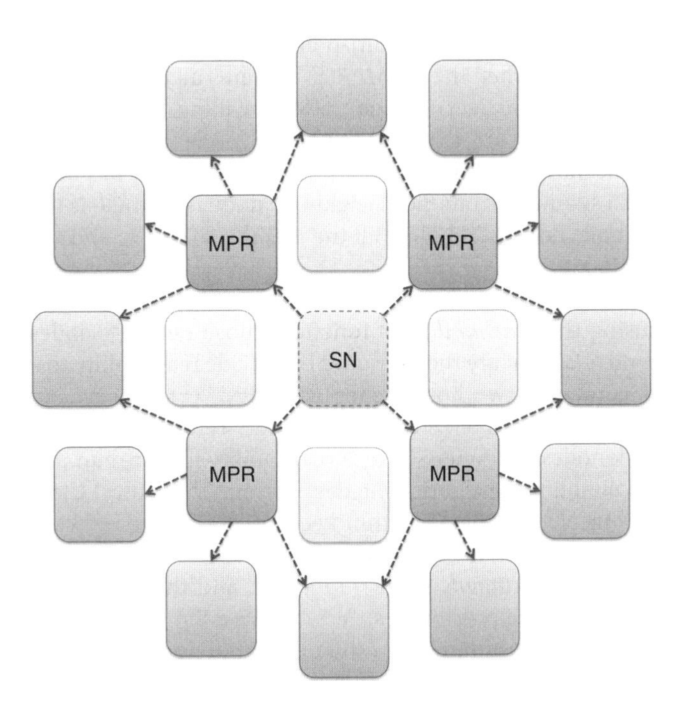

Figure 5.4 Multi-point relay station selection heuristics. Adapted from Qayyum et al. (2002).

7 From the already a bit historical perspective, the Multi-Point Relay Station selection heuristics may be perceived as the main enabler of the autonomic design presented in this book (Wódczak, 2010).

Table 5.1 Neighbourhood type characteristics.

Neighbourhood type	Characteristics
Symmetric one-hop neighbourhood	A set of network nodes which have at least one symmetric link to the source node.
Symmetric two-hop neighbourhood	A set of network nodes, excluding the source node, which have symmetric links to the symmetric one-hop neighbour nodes of that source node.
Strict symmetric two-hop neighbourhood	A set of network nodes, excluding the source node and its neighbour nodes, which have a symmetric link to a symmetric one-hop neighbour node of that source node, characterised by a willingness level different from WILL_NEVER.

where a multi-point relay is understood as a one-hop neighbour node selected by source node x to retransmit all the broadcast messages received from x on condition that, since a retransmitted message cannot be a duplicate, its TTL field must carry a value greater that one (Clausen and Jacquet, 2003). In general, the multi-point relay station selection heuristics is carried out with the aid of both sets of one-hop and two-hop neighbour nodes. For the sake of clarity, one needs to additionally assume that both the sets $N(x)$ and $N^{(2)}(x)$ are formed only of symmetric network nodes that are reachable via bidirectional links, as described in Table 5.1. According to the specification by Clausen and Jacquet (2003), the multi-point relay station selection heuristics is launched by source node x when the set $MPR(x)$ contains all the members belonging to the one-hop neighbourhood $N(x)$ that are always willing to carry and forward traffic. The author additionally assumes that those network nodes are no longer included in the set $N(x)$, and that the degree of each network node n in $N(x)$ has been computed, where the degree of a symmetric one-hop neighbour node is defined as the number of its symmetric neighbour nodes, excluding all the members of $N(x)$ and the source node x itself (Wódczak, 2006).

Examining Algorithm 5.1, adapted by the author from Clausen and Jacquet (2003), the source node x chooses, using the *reachability*(n) function, those network nodes from its one-hop neighbourhood $N(x)$ that are the only ones to provide reachability to a network node $n^{(2)}$ in the strict symmetric two-hop neighbourhood $N^{(2)}(x)$. However, in contrast to the specification, network nodes of zero reachability are removed from the set $N(x)$. Thus, only the network nodes characterised by a reachability higher than one may remain in the set $N(x)$, to simplify the algorithmic description. Then, until there exist uncovered network nodes in $N^{(2)}(x)$, the algorithm keeps selecting the network node in $N(x)$ that has the highest willingness to carry and forward traffic. This operation is performed with the aid of the *max_willingness*($N(x)$) function, and the results are stored in the set W. In the case of multiple possibilities, i.e. when *size*(W) > 1, the candidate is selected through which the highest number of still uncovered nodes in $N^{(2)}(x)$ can be reached. To this end the *max_reachability*(W) function is used, which results in the set R. If it is still impossible to select one network node only, i.e. *size*(R) > 1, the network node of the highest degree is chosen using the function *max_degree*(R). If the problem of multiple choice arises again, i.e. *size*(D) > 1, the author proposes picking one of them at random using the function *random_member*(D). One should note that each time a network node n, which covers a network node $n^{(2)}$, is included in the MPR set, the network node $n^{(2)}$ is removed from the set $N^{(2)}(x)$. Eventually, the network node n is removed from the set $N(x)$, accordingly.[8]

Algorithm 5.1 Modified multi-point relay station selection heuristics.

1: **for all** $n \in N(x)$ **do**
2:   **for all** $n^{(2)} \in N^{(2)}(x)$ **do**
3:     **if** $reachability(n) = 0$ **then**
4:       $N(x) \leftarrow N(x)\backslash\{n\}$
5:     **else if** $reachability(n) = 1$ **then**
6:       $MPR(x) \leftarrow MPR(x) \cup \{n\}$
7:       $N(x) \leftarrow N(x)\backslash\{n\},\ N^{(2)}(x) \leftarrow N^{(2)}(x)\backslash\{neighbour(n)\}$
8:     **end if**
9:   **end for**
10: **end for**
11: $N(x) \leftarrow N(x)\backslash MPR(x)$
12: **while** $N^{(2)}(x) \neq \emptyset$ **do**
13:   $W \leftarrow max\_willingness(N(x))$
14:   **if** $size(W) > 1$ **then**
15:     $R \leftarrow max\_reachability(W)$
16:     **if** $size(R) > 1$ **then**
17:       $D \leftarrow max\_degree(R)$
18:       $n \leftarrow random\_member(D)$
19:     **else**
20:       $n \leftarrow member(R)$
21:     **end if**
22:   **else**
23:     $n \leftarrow member(W)$
24:   **end if**
25:   $MPR(x) \leftarrow MPR(x) \cup \{n\}$
26:   **for all** $n^{(2)} \in N^{(2)}(x)$ **do**
27:     **if** $n^{(2)} = neighbour(n)$ **then**
28:       $N^{(2)}(x) \leftarrow N^{(2)}(x)\backslash\{n^{(2)}\}$
29:     **end if**
30:   **end for**
31: **end while**

Lastly, one should note that, unlike Clausen and Jacquet (2003), the latest version of the specification describing the Optimised Link State Routing protocol provides a new notion of routing multi-point relay (RMPR) stations, as opposed to the legacy ones, currently directly referred to as flooding multi-point relay (FMPR) stations (Clausen

8 Usually only a subset of all the available one-hop neighbour nodes is utilised for the dissemination of control messages. However, there are situations, such as very dynamic changes in the network topology, when it may be necessary to increase the protocol overhead by using redundant MPRs, as indicated by Clausen and Jacquet (2003). Such a special case, which may well pertain to mission-critical (Younis et al., 2009) or tactical (Younis et al., 2010) networks, is yet to be touched upon from a general perspective.

et al., 2014). Essentially, even though, at least at first sight, such a distinction would seem negligible, in fact there appears to be much more to it from the structural and functional perspectives.[9] In particular, looking at the legacy version of the Optimised Link State Routing protocol, the predominant role of the multi-point relay stations was to be 'responsible for forwarding control traffic, intended for diffusion into the entire network' (Clausen and Jacquet, 2003). Most obviously, the capability of multi-point relays was also exploited to provide routing enhancements, yet this was not so much exposed as it is now. In fact, looking at Clausen et al. (2014), one may easily discern that much more emphasis is laid on the routing part, as currently the protocol incorporates the exploitation of additive link metrics different from the previously allowed hop count. There is then no wonder that the FMPRs were given a proper name, while the RMPRs were defined to reflect certain characteristics. In essence, following what was specified by Clausen et al. (2014), on the one hand, while the former need not use any metrics at all, it becomes obligatory for the latter, yet, on the other hand, also quite reasonably, as the former are bound to an interface, the latter are by no means restricted in this respect.

5.2.3 Information Storage Repositories

The Optimised Link State Routing protocol acquires various pieces of information by means of link sensing, neighbour detection, multi-point relay station selection, or topology discovery. Consequently, it is bound to maintain certain information repositories; these will be characterised in this section to prepare for the introduction of the virtual antenna array (VAA) selector tuple which, among other modifications, is to be instrumental in outlining the evolved version of the routing information enhanced algorithm for cooperative transmission (REACT) (Wódczak, 2014a). In fact, examining the repositories in question, one may provide the general classification depicted in Figure 5.5, where four major categories are included: Multiple Interface Association Information Base, Local Link Information Base, Neighbourhood Information Base, and Topology Information Base. Interesting though it may seem, the most recent version of the Optimised Link State Routing protocol, as defined by Clausen et al. (2014), exploits a related, yet modified, group of repositories in the form of the Local Information Base, Interface Information Base,[10] and Neighbour Information Base. Given the fact that the experimentation-targeted version of the Optimised Link State Routing protocol is to be scrutinised in this chapter, any more detailed comparison with its most recent incarnation remains beyond the scope of this book. The four categories will be addressed by the presentation, in turn, of the interface association tuple, the link tuple, the neighbour tuple, two-hop neighbour tuple, MPR selector set, and the VAA selector tuple, to conclude with the topology tuple.

In an effort to provide commentary on the relevant tuples[11] adapted from Clausen and Jacquet (2003), first comes the Multiple Interface Association Information Base

9 The significance in this respect is related to the commercial value of the OLSR protocol; taking into account the experimentation-related dimension addressed in this book, this distinction will not directly influence the workings of the concepts to be addressed in the rest of this chapter.

10 As much as the names are cited directly as in Clausen et al. (2014), the author could not fully justify the plural form behind the term 'Interface Information Bases' found there in select contexts only. Consequently, the singular counterpart is used here.

11 For the sake of introducing the complementary VAA selector tuple, proposed below by the author.

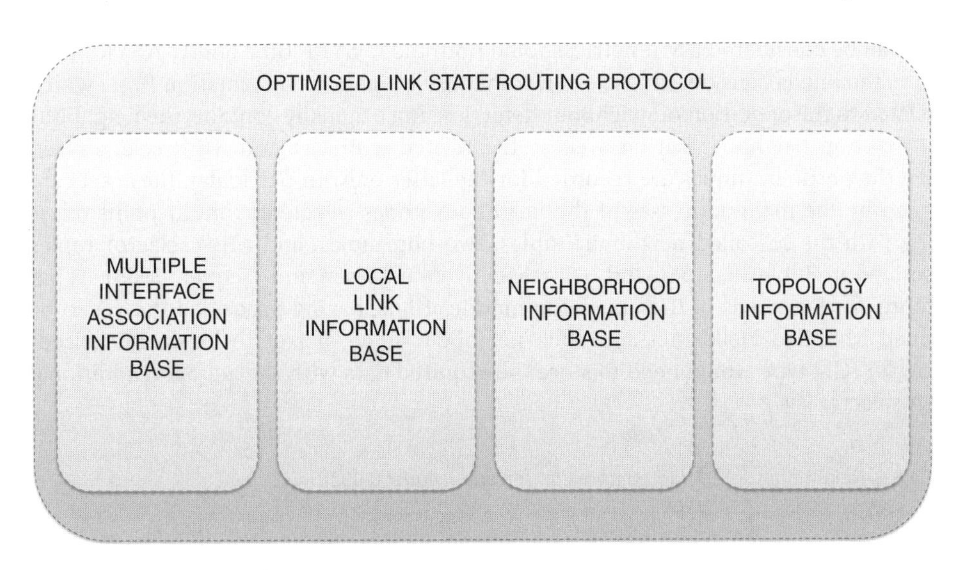

Figure 5.5 OLSR protocol repositories.

containing the interface association set. This set is composed of the so-called interface association tuples, which are stored for every destination node in the network. Such a tuple is composed of the items described in Table 5.2. Next comes the Local Link Information Base, which is related to the operation of link sensing and contains the related link set. This set is formed by the so-called link tuples comprising a number of items, as outlined in Table 5.3. Based on this information, the neighbour interfaces may be declared using Hello messages. One should note that once the L_SYM_time expires, the

Table 5.2 Interface association tuple. Adapted from Clausen and Jacquet (2003).

Item	Description
I_iface_addr	Specifies the address of the interface of a given destination node.
I_main_addr	Specifies the main address of a given destination node.
I_time	Specifies the expiry time at which this tuple will be discarded.

Table 5.3 Link tuple. Adapted from Clausen and Jacquet (2003).

Item	Description
L_local_iface_addr	Specifies the address of the interface of a given local network node.
L_neighbor_iface_addr	Specifies the address of the interface of a given neighbour node.
L_SYM_time	Specifies the time until which a given link shall be considered symmetric.
L_ASYM_time	Specifies the time until which a given interface of a given neighbour node shall be considered heard.
L_time	Specifies the expiry time at which this tuple shall be discarded.

link shall be considered asymmetric; should both the L_SYM_time and L_ASYM_time expire, the link is considered lost. Then comes the Neighbour Information Base, which is related to the operation of neighbour detection and originally contains the neighbour set, two-hop neighbour set, as well as the related MPR set and MPR selector set, while the pertinent tuples are recorded for the latter only. In particular, these sets are formed by the main addresses of the neighbour nodes selected as multi-point relays along with the so-called neighbour tuples, two-hop tuples, and MPR selector tuples described in Tables 5.4, 5.5, and 5.6, respectively (Clausen and Jacquet, 2003). What is more, for the needs of the upcoming modifications, having processed the yet to be defined Modified Hello message, each neighbour node of (yet to be characterised) VAA_NEIGH type would need to store the acquired data with the aid of an additional VAA selector set.

Table 5.4 Neighbour tuple. Adapted from Clausen and Jacquet (2003).

Item	Description
N_neighbor_main_addr	Specifies the main address of a given neighbour node.
N_status	Specifies whether a given neighbour node shall be considered as being connected over a symmetric or nonsymmetric link.
N_willingness	Specifies an integer value within the range 0 to 7 indicating whether a given neighbour node shall be willing to carry and forward[a)] traffic coming from other network nodes.

a) The specification does not explicitly mention forwarding, yet the author assumes that such a meaning is naturally implied in this respect.

Table 5.5 Two-hop neighbour tuple. Adapted from Clausen and Jacquet (2003).

Item	Description
N_neighbor_main_addr	Specifies the main address of a given neighbour node.
N_2hop_addr	Specifies the main address of a given two-hop neighbour node connected to the one denoted by N_neighbor_main_addr with a symmetric link.
N_time	Specifies the expiry time at which this tuple shall be discarded.

Table 5.6 MPR selector tuple. Adapted from Clausen and Jacquet (2003).

Item	Description
MS_main_addr	Specifies the main address of the neighbour node which selected this specific network node as an MPR.
MS_time	Specifies the expiry time at which this tuple shall be discarded.

Table 5.7 VAA selector tuple. Adapted from Wódczak (2014).

Item	Description
VS_main_addr	Specifies the main address of the neighbour node which selected this network node as the element of a VAA.
VS_elem_id	Specifies the VAA element identification number indicating the column according to which the EDSTBE shall process the retransmitted signals.
VS_time	Specifies the expiry time at which this tuple shall be discarded.

Such a VAA selector set, to be maintained in the Neighbour Information Base, as indicated by Wódczak (2014a), would be formed of VAA selector tuples following the format outlined in Table 5.7, so that each network node could determine whether it would be expected to cooperate after receiving a user data packet from any of its neighbours by simply comparing its address with the VS_main_addr. Should there be a match, then the relevant element of the virtual antenna array, also referred to as a VAA_NEIGH network node, would process the signals according to the appropriate column of the code matrix employed by the equivalent distributed space-time block encoder (EDSTBE), as specified by VS_elem_id. Finally, there is the Topology Information Base, which is related to the operation of topology discovery and contains the topology set. This set comprises the so-called topology tuples, which are composed of the items described in Table 5.8. All in all, comparing the structure of and relations between the information storage repositories of the original Optimised Link State Routing protocol with what is offered by its latest incarnation, one may come to the general conclusion that there are two dimensions to be considered. The most recent version of this protocol appears to promote a more consistent view by introducing a reduced number of repositories with a clear distinction between information defined locally and received globally. Moreover, in the experimentation-related version, a given repository or base may contain only a single set, but the latest update provides a much more balanced structure. Regardless of the above advantages, the experimentation track is followed in this book.

Table 5.8 Topology tuple. Adapted from Clausen and Jacquet (2003).

Item	Description
T_dest_addr	Specifies the main address of the neighbour node which may be reached in one hop by the network node denoted by T_last_addr.
T_last_addr	Specifies the network node which typically is the MPR of the destination node denoted by T_dest_addr.
T_seq	Specifies the sequence number.
T_time	Specifies the expiry time at which this tuple shall be discarded.

5.3 Routing Information Enhanced Cooperation

5.3.1 Justification and Algorithmic Outline

Given the background provided in the opening part of this chapter, where certain aspects of relevance for further description, inherently rooted in the Optimised Link State Routing protocol were described with additional commentary, the time has come to advance the analysis towards an evolved version of what was initially devised by the author under the name routing information enhanced algorithm for cooperative transmission (Wódczak, 2012a).[12] As such, the original version of REACT was conceived with the aim of finding a solution that could not only orchestrate, but also enhance the typically link layer driven cooperative relaying based on virtual antenna arrays, where distributed space-time block coding would be instantiated among the network nodes preselected to cooperate. In particular, the observation was made that there is much in common between the multi-point relay station selection heuristics of the Optimised Link State Routing protocol and the manner in which virtual antenna arrays are formed. Then, certain design decisions were made with respect to the workings of the Optimised Link State Routing protocol in order to reuse the routing information readily available at the network layer to enhance the cooperative relaying at the link layer. In the light of the above, the common aspect between the multi-point relay station selection heuristics and the concept of virtual antenna arrays will be detailed, so that it is possible to elaborate on the algorithmic description before the focus shifts to the evolved messaging structure (Wódczak, 2014a).

Keeping in mind the importance of the applied nomenclature, one should note that, even though certain advancement in the naming structure was already indicated, including the notion of virtual antenna arrays to be elevated to the concept of autonomic cooperative sets (ACSs), the author decided that, for the sake of clarity, the original naming pattern employed for REACT should at this stage be generally followed, with additional commentary whenever applicable. This is especially important for the introduction of the common aspect between multi-point relays and virtual antenna arrays, since both of them are soon to be integrated with the concept of the Autonomic Cooperative Node (ACN). Nonetheless, shifting the viewpoint more towards the Optimised Link State Routing protocol itself, one may notice that its inherent mechanisms were designed to allow each of the network nodes to acquire virtually full knowledge about their one-hop and two-hop neighbourhoods. What is more, it is possible to identify the one-hop neighbour nodes in $N(x)$ that can provide connectivity to certain two-hop neighbour nodes in $N^{(2)}(x)$. In fact, this is one of the major reasons why REACT was based on the multi-point relay station selection heuristics, since obviously there exists a common aspect between the two. Generally, only those network nodes are identified as multi-point relays that can provide connectivity to a two-hop neighbour node $n^{(2)}$. As depicted in Figure 5.6, this assumption especially holds true for the network nodes to be preselected as the relay nodes intended to form a virtual antenna array.

Moving into the details, based on additional link state information provided by the Optimised Link State Routing protocol, the proposed solution attempts to assign multi-point relays to specific virtual antenna arrays, as outlined by Wódczak (2011c).

12 One might note that the initial idea behind REACT originated in Wódczak (2006).

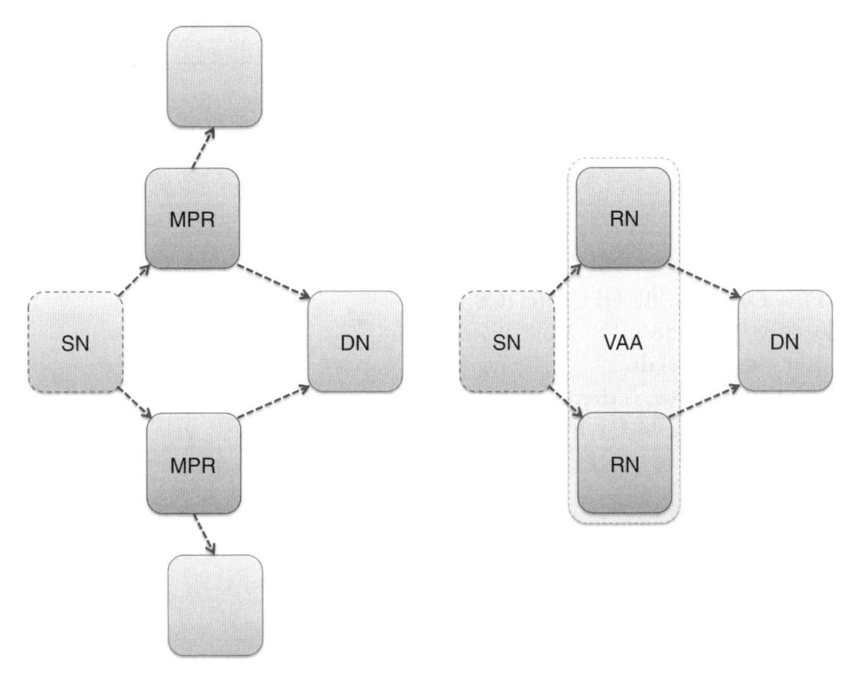

Figure 5.6 Common aspect of MPRs and VAAs.

For the sake of the analysis of Algorithm 5.2, one may recall that the sets $N(x)$ and $N^{(2)}(x)$, containing solely symmetric neighbour nodes reachable over bidirectional links, are formed respectively by the one-hop neighbours and two-hop neighbours of the network node denoted as x. Moreover, the degree of a symmetric one-hop neighbour node is defined as the number of its symmetric neighbour nodes, excluding all the members of $N(x)$ and the network node x itself (Clausen and Jacquet, 2003). Looking into the algorithmic description itself, first, each neighbour node n characterised by zero degree, i.e. $degree(n) = 0$, is removed by the network node x from the set $N(x)$. Then, the classic multi-point relay station selection heuristics is executed iteratively over set $N(x)$, as long as all the potential multi-point relays have been assigned to distinct $MPR^i(x)$ sets. Consequently, each iteration results in an additional set of multi-point relays, i.e. secondary, ternary, and so on, based on which all the neighbour nodes contained therein may become allocated to the most relevant virtual antenna arrays.[13] Such virtual antenna arrays are denoted $VAA(x, n^{(2)})$ and are capable of providing cooperative connectivity between the source node x and the destination node $n^{(2)}$, where $n^{(2)}$ belongs to the two-hop symmetric neighbourhood of x. Given such an approach, any intermediate network node n may become included in more than one virtual antenna array, as indicated by Wódczak (2011a).

One advantage of the proposed routine is that not only do all the intermediate neighbour nodes become preselected, but at the same time additional redundancy is introduced in comparison to the original multi-point relay station selection heuristics. Such

13 At this stage the 'pre-autonomic' nomenclature is still maintained for consistency reasons.

Algorithm 5.2 REACT.

```
1: for all n ∈ N(x) do
2:   if degree(n) = 0 then
3:     N(x) ← N(x)\{n}
4:   end if
5: end for
6: i ← 1
7: while N(x) ≠ ∅ do
8:   MPR^i(x) ← OLSR_MPR_HEURISTICS(N(x))
9:   for all n ∈ MPR^i(x) do
10:     for all n^(2) ∈ N^(2)(x) do
11:       if n = neighbour(n^(2)) then
12:         VAA(x, n^(2)) ← VAA(x, n^(2)) ∪ {n}
13:       end if
14:     end for
15:   end for
16:   N(x) ← N(x)\MPR^i(x)
17:   i ← i + 1
18: end while
```

redundancy may be utilised in the case of any unexpected sudden changes in the topology of the mobile ad hoc network,[14] where these additional $MPR^i(x)$ could be taken into account adaptively, or maybe even autonomically (as will be discussed once the Autonomic Cooperative Networking Protocol has been introduced), in order to provide better coverage, as outlined by Wódczak (2011b). Since, for the Optimised Link State Routing protocol, all the one-hop neighbour nodes are notified about having been chosen as multi-point relays by Hello messages, the same pattern will be followed to inform them about having been assigned to specific virtual antenna arrays. In this way, additional information is conveyed, on receipt of which a given network node n may learn directly, first of all, that it is supposed to take part in the DSTBC-enabled cooperative transmission, and, equally importantly, according to which column of the matrix defining the operation of the EDSTBE it should perform the processing the received signal. Evaluation analysis of the performance of the proposed approach is carried out using the scenario depicted in Figure 5.7, where the mobile ad hoc network is formed by network nodes of the following types: source node (0), relay nodes (1–8), and possible destination nodes (9–24) (Wódczak, 2012a).

Looking at this scenario, one should note that the relay nodes marked 'I' are those that would be selected by the classic multi-point relay station selection heuristics, whereas the ones marked 'II' belong to the redundant secondary set of multi-point relays additionally selected by the routing information enhanced algorithm for cooperative transmission. To reduce the complexity of the simulated configuration, the maximum size of the virtual antenna array is limited to 2, so that the full-rate G_2 code matrix is applicable, originally proposed by Alamouti (1998); other matrices would still be applicable (Tarokh

14 Apart from the aforementioned emergency-related use cases, one may also come across a growing role for social-networking-driven deployments (Chessa et al., 2016).

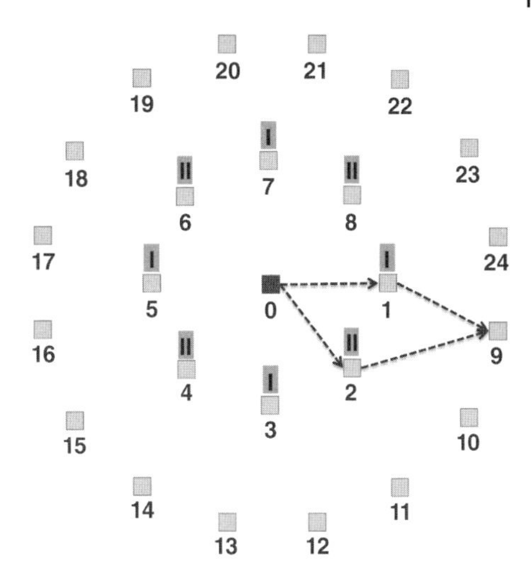

Figure 5.7 REACT scenario.

et al., 1999). Consequently, once the routing information enhanced algorithm for cooperative transmission has been activated, the primary $MPR^1(0) = \{1, 3, 5, 7\}$ and secondary $MPR^{(2)}(0) = \{2, 4, 6, 8\}$ sets are created. In fact, this is where the first indication of the readiness of the proposed solution for incorporation into the Autonomic Cooperative Networking Architectural Model becomes clearly visible, as the latter set may be used autonomically, should an increase in the control overhead become necessary. Based on both these sets, as well as on the assumption that the data stream originates from source node 0 and is destined for destination node number 9, $VAA(0, 9) = \{1, 2\}$ is created at the beginning of the simulation, as indicated in Figure 5.7. However, since the relay nodes and the destination node are assumed to be mobile and moving around at a speed of 5 km h^{-1}, the assignment of relay nodes to $VAA(0, 9)$ will be dynamic during the entire course of the simulation analysis, as outlined by Wódczak (2014a).

Although the devised simulation environment supported switching between single-path relaying and the routing information enhanced algorithm for cooperative transmission mode, it was guaranteed that at least two relay nodes would be available in the region of interest, so that DSTBC-enabled cooperative transmission was continuous. As such, the analysis was carried out on the downlink, where single-input multiple-output, at the first hop, and multiple-input single-output, at the second hop, block Rayleigh channels were used, as described by Wódczak (2012a). The channel coefficients for the links between distinct pairs of network nodes were generated according to the previously described formulas, as originally proposed by Zheng and Xiao (2003). The total transmitted power, either by a single network node or a virtual antenna array, was always normalised to unity. Also, the quadrature phase-shift keying (QPSK) modulation scheme was used, and the signal was perturbed by additive white Gaussian noise with zero mean and $N_0/2$ variance per dimension. Given such a simulation configuration, both the single-path relaying mode, where the transmission was assisted by one relay node only, and the routing information enhanced algorithm for cooperative transmission mode, exploiting two relay nodes, were investigated. The corresponding results are presented in Figure 5.8, where the numbers placed in the legend next to the names of the specific system configurations are used to identify the

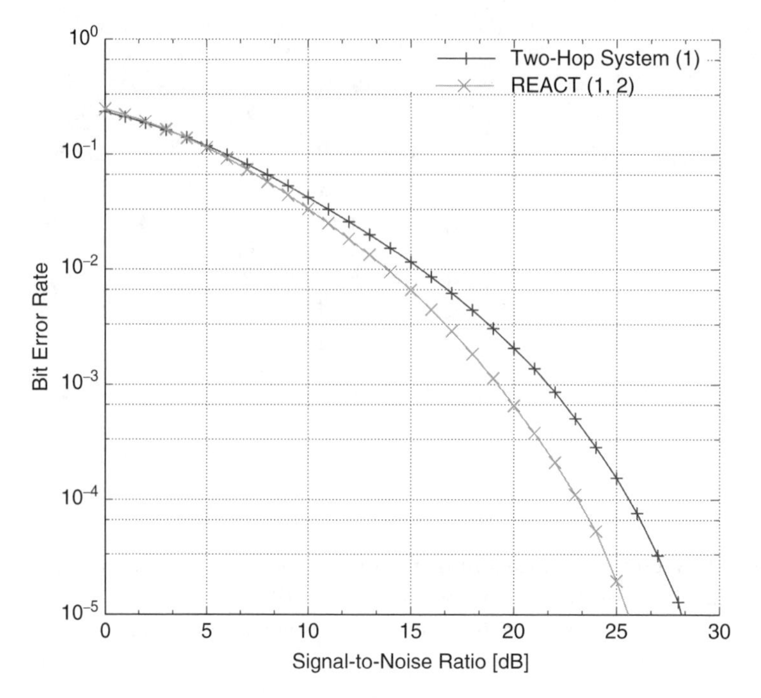

Figure 5.8 Performance of REACT.

next hop neighbour node or neighbour nodes. Next, the focus is to be shifted more towards the complementary evolved messaging structure (Wódczak, 2014a).

5.3.2 Evolved Messaging Structure

Attempting to understand the limitations of the Optimised Link State Routing protocol, one should note that too rough a Link Type classification it offers could make the entire process of virtual antenna array preselection rather ineffective, thereby spoiling the expected gain from cooperative transmission. Simply because such a generic Link Type allocation may seem justified in the context of mobile ad hoc networks, where knowledge of whether a link is symmetric or not suffices for the orchestration of the communications, the cooperative transmission approach calls for much more detailed feedback regarding the power level of the signal between neighbour nodes. As the Optimised Link State Routing protocol implies that each network node may group all the Neighbour Interface Addresses characterised by both the same Link Type and Neighbour Type in a single Link Message, the imprecise Link Type information becomes advantageous, since it guarantees that a number of Neighbour Interface Addresses are assigned to the same Link message, whereas increasing the accuracy could reduce the effectiveness of the protocol in terms of its control overhead (Wódczak, 2012a). While the achievable trade-off in terms of accuracy and overhead is yet to be scrutinised in detail, the evolution of the messaging structure is presented below. First, the emphasis will be laid on the Link Type and Neighbour Type in order to justify the modifications to follow. Then the Modified Hello message structure will be presented along with its respective Extended Link Code (ELC) and Extended Link Mask (ELM). Finally, following Wódczak (2012a), the format of the Generalised Hello message will be outlined.

Leveraging the evolved messaging structure (EMS), as outlined by Wódczak (2012a), for the needs of this book, it seems apparent that the Link Code of the Hello message becomes instrumental in the introduction of all the relevant modifications. This is mostly because of its internal structure, as depicted in Figure 5.9, which may be both an indicator of the existing limitations of and the areas for improvement with regard to the Optimised Link State Routing protocol. Essentially, as adapted in Table 5.9, this protocol is able to collect merely very generic parameters for the radio connections between the neighbour nodes, which results in the assignment of specific links to four distinct groups: UNSPEC_LINK, ASYM_LINK, SYM_LINK, and LOST_LINK (Clausen and Jacquet, 2003). While such a classification may be sufficient for the functioning of the said mobile ad hoc network, the enabling mechanism for cooperative transmission requires much more precise parameters, as indicated by Wódczak (2014a). What is more, the selection of the complementary Neighbour Types also appears to be somewhat restricted, as only the values NOT_NEIGH, SYM_NEIGH, and MPR_NEIGH are available. Yet, not too much effort will be required to arrange for an expansion in this respect, as the existence of an additional and not yet allocated Neighbour Type makes the introduction of the new value of VAA_NEIGH fairly straightforward, as described in Table 5.10 (Clausen and Jacquet, 2003). Thus it becomes feasible to produce the first stage of the concept behind enabling cooperative transmission between or among multi-point relay stations using virtual antenna arrays (Wódczak, 2014a).[15]

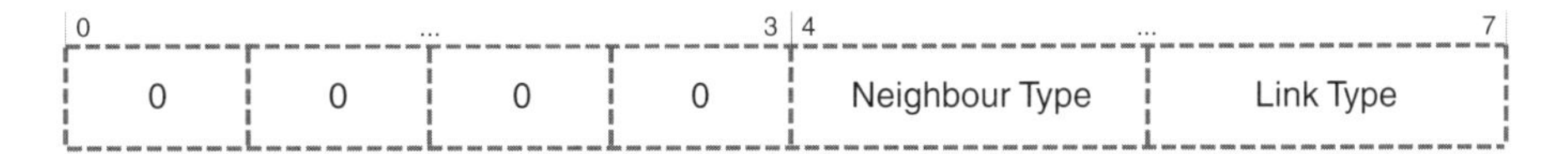

Figure 5.9 Link Code. Adapted from Clausen and Jacquet (2003).

Table 5.9 Link types. Adapted from Clausen and Jacquet (2003).

Link Type	Value	Description
UNSPEC_LINK	0	Indicates that no information about a given link is specified.
ASYM_LINK	1	Indicates that a given link is asymmetric, which means that it is only heard.
SYM_LINK	2	Indicates that a given link is symmetric and heard bidirectionally in both directions.
LOST_LINK	3	Indicates that a given link has been lost.

15 For the sake of clarity, one might note that compared to the specification by Clausen and Jacquet (2003), where both the Link Types and Neighbour Types are described from the perspective of the relation between a given network node and all its neighbour nodes, the approach assumed in Tables 5.9 and 5.10 is more of a point-to-point nature, which results from a slightly different perception exercised by the author with regard to the concept.

Table 5.10 Neighbour types. Adapted from Clausen and Jacquet (2003), as well as from Wódczak (2014).

Neighbour Type	Value	Description
NOT_NEIGH	0	Indicates that a given network node is no longer considered as or has not yet become a symmetric neighbour of a specific network node.
SYM_NEIGH	1	Indicates that there exists at least one symmetric link between this network node and a given neighbour node.
MPR_NEIGH	2	Indicates that there exists at least one symmetric link between this network node and a given neighbour node, additionally selected as an MPR.
VAA_NEIGH	3	Indicates that there exists at least one symmetric link between this network node and a given neighbour node, additionally selected as an element of a given VAA.

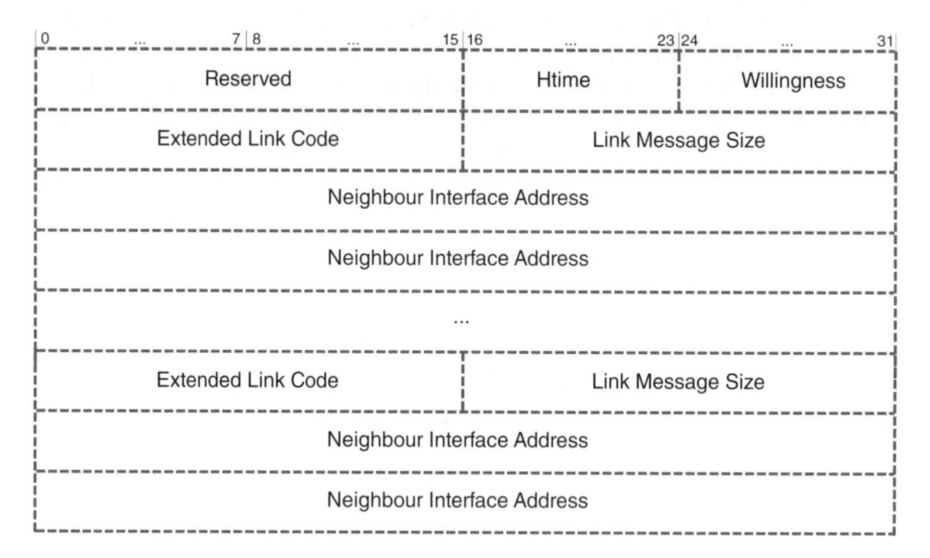

Figure 5.10 Modified Hello message.

In order to expand the capability of the Optimised Link State Routing protocol in terms of enabling more accurate link quality collection, a Modified Hello message[16] format is introduced, as depicted in Figure 5.10, where a 16 bit Extended Link Code field is allocated to replace the originally available Link Code and Reserved fields (Wódczak, 2012a). The structure of such an Extended Link Code is outlined in Figure 5.11. In particular, it is suggested that the four MSBs of the Link Code field be used along with eight bits of the Reserved field in order to constitute the Power Level field of total length 12 bits, so that additional information on the power level of the radio signal between neighbouring network nodes can be conveyed. As a result, the network node x will

16 This is, in fact, where the modifications inherent in EREACT start being rolled out, as originally outlined by Wódczak (2007). As such they will become instrumental in the introduction of the Autonomic Cooperative Networking Protocol later in this chapter.

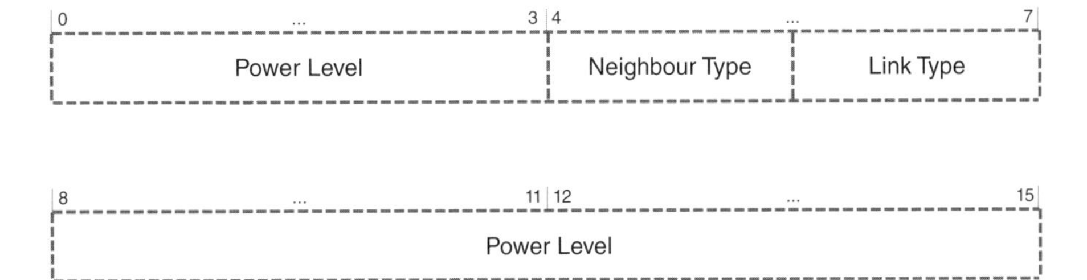

Figure 5.11 Extended Link Code.

be able to find out whether its one-hop neighbour *n* can hear the signals it transmits, and also learn what the precise power level of such a signal would be. Moreover, as the Hello messages periodically disseminated by each network node *n* usually contain similar information pertaining to its one-hop neighbourhood, potentially forming the two-hop neighbourhood of network node *x*, the latter could acquire a far more concrete overview of the link parameters in its entire one-hop and two-hop neighbourhoods, especially if radio channel reciprocity is assumed.[17] However, this modification to the Hello message format would no longer be completely transparent to the workings of the Optimised Link State Routing protocol. In other words, unlike the initial modification, where it was sufficient to make the protocol aware of the new VAA_NEIGH type, here the situation appears more demanding.

Now, not only is the Reserved field to be utilised, which, according to the specification, should remain unaltered, but also the aforementioned four MSBs are to be exploited, which, despite being meant for future extensions, are still not straightforwardly applicable to the contemplated modification (Wódczak, 2012a). Therefore, in order to overcome any such issues, a compromise attempt guaranteeing backward compatibility is needed. In essence, should a Hello message be processed for the purposes of performing the classic protocol operations, the newly introduced Extended Link Code field would need to be masked by application of the Extended Link Mask, as depicted in Figure 5.12. Going further, it would mean that any specific implementation of an accordingly modified Optimised Link State Routing protocol would need to exploit the Extended Link Mask to perform the operation of logical conjunction over all the Extended Link Codes included in a specific Modified Hello message.[18] The

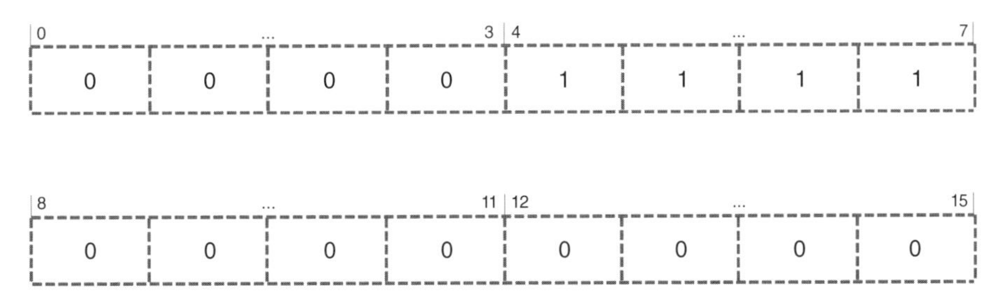

Figure 5.12 Extended Link Mask.

---

17 One could also consider the possibility of employing a still suboptimal approach consisting in the sharing of the available Power Level bits between the incoming and outgoing direction for each of the links.

18 For the wider context of the protocol stack, the reader is referred to Wódczak (2012c).

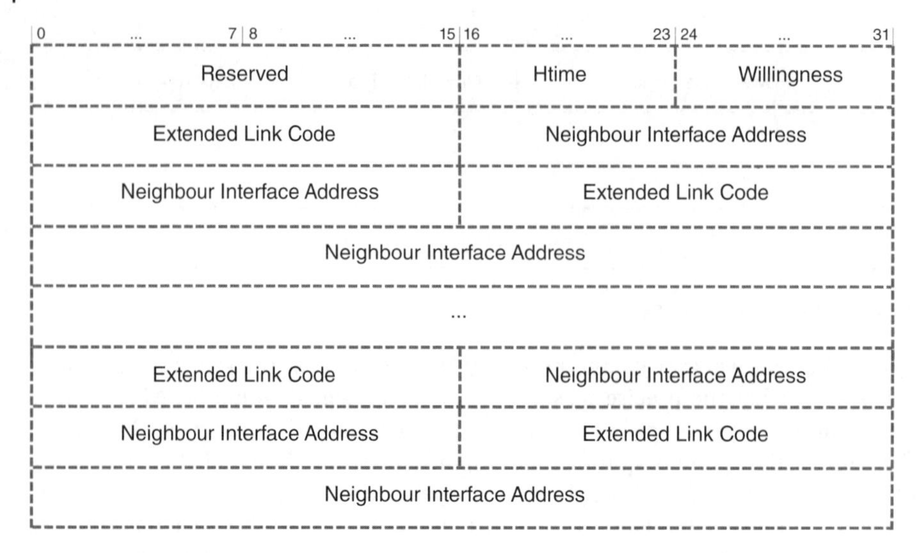

Figure 5.13 Generalised Hello message.

only exception to this rule would hold when the Power Level data should be accessed for arranging for and orchestrating cooperative transmission. Even though such an approach would seem appropriate from the backward compatibility perspective, one could also consider the introduction of a Generalised Hello message, as depicted in Figure 5.13. However, its format would be bound to differ from both the structure of the original and the Modified Hello message, as it lacks the Link Message Size field to allow overhead optimisation by avoiding Link messages that would contain a single Neighbour Interface Address only (Wódczak, 2012a).

5.3.3 Address Auto-Configuration and Duplication

Attempting to propose certain cooperative transmission related extensions in the form of the above-mentioned REACT to be contained under the umbrella of the Generic Autonomic Network Architecture, one needs to take into consideration a much wider context of network discovery and configuration, going beyond what may be covered by the Optimised Link State Routing protocol alone. On the one hand, the Generic Autonomic Network Architecture is fully integrated with Internet Protocol version 6 (IPv6) which provides certain mechanisms of relevance, such as the Neighbour Discovery (ND) protocol. Among others, as outlined by Narten et al. (2007), Neighbour Discovery is exploited by network nodes sharing the same link to discover the presence of each other or one another. On the other hand, however, apart from the discovery aspects, there may appear a highly sensitive configuration-related issue, especially likely to be encountered in mobile ad hoc networks, where address duplication could take place under certain circumstances. In essence, one may think of an unlikely, yet possible, situation where a lack of direct visibility between network nodes and the stateless approach to address auto-configuration (AAC) would result in such address duplication. Since both the above-mentioned aspects are vital for networked setups of the scale under consideration for the Autonomic Intelligence Evolved Cooperative Networking, where the design principles of the Generic Autonomic Network Architecture

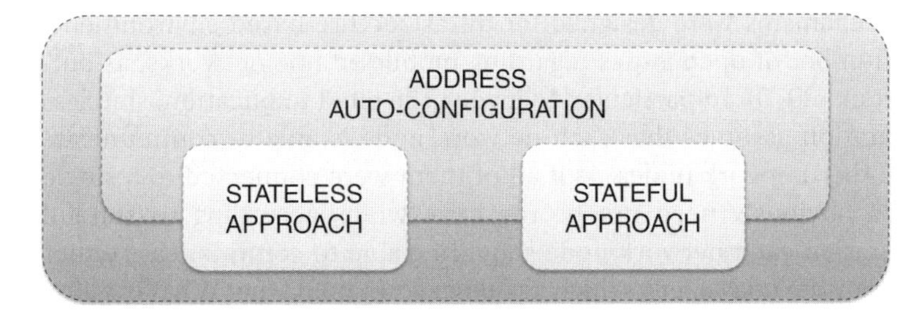

Figure 5.14 Address auto-configuration.

hold, below its characteristics related to both address auto-configuration and duplicate address detection are described.

Most generally, thanks to the mechanism of Neighbour Discovery, a group of network nodes should be able not only to auto-discover themselves in a possibly new context, but in the event that a given neighbour node of theirs becomes no longer available, they could actively search for relevant up-and-running alternatives. Moving into the details, following Narten et al. (2007), should a network interface be activated, a given network node becomes entitled to commence the distribution of Router Solicitation messages in order to stimulate other neighbour nodes to generate their Router Advertisement messages immediately. In general, based on the information acquired in this way, a given host[19] may start address auto-configuration. Then, routers may use the Router Advertisement messages in order to indicate whether a given host should proceed with the stateless or stateful process, as classified in Figure 5.14. In the case of stateless address auto-configuration (SLAAC), as outlined by Thomson et al. (2007), each host can generate its address as a combination of subnetwork and interface identifiers, which appears very convenient, as long as they are unique and properly routable, while there are no requirements as to their specific properties. Following the design guidelines given by Thomson et al. (2007), it is assumed that manual configuration shall be not required, while neither small networks, where the same link is shared, nor large networks, composed of subnetworks, shall require stateful address auto-configuration (SFAAC). While in the former case all the addresses will share the same prefix, in the latter each subnetwork will be assigned its own prefix. Lastly, the possibility of renumbering provision shall be guaranteed in any of the cases.

Stateful address auto-configuration is characterised by the existence of Dynamic Host Configuration Protocol (DHCP) servers delivering configuration parameters to network nodes. Such parameters are by no means limited to IPv6 addresses and may include other information carried as options. The DHCP clients, in turn, transmit their messages to a reserved multicast address of link scope, so they do not need to be configured with the address or addresses of the DHCP servers, as indicated by Droms et al. (2003). In the case that a given network node and a given DHCP server are not connected to the same link, a DHCP relay agent acts as an intermediary network node to take care of proper delivery, at the same time operating transparently from the perspective of the network node itself. Once a given network node determines the address of the DHCP server, it might contact this server directly in some cases. Most

19 One should note that, following Narten et al. (2007) strictly, the notion of a host shall be understood as 'any node that is not a router'.

importantly, these mechanisms were designed for infrastructure-based environments and there are still a number of open issues regarding mobile ad hoc networks, as outlined by Bernardos et al. (2010). In particular, following the cited publication, stateless address auto-configuration assumes that each network node is able to communicate directly with all the other network nodes, as if all of them were connected to a single multicast link, which obviously need not be the case, while, according to stateful address auto-configuration each network node should be able to communicate either with the DHCP server or the relay agent, which again need not hold true. What is more, one should keep in mind the pertinent issues related to multi-hop support and network merging and partitioning (Bernardos et al., 2010).[20]

Shifting from address auto-configuration to duplicate address detection (DAD), it was previously claimed that, under certain circumstances, especially inherent in mobile ad hoc networks, the issue of address duplication could take place even in a system characterised by a correct initial configuration. In other words, it cannot be excluded that physical separation, for example, as depicted in Figure 5.15, might result in two network nodes being assigned the same IPv6 address. One should note, however, that in some cases such duplication may take place without any physical obstacles, as it all boils down to the workings of a specifically applied protocol. In particular, analysing the duplicate address avoidance (DAA) aspects of the specification of Neighbour Discovery in IPv6 networks by Narten et al. (2007), both neighbour solicitation and proper seed calculation appear to be implied as the preferred countermeasures against this cumbersome phenomenon. Regardless of the way the address duplication took place, there appear to exist a number of methods for coping with it in different ways, as classified in a very detailed survey by Bernardos et al. (2010). Given the scope of this book, the first part of the survey appears of highest relevance because it is related to standalone mobile ad hoc network scenarios. However, taking into account that the major emphasis here is on the Optimised Link State Routing protocol concepts, the three most pertinent approaches, as described in Table 5.11, are summarised below in a more detailed way;

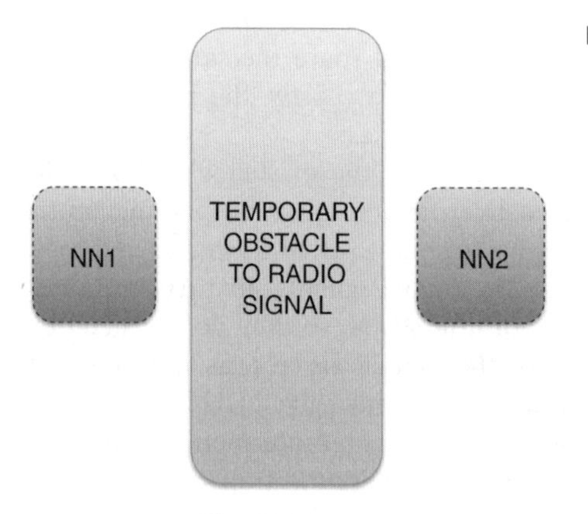

Figure 5.15 Address duplication scenario.

20 Pertinent to the issue of merging and partitioning is the question of maintaining a proper network hierarchy. For additional context in this respect, the reader may also refer to Villasenor-Gonzalez et al. (2005), for instance.

Table 5.11 Duplicate address detection for the OLSR protocol.

Approach	Characteristics
No overhead auto-configuration OLSR	Passive duplicate address detection approach is used together with the OLSR protocol. It is based on the observation that some protocol events occur generally for duplicate addresses and rarely for unique ones.
Passive duplicate address detection for OLSR	Algorithmic approach aiming to detect duplicate addresses through different parameters contained in Hello and TC messages, as well as the addresses of OLSR protocol messages and IP headers.
Address auto-configuration in OLSR	Network nodes periodically disseminate their addresses and a randomly generated sequence of bits of a fixed length. Duplicate addresses exist if the identifiers for a given address do not match.

for a thorough analysis containing all the references, the reader is pointed directly to Bernardos et al. (2010).

Examining the approaches most suited to the major theme of this book, first comes the no overhead auto-configuration OLSR (NOA-OLSR) extension proposed by Mase and Adjih (2006). In particular, this approach is intended for both the initial address auto-configuration and duplicate address detection, while keeping the Optimised Link State Routing protocol operating in the usual manner. One should note, however, that a network node may only be allowed to fully run the Optimised Link State Routing protocol after it has been confirmed that its address is unique. In the meantime, the network node goes through different states and is considered by its neighbour nodes as not fully reliable. As such, the process of address auto-configuration involves three stages in the form of address generation, progressive duplicate address detection, and gradual admittance to the network connected with avoidance of routing table contamination (Mase and Adjih, 2006). Essentially, during the first stage, a given network node monitors the message flow in the network to create a list of addresses in use. Based on this, it becomes entitled to select a still tentative address not in the list, yet the specific procedure is not defined in detail. In the second stage the process of duplicate address detection is commenced, which consists in checking for inconsistencies in both the Hello and TC messages, and also in the sequence numbers. With the third stage, the network node is gradually admitted to the network so that an increasing number of other network nodes can use its messages, while it is still verified if it has passed the duplicate address detection procedure in order to become included in the routing table.

Then there is the passive duplicate address detection OLSR (PDAD-OLSR) extension, which is based on a set of algorithms designed to allow the network nodes to detect conflicts in the network by observing protocol anomalies, as outlined by Mase and Weniger (2006). In general, such an approach may be translated into the assumption that some protocol-related events may clearly occur should duplicate addresses exist, while being extremely unlikely in the case of unique assignments, as explained by Baccelli et al. (2005). In particular, a network node running PDAD-OLSR may obtain information about the state of the entities governing the behaviour of the routing protocol running at other network nodes on the basis of analysis of the incoming messages. What is more, the network node performing such an analysis must be aware of the exact time at which the message was dispatched. Should this not be the case, some vital decisions could become immediately outdated, as the state of the routing protocol is, most obviously, undergoing continual changes with the passage of time. Thus, the above-mentioned algorithms used by network nodes for the purposes of duplicate address detection make use of different parameters of the Hello and TC messages, such as link states, link codes, message sequence numbers, as well as the addresses contained in the Optimised Link State Routing protocol messages and the IP headers. Consequently, it is possible to ensure that the multi-point relay station selection heuristics may remain unaffected (Mase and Weniger, 2006).

Last, but not least, comes address auto-configuration OLSR (AAC-OLSR),[21] which is targeted at guaranteeing address uniqueness in very demanding situations, when different mobile ad hoc networks are about to undergo the process of merging. The method proposed by Adjih et al. (2005) assumes a distributed approach, where each network node is expected to periodically send out both the list of all its addresses and its identifier using so-called Multiple Address Declaration (MAD) messages. In particular, first of all, an incoming network node assigns itself an address by means of random selection, acquisition of an address advertised by another network node as unused, or exchange of control messages. What follows is duplicate address detection performed by means of analysing the MAD messages, so that when a given network node finds its own address on the list yet characterised by a different identifier, it chooses a new address for itself. Such a problem typically occurs when networks merge, which means that other network nodes also detect the related addressing conflict, and should all of them start announcing it, there could be a significant control overhead increase. This is why it is suggested that the MAD messages are allowed to reach all the network nodes in the network before the conflicting situation is announced to the network nodes, since, otherwise, the network could inevitably find itself in the situation referred to as a 'broadcast storm' (Adjih et al., 2005). As such, the approach does not exclude optional routines such as the avoidance of routing table contamination or passive duplicate address detection.

5.4 Node Level Overlay Logic

5.4.1 Autonomic Cooperative Networking Protocol

Following the prior analysis of the rationale behind the concept of the Autonomic Cooperative Node, the incremental description towards the ultimate Autonomic Cooperative Networking Architectural Model is continued with the next major component in the form of the Autonomic Cooperative Networking Protocol (ACNP) (Wódczak, 2014a). Essentially, the Autonomic Cooperative Networking Protocol fulfils the classification of a cross-layer solution by not only incorporating the workings of the Optimised Link State Routing protocol residing at the network layer, but also being profoundly rooted in the distributed space-time block coding inherent in the link layer,

21 For the sake of clarity, the acronym AAC-OLSR is applied in this book only for naming consistency reasons with regard to the previously described NOA-OLSR and PDAD-OLSR.

as well as indirectly reaching the physical layer to orchestrate the virtual multiple-input multiple-output (VMIMO) radio channel. In other words, remaining in the dimension of the Open Systems Interconnection Reference Model, one could justly say that the Autonomic Cooperative Networking Protocol capitalises directly on the workings of the extended routing information enhanced algorithm for cooperative transmission (EREACT), as outlined by Wódczak (2007). There is, however, profound added value differentiating the two, as all the building blocks of the Autonomic Cooperative Networking Protocol make it fully integrated into the perpendicular or orthogonal dimension of the Generic Autonomic Network Architecture. Given such a context, first of all, the emphasis is to be laid on the elevated building blocks, especially involving the consolidated notion of the Autonomic Cooperative Behaviour (ACB). Then, the incorporation of the already discussed evolved messaging structure will be justified, along with the introduction of a properly enabled routing table design.

In particular, analysing the transition[22] from EREACT to the Autonomic Cooperative Networking Protocol, one should note that, on the one hand, as described earlier in this chapter, certain parts of the former were already outlined on the occasion of introducing the evolved messaging structure on the basis of the original REACT. On the other hand, the upgraded algorithmic description driving the behaviour of EREACT will not be outlined until the logic behind the cooperation management decision element (CMDE) has been presented in the next section. Such an approach should by no means be read as restrictive, because at this stage those will be mostly the elevated building blocks to be upgraded from the terminological perspective. Essentially, one should note that, as long as the Autonomic Cooperative Networking Protocol has been designed to fuse certain workings of both the link layer and the network layer, its role is to orchestrate Autonomic Cooperative Behaviour instantiated between or among Autonomic Cooperative Nodes, forming an elevated version of VAAs, presently referred to as ACSs. So, much as the Autonomic Cooperative Nodes may be perceived as an upgraded incarnation of relay nodes or multi-point relays, depending on the context,[23] the concept of Autonomic Cooperative Behaviour, described in the previous chapter, brings a completely new value to what was previously known as EREACT.[24]

Key to the analysis appears to be the integration between the MPRs and ACNs, as depicted in Figure 5.16, which stems directly from and builds upon the rationale behind the concept of EREACT, where the multi-point relay station selection heuristics was similarly employed to facilitate VAA-aided cooperative transmission, as introduced by Wódczak (2012a). Even though the idea of VAAs was elevated by Wódczak (2014a) to institute the notion of virtual cooperative sets (VCSs), or using the latest nomenclature, the said ACSs, while the already legacy relay nodes were replaced with the latest Autonomic Cooperative Nodes, the major assumption still consists in the execution of the multi-point relay station selection heuristics in an iterative manner in order to identify those Autonomic Cooperative Nodes that could act together as ACSs and, thus, expose Autonomic Cooperative Behaviour through the DSTBC-driven orchestration of the virtual multiple-input multiple-output radio channel. In fact, as the Optimised Link

22 Most obviously, the transition is done with the OLSR protocol playing the role of a synchronising entity.
23 Referring to the original nomenclature inherent in EREACT, only carefully preselected MPRs may form VAAs, while those MPRs are chosen from among relay nodes.
24 In the light of the upcoming respective context of emergency communications networks, the reader may also refer to Wódczak (2014b).

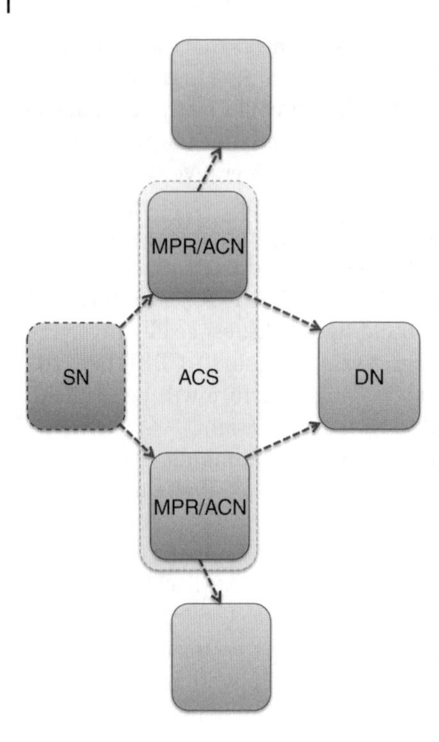

Figure 5.16 Integration of MPRs and ACNs.

State Routing protocol belongs to the network layer, each of the Autonomic Cooperative Nodes is not only entitled to acquire the knowledge of its one-hop and two-hop neighbourhoods, but may equally well identify such one-hop neighbour nodes as would be able to provide connectivity to some preselected two-hop neighbour node, in turn. From these, solely neighbour nodes, or rather relay nodes,[25] may be identified as Autonomic Cooperative Nodes which help minimise the protocol control overhead.

Algorithm 5.3 Preselection of ACNs.

1: **for all** $n \in N(x)$ **do**
2:   **while** $j \geq 0$ **and** $P_x^{ACS(x,n^{(2)})[j]} < P_x^n$ **do**
3:     $ACS(x, n^{(2)})[j+1] \leftarrow ACS(x, n^{(2)})[j]$
4:     $j \leftarrow j - 1$
5:   **end while**
6: **end for**
7: $ACS(x, n^{(2)})[j+1] \leftarrow n$

Moving towards the justification for the introduction of the previously outlined evolved messaging structure, one might recall Figure 5.8, where the bit error rate curves almost overlapped in the region of low signal-to-noise ratio (SNR) values. This was undoubtedly related to the fact that the first-hop errors were propagated further

25 This is so since, using the original nomenclature for reference purposes, a neighbour node qualifying to become an MPR needs to expose its willingness to carry and forward traffic, which, to some extent by default, makes one think of such a network node as if it were a relay node from the very outset.

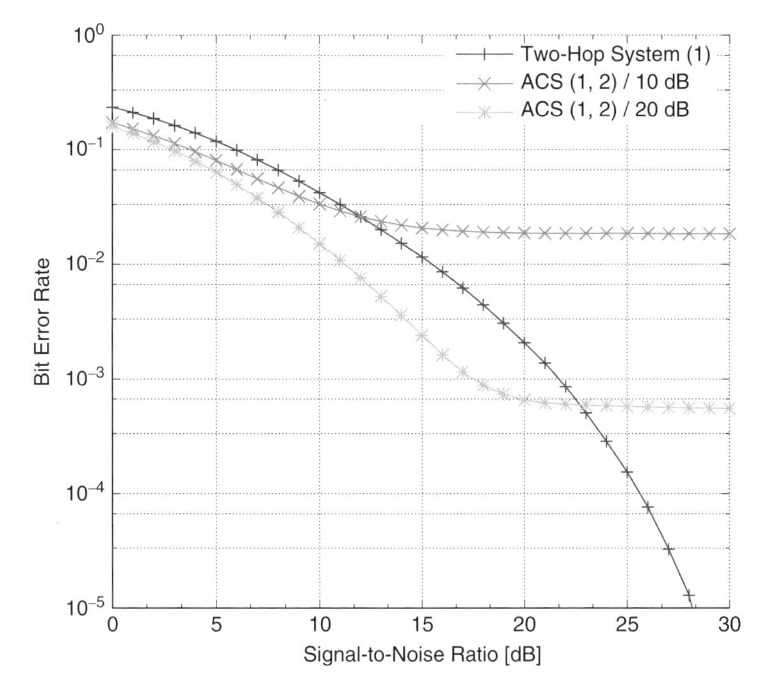

Figure 5.17 Performance for first-hop 10 dB or 20 dB SNR.

during the second hop, as analysed by Wódczak (2012a). It was so since the first-hop transmission, where the source node feeds the preselected relay nodes, or, using the present nomenclature, the Autonomic Cooperative Nodes, with the aid of single-input single-output radio links, is less robust to radio channel impairments when compared to the DSTBC-enhanced cooperative transmission at the second hop, where the Autonomic Cooperative Nodes, forming an Autonomic Cooperative Set, feed the destination node. The question then arises of what gain could be achievable should a better preselection be made on the basis of more accurate link parameters at the first hop only. In order to quantify their influence, the logic outlined in Algorithm 5.3 is validated, where the entire set $N(x)$ is explored to find and promote these potential Autonomic Cooperative Nodes, which appear to be characterised by the highest received power level expressed as P_x^n. In particular, two different cases are evaluated, where the SNR value at the first hop is maintained at the level of either 10 dB or 20 dB, as outlined by Wódczak (2012a). The results are depicted in Figure 5.17, where the numbers in the legend denote either the first hop neighbour node or Autonomic Cooperative Nodes, forming an Autonomic Cooperative Set, following the original notation of Figure 5.7.

The above analysis helps to answer the question of what bit error rate one could expect in the case of an equivalent dynamic system, where the source node could preselect such Autonomic Cooperative Nodes that would observe the received power level P_0^n remaining, respectively, 10 dB or 20 dB higher than the mean noise power N.[26] Indeed, the expected gain in bit error rate appears observable, becoming higher for higher

26 In other words, the performance of such a system would be not worse than is indicated by the curves in Figure 5.17.

1	R_dest_addr	R_next_addr	R_dist	R_iface_addr
2	R_dest_addr	R_next_addr	R_dist	R_iface_addr
3	R_dest_addr	R_next_addr	R_dist	R_iface_addr
...	...	...	...	...
K	R_dest_addr	R_next_addr	R_dist	R_iface_addr

Figure 5.18 ACNP routing table.

theoretically guaranteed SNRs. Most importantly, such a system begins to saturate at values close to the assumed 10 dB or 20 dB, thereby additionally showing that the first hop is critical in this respect, which justifies the incorporation of the Extended Link Code. In particular, the availability of additional power level information conveyed within such an Extended Link Code of the Modified Hello message makes it possible to place the Autonomic Cooperative Nodes in $ACS(x, n^{(2)})$ at positions derived from the strength of the signal heard by the destination node.[27] In general, should a network layer entity be about to send a packet to a destination node at R_dest_addr,[28] it would need to use R_next_addr and request the link layer to handle the transmission to the medium access control (MAC) address[29] corresponding to R_next_addr. However, in the case of the Autonomic Cooperative Networking Protocol, such a packet would need to be transmitted concurrently over all the Autonomic Cooperative Nodes belonging to a given ACS (Wódczak, 2012a). This issue could be addressed by associating an additional routing table with each column of the space-time block coding matrix, which leads to a multidimensional form thereof, as depicted in Figure 5.18.

5.4.2 Cooperation Management Decision Element

Progressing even further towards the perpendicular or orthogonal dimension determined by the Generic Autonomic Network Architecture one immediately comes across the cooperation management decision element (CMDE), whose inner logic is directly programmed to orchestrate the composition of Autonomic Cooperative Behaviour with the multi-point relay station selection heuristics of the Optimised Link State Routing protocol. As such, the heuristics is executed iteratively over the set of one-hop neighbour nodes, $N(x)$, of a given source node, until all the Autonomic Cooperative Nodes potentially meeting the selection criteria have been assigned to distinct $MPR^i(x)$ subsets. Looking more into the details, one should note that the routine being outlined very much mirrors the workings of EREACT. On top of this, it is also assumed that all the

27 Consequently, the first ACNs in an ACS are those characterised by the ability to provide the best-quality cooperative transmission towards the destination node. As such, cooperative transmission enabled in this way would definitely require proper packet routing.
28 Denoting the IP address of this destination node.
29 Resolved with the aid of the Address Resolution Protocol (ARP), as outlined by Plummer (1982).

Autonomic Cooperative Nodes can expose the said Autonomic Cooperative Behaviour, and thus could be perceived as being capable of forming Autonomic Cooperative Sets. Consequently, each iteration will similarly result in redundant subsets of multi-point relays, i.e. secondary, ternary, and so on, denoted as $MPR^i(x)$, thereby making it possible to distribute all the Autonomic Cooperative Nodes, on the basis of additional power level information, among the most relevant Autonomic Cooperative Sets. In the light of the above, the algorithmic description defining the logic behind the CMDE will be outlined first, so that the relevant scenario overview and performance results may follow, to be complemented with an analysis of the potential control overhead.

Essentially, as adapted in Algorithm 5.4,[30] one should note that, similarly to VAAs, the ACSs, denoted as $ACS(x, n^{(2)})$, are designed to support cooperative transmission on the two-hop path between the source node x and the destination node $n^{(2)}$. What is more, the general outline of the proposed logic resembles the previously analysed Algorithm 5.2 to such an extent that again each network node n characterised by a zero degree, i.e. $degree(n) = 0$, is first removed by the source node x from the one-hop set of neighbour nodes $N(x)$. What becomes clear after additional inspection, however, is that it also incorporates certain aspects of Algorithm 5.3, where the awareness of the power level issue was investigated. In fact, the extended logic extensively exploits any additional information regarding the power level of the signals received by a given network node from its one-hop neighbour nodes, as collected by the relevantly modified version of the Optimised Link State Routing protocol, where the evolved messaging structure described earlier is employed. In particular, such information is stored in the Power Level field of the Extended Link Code, so that, based on the value it carries, the placement of a specific neighbour node of the source node in $ACS(x, n^{(2)})$ may be adjusted accordingly, allowing it to take the position corresponding directly to the power level at which it was heard by the destination node. This means that, again, the Autonomic Cooperative Nodes taking the most elevated positions in a given Autonomic Cooperative Set will, at least theoretically, provide the best connectivity to the said destination node, thereby additionally optimising the process of preselection.

The operation of a CMDE defined in this way is verified in a low-mobility and high-density hot-spot scenario, where mobile Autonomic Cooperative Nodes are taken into account, moving around at velocities in the range 0–5 km h^{-1} (equivalent to 0–1.4 m s^{-1}). In particular, following Dottling et al. (2009), a line-of-sight channel model is assumed, characterised by the path loss $L(d)$, defined as

$$L(d) = 13.4 \log_{10}(d) + 36.9 \ \text{[dB]}, \tag{5.4}$$

where d represents the distance, expressed in meters, $5 < d < 29$ m, while the standard deviation of the shadow fading σ is 1.3 dB. What is more, the source node, taking the form of a fixed base station (BS) in this scenario, is assumed to transmit with an average power of 200 mW and is equipped with an antenna characterised by a gain of 8 dBi. The mobile relay nodes and the destination node are all embodied by user terminals (UTs), characterised by the corresponding parameters being equal to 200 mW and 0 dBi, respectively.[31] In particular, the downlink is investigated and QPSK modulation

30 For the previous version thereof the reader is referred to Wódczak (2014a).

31 One should note that having the BS equipped with an antenna of a higher gain makes it possible to verify the previously discussed influence of the first-hop errors on the overall performance of the cooperative transmission.

Algorithm 5.4 Logic of CMDE.

1: **for all** $n \in N(x)$ **do**
2:   **if** $degree(n) = 0$ **then**
3:     $N(x) \leftarrow N(x) \backslash \{n\}$
4:   **end if**
5: **end for**
6: $i \leftarrow 1$
7: **while** $N(x) \neq \emptyset$ **do**
8:   $MPR^i(x) \leftarrow$ OLSR_MPR_HEURISTICS($N(x)$)
9:   **for all** $n \in MPR^i(x)$ **do**
10:     **for all** $n^{(2)} \in N^{(2)}(x)$ **do**
11:       **if** $n = neighbour(n^{(2)})$ **then**
12:         $j \leftarrow size(ACS(x, n^{(2)})) - 1$
13:         **while** $j \geq 0$ **and** $P^{n^{(2)}}_{ACS(x,n^{(2)})[j]} < P^{n^{(2)}}_{n}$ **do**
14:           $ACS(x, n^{(2)})[j+1] \leftarrow ACS(x, n^{(2)})[j]$
15:           $j \leftarrow j - 1$
16:         **end while**
17:         $ACS(x, n^{(2)})[j+1] \leftarrow n$
18:       **end if**
19:     **end for**
20:   **end for**
21:   $N(x) \leftarrow N(x) \backslash MPR^i(x)$
22:   $i \leftarrow i + 1$
23: **end while**

is assumed, while the noise figure introduced by the radio frequency chain of a UT is assumed to be 7 dB. Essentially, the analysed network topology, as presented in Figure 5.19, very much resembles a mobile ad hoc network configuration in the access part, except for the BS. The destination UT (18) is located 29 m from the BS (0), while the distances between the BS and each UT belonging to the first (1–6) and second groups (7–17) are 10 m and 24 m, respectively.[32]

The related computer-assisted simulations were performed in the presence of a zero-mean additive white Gaussian noise characterised by power level N expressed in dBW, where the transmission was originated by the BS (0) and destined for UT (18). First of all, although it was assumed that the destination UT is conceptually not a neighbour node of the BS, the performance of the one-hop link towards this UT was evaluated as the worst, reference, case. Then two different configurations of conventional relaying were tested, where the transmission was assisted either by UT (3) or UT (12). An advantage was observable, especially in the latter case, where the relaying UT was

32 One should keep in mind that the BS and the UTs are equipped with antennas offering different gains. Consequently, even if the power level of the signal received from the BS by the destination UT (18) is acceptable, it need not necessarily hold true in the opposite direction. This is to some extent in contrast with mobile ad hoc network environments, for which the OLSR protocol was originally designed, as indicated by Wódczak (2011d). In other words, in the case of the OLSR protocol, if two network nodes hear each other, a symmetric link can be established without any additional considerations regarding the corresponding power levels, since, in general, such network nodes are perceived as homogeneous (Doppler et al., 2007).

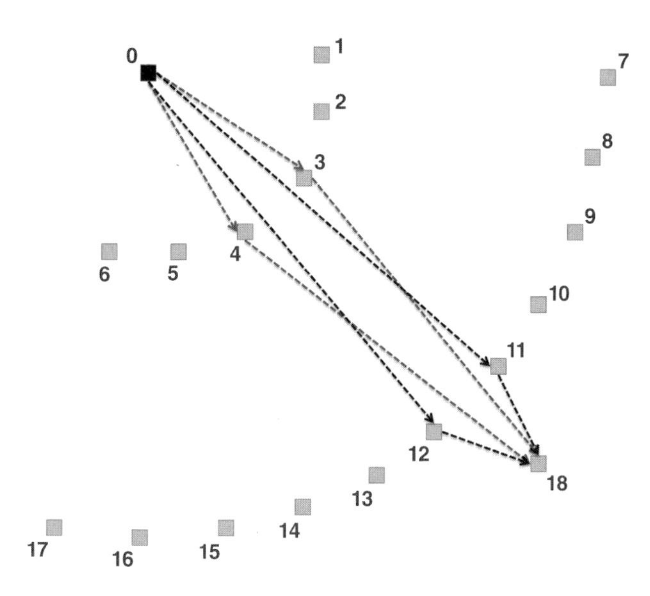

Figure 5.19 Evaluation scenario.

situated closer to the destination node so that the antenna gain of the BS could be more efficiently exploited. Finally, both REACT and ACNP were validated, where, in the case of the former, ACN (3) and ACN (4) were selected to form $ACS(0, 18) = \{3, 4\}$, while, in the case of the latter, the Autonomic Cooperative Nodes were preselected which were closer to the destination node, consequently forming $ACS(0, 18) = \{11, 12\}$. While the former configuration provided a significant improvement in performance, the latter setup offered, as expected, an even more visible gain. For a detailed comparison of the results the reader is referred to Figure 5.20, where the numbers placed in the legend next to the labels defining specific system configurations are meant to denote the next hop neighbour node or neighbour nodes (Wódczak, 2012a).

As previously suggested, one could also expect increased control overhead related to the introduction of the Modified Hello message containing the Extended Link Code field. In fact, the Optimised Link State Routing protocol may typically differentiate between four distinct Link Types and three distinct Neighbour Types, giving 12 combinations and meaning that the Neighbour Interface Addresses may become assigned to 12 groups, i.e. Link Messages, at most. Taking into consideration the proposed modifications to the Optimised Link State Routing protocol, due to the type of information required by the multi-point relay station selection heuristics, as well as by the resulting Autonomic Cooperative Networking Protocol, it suffices to focus on SYM_LINK along with SYM_NEIGH and MPR_NEIGH. This would limit the number of the theoretically available 16 combinations, enlarged with the introduction of VAA_NEIGH, merely to two. Consequently, as indicated by Wódczak (2012a), the main factor influencing the overhead would clearly stem from the size of the Power Level field in question. As such, the length of this field, amounting to 12 bits, would imply that, in theory, there could be 4096 values allowed. Multiplied further by the aforementioned two combinations, this would result in 8192 possibilities. What is more, taking into account that, in the worst case, solely singular interfaces could become characterised by a given Power Level, one

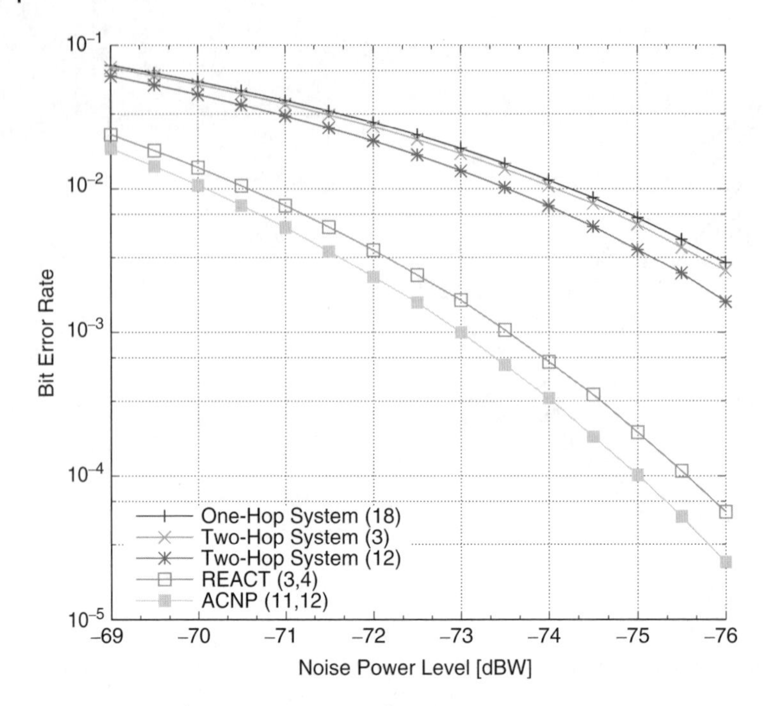

Figure 5.20 Performance comparison for ACNP.

could expect to end up with as many as 8192 Link messages, each accompanied by a header of length 32 bits. This, of course, would be the worst option, while, as originally shown by Wódczak (2012a) and depicted in Figure 5.21, the observed overhead could be almost diminishable for as few as six bits (Wódczak, 2014a).[33]

5.4.3 Architectural Integration Aspects

Touching upon the architectural integration aspects once again, as well as keeping in mind all the developments rolled out in this chapter, one might expect some extensive analysis, especially taking into account the ever increasing complexity of the interaction taking place on the verge of a conceptual junction between the perpendicular or orthogonal positioned entities of both the Open Systems Interconnection Reference Model and the Generic Autonomic Network Architecture.[34] Since this junction appears to be multidimensional, a few different perspectives will have to be taken to provide a sufficiently detailed insight into all the architectural aspects. In particular, on the one hand, there are all the components revolving around the multi-point relay station selection heuristics, which not only brings in the relevantly modified Optimised Link State Routing protocol along with its evolved messaging structure,

33 For the wider context of performance evaluation methods for mobile ad hoc networks, the reader is also referred to Papadopoulos et al. (2016).

34 One should keep in mind that the conceptual analysis pertains to mobile ad hoc network environments only. For a complementary mesh-enabled architectural framework, the reader is referred to Szott et al. (2012a,b).

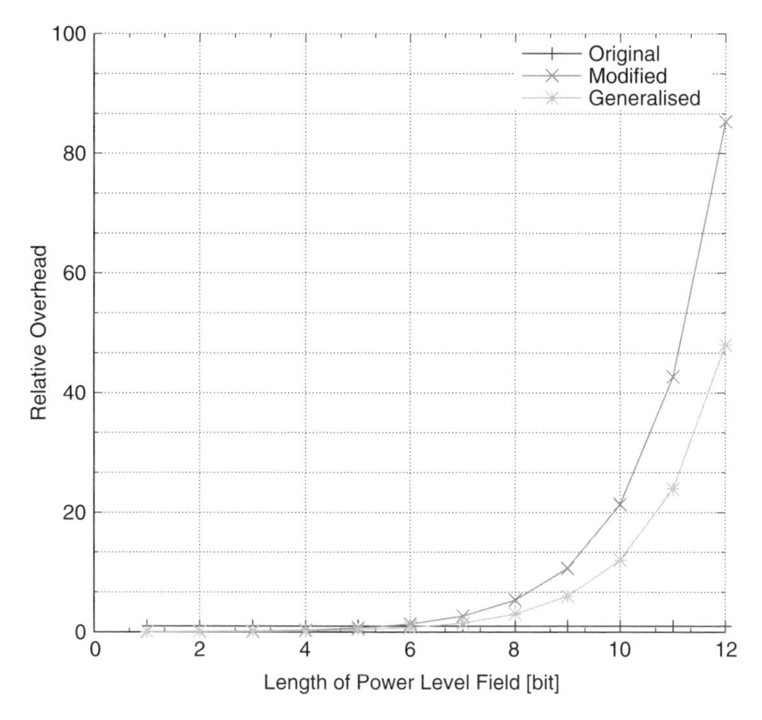

Figure 5.21 Overhead of Modified Hello message and Generalised Hello message.

but also incorporates VAA-based and DSTBC-enabled cooperative transmission using the equivalent virtual multiple-input multiple-output radio channel. On the other hand there emerges the Autonomic Cooperative Behaviour, being a tightly fused and integral part of the Autonomic Cooperative Networking Protocol, also involving the notion of ACSs, where the related Autonomic Cooperative Nodes are intended to reside. In order to provide a relevant background, first, the roots of the Autonomic Cooperative Networking Protocol will be explained, in order to move into analysis of certain necessary conceptual transitions, so that the dependencies between various routines and the architectural relations may be described.[35]

In investigating the roots of the Autonomic Cooperative Networking Protocol, one needs to commence the analysis with the initial solution conceived by Wódczak (2012a) under the name of the routing information enhanced algorithm for cooperative transmission (REACT). As such, REACT was intended merely to enable the performance of DSTBC-based cooperative transmission at the link layer with the aid of additional routing information, readily available at the network layer, through an integration entity in the form of multi-point relay station selection heuristics. Because of certain limitations of such an approach, a more advanced concept in the form of the extended routing information enhanced algorithm for cooperative transmission (EREACT) was introduced, along with its evolved messaging structure. In fact, this latest development was advanced enough to become elevated to yet another level with the conception of the target Autonomic Cooperative Networking Protocol, thereby becoming one of the

35 In the light of the upcoming closing chapter, the reader may refer to Wódczak (2014c, 2017) for encyclopaedic references leading to the most up-to-date concept.

major building blocks of the entire Autonomic Cooperative Networking Architectural Model. One should note, however, that as much as the described transition path may appear straightforwardly incremental, it also carries certain flavour of shifting the evolved concept from an approach rooted only in the Open Systems Interconnection Reference Model towards a Generic Autonomic Network Architecture driven design, as depicted in Figure 5.22. While such a perception cannot be denied, one may also claim that the Autonomic Cooperative Networking Protocol bridges both the above-mentioned ends of the transition path.

Whether the notion of the Autonomic Cooperative Networking Protocol should be perceived as being more of a bridging or more of a synergetic nature appears not so obvious, especially when Figure 5.23 is considered, where certain conceptual transitions

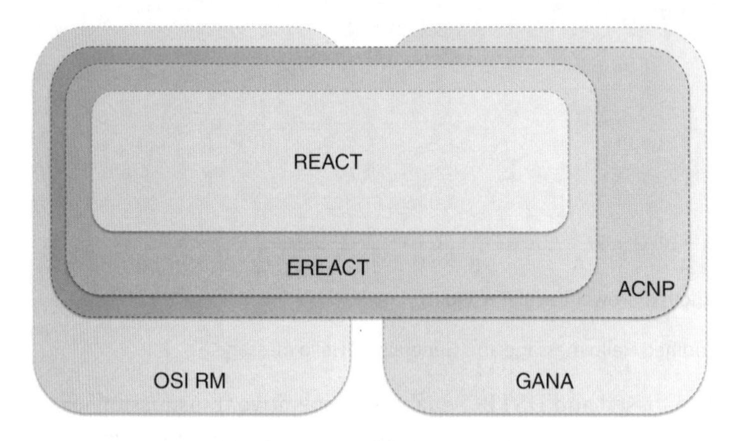

Figure 5.22 Roots of the ACNP.

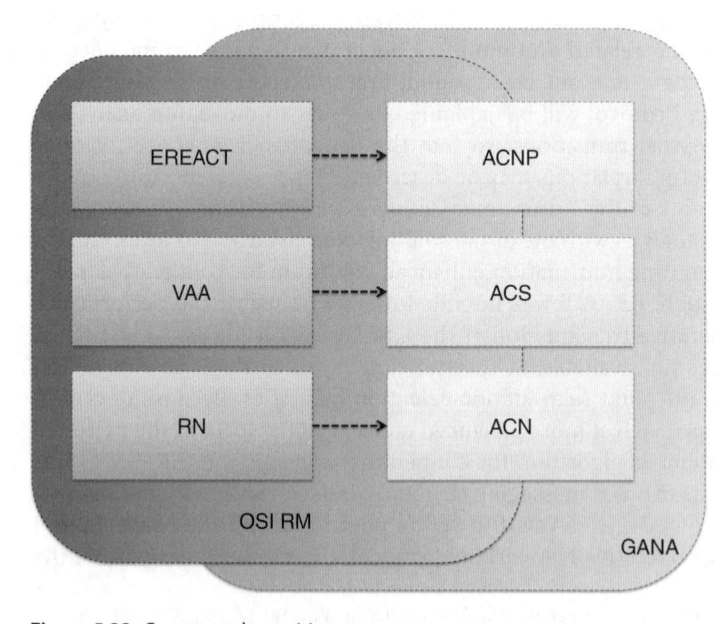

Figure 5.23 Conceptual transitions.

are outlined. One should note that, once again, there exist backgrounds created by both the Open Systems Interconnection Reference Model and the Generic Autonomic Network Architecture, yet, despite their mutual perpendicularity or orthogonality, the present depiction should be read to way more emphasise the fact that even though the latter may stem from the former, there is by no means a full overlap between the two. Given such a context, as well as assuming the usually followed bottom-up approach to analysis, first comes the relay node which, keeping in mind the aforementioned transition path, becomes elevated to the Autonomic Cooperative Node. This upgrade should be viewed as maintaining the roots of the Open Systems Interconnection Reference Model, while adding the extra flavour brought by the Generic Autonomic Network Architecture. In fact, moving upwards, one immediately comes across a fairly similar transformation, where what was originally referred to as a VAA becomes upgraded to an ACS, intended to virtually arrange Autonomic Cooperative Nodes. Last comes the already discussed transition from EREACT towards the Autonomic Cooperative Networking Protocol related to the orchestration of the ACSs.

In fact, this orchestration is done under the umbrella of the Autonomic Cooperative Behaviour, whose role and scope enlarges substantially as the conceptual plot develops. In fact, attempting to make a comparison between what was outlined in the previous chapter in Figure 4.34 and what is presently proposed in Figure 5.24, one may easily discern that the changes between the link layer and the network layer are rather substantial. The discrepancy results from the fact that previously the network layer was incorporated into the global picture of the proposed design to create a fully-fledged appearance of the notion of Autonomic Cooperative Behaviour. This way it was sufficient to refer to the Optimised Link State Routing protocol only, with the multi-point relay station selection heuristics playing more the role of a gluing entity.

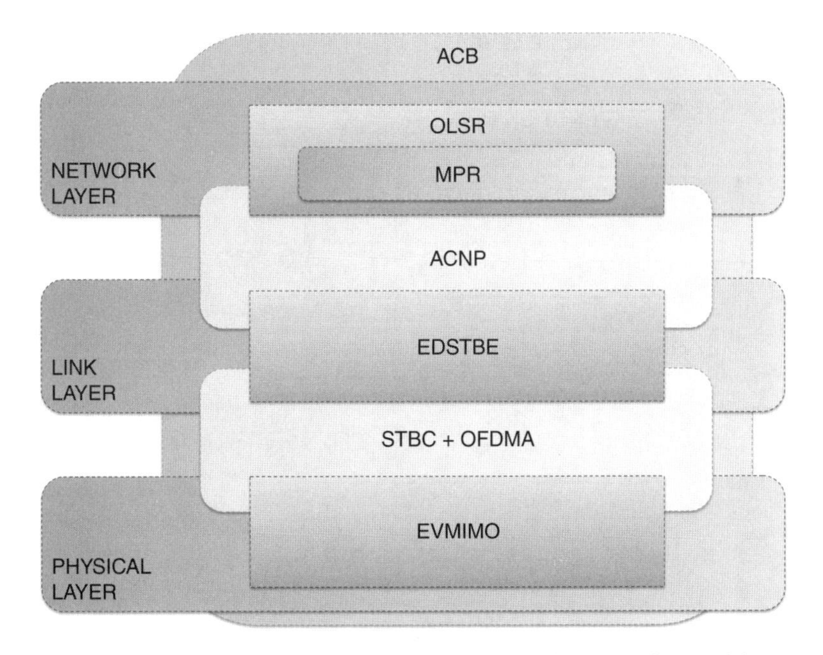

Figure 5.24 Dependencies among the physical layer, link layer, and network layer routines.

However, once the workings of the network layer started becoming substantial in this chapter, it also became apparent that the multi-point relay station selection heuristics should be emphasised much more as an inherent element of the Optimised Link State Routing protocol, while the interaction with the EDSTBE should be performed over an interfacing entity in the form of the Autonomic Cooperative Networking Protocol. In essence, analysing the incorporation of the multi-point relay station selection heuristics into the Autonomic Cooperative Networking Protocol, one should not forget the need for an intermediary step in this respect, taking the form of EREACT (Figure 5.23).

Given the above context, one comes to the most fully-fledged depiction of the architectural relations stemming from the interworking of both the link layer and the function level and their adjacent entities. In particular, once again making a comparison between what was presented in the previous chapter in Figure 4.35 and what is outlined in Figure 5.25, one may clearly discern that while the former concept did not go beyond the notion of Autonomic Cooperative Behaviour, mostly due to the stage of conceptual progression on the side of the Open Systems Interconnection Reference Model itself, this time, with the introduction of the Autonomic Cooperative Networking Protocol, the entire picture of the Autonomic Cooperative Networking Architectural Model, being the ultimate enabler of the Autonomic Intelligence Evolved Cooperative Networking, has become substantially advanced. This is thanks to the fact that the mechanism of Autonomic Cooperative Behaviour, as expressed separately by many ACSs, has now become orchestrated by the CMDE at the node level. One should note, however, that

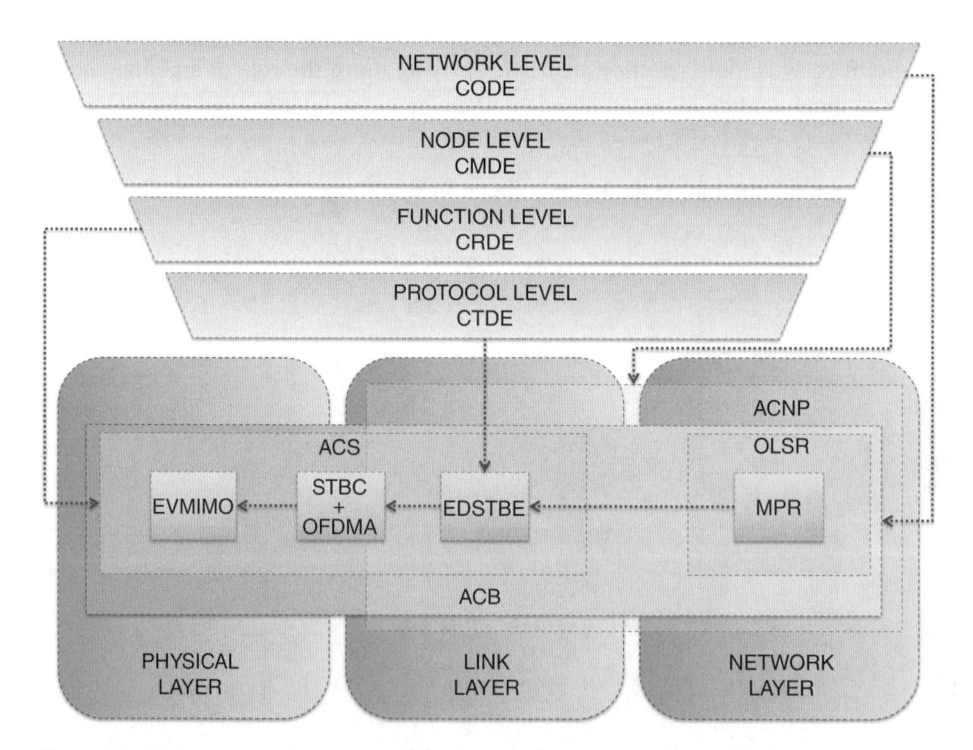

Figure 5.25 Architectural relations stemming from the link layer and function level.

the scope of Autonomic Cooperative Behaviour understood in a more general sense is by no means limited by the routines of the Autonomic Cooperative Networking Protocol. In fact, as can be seen from Figure 5.24, in the final version of the still evolving design, the Autonomic Cooperative Behaviour is supposed to interact directly with the cooperation orchestration decision element of the network level, thereby surpassing the Autonomic Cooperative Networking Protocol that it has so far been merely a part of.

5.5 Conclusion

In this chapter, first of all the workings of the experimentation-related version of the Optimised Link State Routing protocol were discussed, with special emphasis on its functional and structural characteristics related to the field of applicability and the assumed messaging structure. Apart from the proactivity-driven relevance to mobile ad hoc network scenarios, special stress was put on the multi-point relay station selection heuristics, with the incorporation of certain small alignments thereto. Additionally, the information storage repositories were scrutinised, to provide the required context for further developments and to introduce new elements in the form of the VAA selector set and its related VAA selector tuples, intended to become the enablers of the target concept of routing information enhanced cooperative transmission. What followed directly were the developments originating from the routing information enhanced algorithm for cooperative transmission, which was conceived by the author as a method for applying the additional information collected by the Optimised Link State Routing protocol inherent in the network layer, and its modified version in particular, to enable and orchestrate the cooperative transmission at the link layer. To this end the justification for the introduction of the routing information enhanced algorithm for cooperative transmission was given, with special attention paid to the relevant algorithmic description, additionally assuming certain elements and nomenclature of the Optimised Link State Routing protocol, predominantly because of the direct usage of the outcome of the multi-point relay station selection heuristics.

Then, an elevated concept in the form of the extended routing information enhanced algorithm for cooperative transmission was outlined, along with the evolved messaging structure, in order to lay the ground for the target Autonomic Cooperative Networking Protocol. In this respect, the vital topics of address auto-configuration and duplicate address detection were brought up, before the focus shifted towards the umbrella formed by the function level overlay logic. Thus the workings of the Autonomic Cooperative Networking Protocol were outlined, covering the role of the extended routing information enhanced algorithm for cooperative transmission in its conception, and also the justification of the role of the evolved messaging structure in the process of Autonomic Cooperative Node preselection, along with the layout of and the reasoning for the specific design of the routing table. Given such a context, the extended algorithmic description defining the logic of the cooperation management decision element was scrutinised in reference to what was previously outlined for the original routing information enhanced algorithm for cooperative transmission. Based on this, it became possible not only to evaluate its advantages by simulation analysis, but also to address the overhead aspects of the evolved messaging structure. Last, but not least, the entire analysis was elevated once again to be concluded with the aspects of architectural integration covering the roots of the Autonomic Cooperative Networking Protocol and involving the pertinent conceptual transitions, as well as the related dependencies among its architectural entities.

References

Abolhasan M, Wysocki T, and Lipman J 2005 Performance investigation on three classes of MANET routing protocols. *Asia-Pacific Conference on Communications*, pp. 774–778.

Adjih C, Baccelli E, and Jacquet P 2003 Link state routing in wireless ad-hoc networks. *IEEE Military Communications Conference, MILCOM*, pp. 13–16.

Adjih C, Boudjit S, Jacquet P, Laouiti A, Muhlethaler P, and Mase P 2005 *Address Autoconfiguration in Optimized Link State Routing Protocol*. Internet-Draft.

Alamouti S 1998 A simple transmit diversity technique for wireless communications. *IEEE Journal on Selected Areas in Communications* **16**(8), 1451–1458.

Andrews J, Shakkottai S, Heath R, Jindal N, Haenggi M, Berry R, Guo D, Neely M, Weber S, Jafar S, and Yener A 2008 Rethinking information theory for mobile ad hoc networks. *IEEE Communications Magazine* **46**(12), 94–101.

Baccelli E, Clausen T, and Garnier J 2005 *OLSR Passive Duplicate Address Detection*. Internet-Draft.

Bernardos C, Calderon M, and Moustafa H 2010 *Survey of IP Address Autoconfiguration Mechanisms for MANETs*. Internet-Draft.

Chessa S, Corradi A, Foschini L, and Girolami M 2016 Empowering mobile crowdsensing through social and ad hoc networking. *IEEE Communications Magazine*, **54**(7), 108–114.

Clausen T and Jacquet P 2003 *Optimised Link State Routing Protocol (OLSR)*. RFC 3626.

Clausen T, Dearlove C, Jacquet P, and Herberg U 2014 *The Optimized Link State Routing Protocol Version 2*. RFC 7181.

Conti M and Giordano S 2014 Mobile ad hoc networking: Milestones, challenges, and new research directions. *IEEE Communications Magazine*, **52**(1), 85–96.

Doppler K, Redana S, Wódczak M, Rost P, and Wichman R 2007 Dynamic resource assignment and cooperative relaying in cellular networks: Concept and performance assessment. *EURASIP Journal on Wireless Communications and Networking* **2009**(475281), 1–14.

Dottling M, Mohr W, and Osseiran A 2009 *Radio Technologies and Concepts for IMT-Advanced*. Wiley.

Droms R, Bound J, Volz B, Lemon T, Perkins C, and Carney M Jul. 2003 *Dynamic Host Configuration Protocol for IPv6 (DHCPv6)*. RFC 3315.

Mase K and Adjih C 2006 *No Overhead Autoconfiguration OLSR*. Internet-Draft.

Mase K and Weniger K 2006 *PDAD-OLSR: Passive Duplicate Address Detection for OLSR*. Internet-Draft.

Narten T, Nordmark E, Simpson W, and Soliman H 2007 *Neighbor Discovery for IP Version 6 (IPv6)*. RFC 4861.

Papadopoulos G, Kritsis K, Gallais A, Chatzimisios P, and Noel T 2016 Performance evaluation methods in ad hoc and wireless sensor networks: A literature study. *IEEE Communications Magazine* **54**(1), 122–128.

Plesse T, Lecomte J, Adjih C, Badel M, Jacquet P, Laouiti A, Minet P, Muhlethaler P, and Plakoo A 2004 OLSR performance measurement in a military mobile ad-hoc network.

24th International Conference on Distributed Computing Systems Workshops, pp. 704–709.
Plummer DC 1982 *An Ethernet Address Resolution Protocol*. RFC 826.
Postel J 1980 *User Datagram Protocol*. RFC 768.
Qayyum A, Viennot L, and Laouiti A 2002 Multipoint relaying for flooding broadcast messages in mobile wireless networks. *35th Annual Hawaii International Conference on System Sciences, HICSS*.
Sholander P, Yankopolus A, Coccoli P, and Tabrizi S 2002 Experimental comparison of hybrid and proactive MANET routing protocols. *IEEE Military Communications Conference, MILCOM*, pp. 513–518.
Szott S, Wódczak M, Chaparadza R, Meriem TB, Tsagkaris K, Kousaridas A, Radier B, Mihailovic A, Natkaniec M, Loziak K, Kosek-Szott K, and Wagrowski M 2012a Standardization of an autonomicity-enabled mesh architecture framework, from ETSI-AFI Group perspective: Work in progress (Part 1 of 2). *Fourth IEEE International Workshop on Management of Emerging Networks and Services (IEEE MENS 2012) at IEEE GLOBECOM 2012*, Anaheim, California, USA.
Szott S, Wódczak M, Chaparadza R, Meriem TB, Tsagkaris K, Kousaridas A, Radier B, Mihailovic A, Natkaniec M, Loziak K, Kosek-Szott K, and Wagrowski M 2012b Standardization of an autonomicity-enabled mesh architecture framework, from ETSI-AFI Group perspective: Work in progress (Part 2 of 2). *Fourth IEEE International Workshop on Management of Emerging Networks and Services (IEEE MENS 2012) at IEEE GLOBECOM 2012*, Anaheim, California, USA.
Tarokh V, Jafarkhani H, and Calderbank AR 1999 Space-time block coding for wireless communications: Performance results. *IEEE Journal on Selected Areas in Communications* **17**(3), 451–460.
Thomson S, Narten T, and Jinmei T 2007 *IPv6 Stateless Address Autoconfiguration*. RFC 4862.
Villasenor-Gonzalez L, Ge Y, and Lament L 2005 HOLSR: A hierarchical proactive routing mechanism for mobile ad hoc networks. *IEEE Communications Magazine* **43**(7), 118–125.
Wódczak M 2006 *On Routing information Enhanced Algorithm for space-time coded Cooperative Transmission in wireless mobile networks*. PhD thesis Faculty of Electrical Engineering, Institute of Electronics and Telecommunications, Poznań University of Technology, Poland.
Wódczak M 2007 Extended REACT: Routing information Enhanced Algorithm for Cooperative Transmission. *16th IST Mobile & Wireless Communications Summit 2007*, Budapest, Hungary.
Wódczak M 2010 Future autonomic cooperative networks. *Second International ICST Conference on Mobile Networks and Management*, Santander, Spain.
Wódczak M 2011a Aspects of cross-layer design in autonomic cooperative networking. *IEEE Third International Workshop on Cross Layer Design*, Rennes, France.
Wódczak M 2011b Autonomic cooperation in ad-hoc environments. *Fifth International Workshop on Localised Algorithms and Protocols for Wireless Sensor Networks (LOCALGOS) in conjunction with IEEE International Conference on Distributed Computing in Sensor Systems (DCOSS)*, Barcelona, Spain.
Wódczak M 2011c Autonomic cooperative networking for wireless green sensor systems. *International Journal of Sensor Networks (IJSNet)* **10**(1–2), 83–93.

Wódczak M 2011d Convergence aspects of autonomic cooperative networking. *IEEE Fifth International Conference on Next Generation Mobile Applications, Services and Technologies*, Cardiff, Wales, UK.

Wódczak M 2012a *Autonomic Cooperative Networking*. Springer.

Wódczak M 2012b Optimised link state routing protocol as enabler of cooperative transmission for emergency communications. *Eighth IEEE International Symposium on Instrumentation and Control Technology (IEEE ISICT),* University of Westminster in London, UK.

Wódczak M 2012c Simulation environment for autonomic cooperative networking in indoor scenario. *14th International Conference on Modelling and Simulation (UKSim).*

Wódczak M 2014a *Autonomic Computing Enabled Cooperative Networked Design.* Springer.

Wódczak M 2014b Autonomic Cooperative Behaviour enabled emergency system design. *The Mediterranean Journal of Computers and Networks,* **10**(1).

Wódczak M 2014c Autonomic cooperative networking. In *Encyclopaedia of Information Science and Technology,* third edition, ed. Khosrow-Pour, M. IGI Global.

Wódczak M 2017 Autonomic cooperative communications. In *Encyclopaedia of Information Science and Technology,* fourth edition, ed. Khosrow-Pour, M. IGI Global.

Younis O, Kant L, Chang K, Young K, and C G 2009 Cognitive MANET design for mission-critical networks. *IEEE Communications Magazine* **47**(10), 64–71.

Younis O, Kant L, McAuley A, Manousakis K, Shallcross D, Sinkar K, Chang K, Young K, Graff C, and Patel M 2010 Cognitive tactical network models. *IEEE Communications Magazine* **48**(10), 70–77.

Zheng YR and Xiao C 2003 Simulation models with correct statistical properties for Rayleigh fading channels. *IEEE Transactions on Communications* **51**(6), 920–928.

6

Network Level System Orchestration

6.1 Introduction

In the previous chapter the notion of node level routing mechanisms was analysed, starting with the background formed by the experimentation version of the Optimised Link State Routing protocol and certain aspects of its standardised upgrade, advanced with the rationale behind the related extended routing information enhanced algorithm for cooperative transmission, to eventually prepare the ground for the introduction of the cooperative transmission aware approach, where the Autonomic Cooperative Networking Protocol plays the role of the ultimate wrapping entity. The time has come to progress to the final stage of the discussion, related to the network level system orchestration. Essentially, the description is opened with the introduction of a standardisation orientated design, where first of all a research and investment driven perspective is assumed. Such an approach allows to understand where the Autonomic Cooperative Networking Architectural Model stems from – it touches upon issues related to standardisation of the Open Systems Interconnection Reference Model, and emphasises the role of prestandardisation related to the Generic Autonomic Network Architecture itself. What naturally follows is a description of the staged instantiation of the Generic Autonomic Network Architecture Reference Model, depicting the progression of various levels of abstraction in an incremental manner. The introductory part is concluded with certain cross-specification-related considerations involving selected concepts, such as software-defined networking, machine-to-machine communications, and intelligent transport systems.

Once the relations between the, to some extent, peer concepts of software-defined networking and Generic Autonomic Network Architecture have been discussed, most naturally extending to encompass the ultimate Autonomic Cooperative Networking Architectural Model, as well as the instantiation options of the latter for machine-to-machine communications and intelligent transport systems, the focus shifts to yet another, highly practical, deployment scenario based on an emergency communications network. This scenario becomes especially appealing because of a combination of specifically tailored emergency-related requirements, where safety takes precedence even before the latest technological advancements. In particular, it is necessary that the system operation follows exactly the hierarchy between chief first responders and their respective first responders originating directly from human established relations. Given such a context, the preferred, or rather imposed, network topologies are discussed, and certain supportive configurations involving a pair of chief first responders are

Autonomic Intelligence Evolved Cooperative Networking, First Edition. Michał Wódczak.

scrutinised to prepare for the further incorporation of the autonomic overlay under consideration. Once the cooperative mode of operation has been introduced, its instantiation is presented along with the proactive and reactive resiliency process, so that it is possible to move to the integration of the emergency communications network into the Autonomic Cooperative Networking Architectural Model. Lastly, the cooperative enhancement is justified with the aid of performance evaluation analysis.

In the light of the above, the network level overlay logic is then brought up to complement the overall picture in such a way that the remaining internals of the Autonomic Cooperative Networking Architectural Model are scrutinised in the first place. Thus the mutual relation between the Autonomic Cooperative Networking Protocol and Autonomic Cooperative Behaviour is presented from the perspective of the actual, and so far not yet comprehensively discussed, precedence between the two, on the grounds of their being inherent in the orthogonal dimensions of the Open Systems Interconnection Reference Model and the Generic Autonomic Network Architecture. Based on this, the notion of the cooperation orchestration decision element is introduced by example, as it becomes possible to show more tangibly in what cases the Autonomic Cooperative Behaviour may be prioritised over the Autonomic Cooperative Networking Protocol. In particular, the already analysed relay-enhanced cell scenario is revisited under certain additional assumptions related to removing the base station and feeding a bigger mesh-like setup of a grid of Autonomic Cooperative Nodes over wired connections for a more accurate evaluation of the second hop. All in all, essentially to summarise, finally the architectural integration aspects are raised for the last time to account for any still outstanding questions related to the mutual operation of all the decision elements discussed. Thus additional synergy is introduced, helping to provide further bindings to the already comprehensive picture of the Autonomic Cooperative Networking Architectural Model.

6.2 Standardisation Driven Design

6.2.1 Research and Investment Perspective

Looking back already from a little bit historical perspective, it transpires that ever since the role of standardisation in the realm of the future internet (FI) has started growing more and more profoundly, the related need for consensus became sought after not only by various related research-driven, but also investment-orientated parties thereto. The ever increasing and substantial interest in this respect may be, most likely, attributed to the fact that the reference to the notion of the 'future' in the very name thereof, makes it appear to be continually evolving in an open-ended manner, thereby, definition-wise, possible to be captured at certain time instances only, but rather not in general. What is more, aside from such evolutionary development, one could also assume an even wider perspective, where more innovative, or even disruptive, approaches permanently stimulate continuous conception of a wide variety of different incarnations of the future internet in parallel. As much as there is little wonder that such a situation may be attributable to today's highly competitive market of new technologies, there also appears to exist a substantial stimulus thereto, especially given the fact that the foreseen number of devices to be interconnected worldwide is expected to grow even more drastically than ever planned. Such a context is directly pertinent to one of the said incarnations, in the form of the Generic Autonomic Network Architecture, advocating virtually full self-manageability of the future internet. Given the above mood, following the mutual relations among the stages of research, standardisation, and investment, are going to be analysed in the first place. Based on this, the specifics of the Generic Autonomic Network Architecture case will be scrutinised and the standardisation ecosystem will be reviewed.

The major effort in the standardisation of the Generic Autonomic Network Architecture was undertaken by the Industry Specification Group (ISG) on Autonomic network engineering for the self-managing Future Internet (AFI) functioning under the auspices of the European Telecommunications Standards Institute (ETSI), as outlined by Chaparadza et al. (2009a).[1] In fact, the ETSI ISG AFI was established in response to a commonly prevailing consensus across industry and academia that the prior developments and conceptual works, as well as numerous follow-up efforts materialising at that very time, which, otherwise, should have been progressing as a joint effort, were, in fact, advanced, but rather separately, if not competitively, despite the very identical objective of achieving a comprehensive architectural design for the emerging concept of autonomic networking. For that reason, keeping in mind the history of the known standardisation deficiencies related to the Open Systems Interconnection Reference Model, a relevantly urgent harmonisation was required, especially given the rapid progress in the realm of the future internet, as indicated by Wódczak et al. (2011). In fact, the historical perspective seems to have played a key role in this respect, as the process of standardisation of new concepts cannot be done effectively and efficiently if detached from the technology and business-related background, as explained by Tanenbaum and Wetherall (2011). In other words, it may be claimed that, aside from standardisation per se, it is also necessary to take into account the phases of research and investment, acting as triggers for and delimiters of the entire process.

In fact, by no means should the above-mentioned explanation by Tanenbaum and Wetherall (2011), also known as 'the apocalypse of two elephants', be considered a mere claim. Essentially, there exists conspicuous and 'living' proof in the form of the Open Systems Interconnection Reference Model, since the standardisation cycle of its protocol stack was obstructed by certain factors, such as wrong timing, inadequate technology, unsatisfactory implementations, and impaired politics (Tanenbaum and Wetherall, 2011). Consequently, the OSI RM was prepared before the actual protocols were devised, quite the contrary to TCP/IP, where the protocols were delivered first, while the Reference Model simply served as their description. It then becomes clear that the research phase is inevitable in the standardisation scenario, yet proper timing is maybe even more important. As can be seen in Figure 6.1, should the related standards be rolled out prior to the research phase, the resulting specifications would be likely to remain immature. If the opposite held true, where the specifications are published with too great a delay, the market would be unlikely to wait and so investments in proprietary solutions could prevail, resulting in a lack of orchestration. Nowadays, although it may still be necessary to arrange for the proper timing, the level of awareness of the parties

---

1 For the sake of clarity, it might be reasonable to underline that the account related to the standardisation-driven approach is supported by the profound hands-on experience of the author who, as well as being co-founder, actively served as Vice Chairman and Rapporteur of the ETSI ISG AFI.

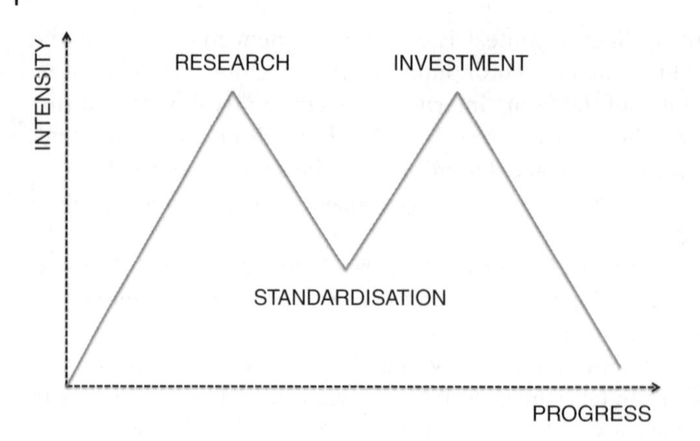

Figure 6.1 Standardisation cycle. Adapted from Tanenbaum and Wetherall (2011).

involved appears to have become sufficiently high, given the lessons learned in the past. Nonetheless, taking into account the scale of investment needed, it similarly became of prime importance for the standardisation[2] of autonomic networking to be agreed upon properly.

In the light of the above background and reasoning, it was decided that the structure of the ETSI ISG AFI be arranged according to specific Work Items (WIs), intended to form the basis for the preparation and publication of the relevant Group Specifications (GSs), at the same time keeping the option of having them upgraded to Technical Specifications (TSs) once the effort has matured even further, as outlined by Wódczak et al. (2011).[3] Consequently, from a time perspective, it is possible to emphasise that the standardisation work of the ETSI ISG AFI resulted in the finalisation of two preparatory WIs, so that the scenarios, use cases, and requirements for autonomic and self-managing future internet were outlined in the ETSI-GS-AFI-001 (2011) GS, while the aforementioned architectural reference model for autonomic networking, cognitive networking, and self-management as required by autonomic network engineering for the self-managing future internet was defined in the ETSI-GS-AFI-002 (2013) GS. Following the completion of these WIs, the emphasis moved to the standardisation process in the area of autonomic reference architectures, analysis of requirements, and specification of implementation-orientated solutions (Wódczak et al. 2011). To this end, the third WI was created and initially subdivided into four complementary branches with the following scopes and naming patterns: 'Autonomicity-enabled NGN Reference Architecture (Fixed/Wired Networks)', 'Autonomicity-enabled Broadband Forum (BBF) Reference Architecture', 'Autonomicity-enabled Mobile Network Architecture (3GPP and Non-3GPP)', and 'Autonomicity-enabled Wireless Ad-Hoc/Mesh/Sensor Network Architecture'.

As the standardisation effort carried out by the ETSI ISG AFI started becoming more and more widespread, the body agreed to openly establish a methodology consisting of an outer and inner process, as outlined by Wódczak et al. (2011). In particular, as depicted in Figure 6.2, the overall outer process became based on the assumption

2 For a more generic view on standardisation the reader may also refer to Gelonch-Bosch et al. (2017).
3 For a broader background, the reader is also referred to Chaparadza et al. (2009b).

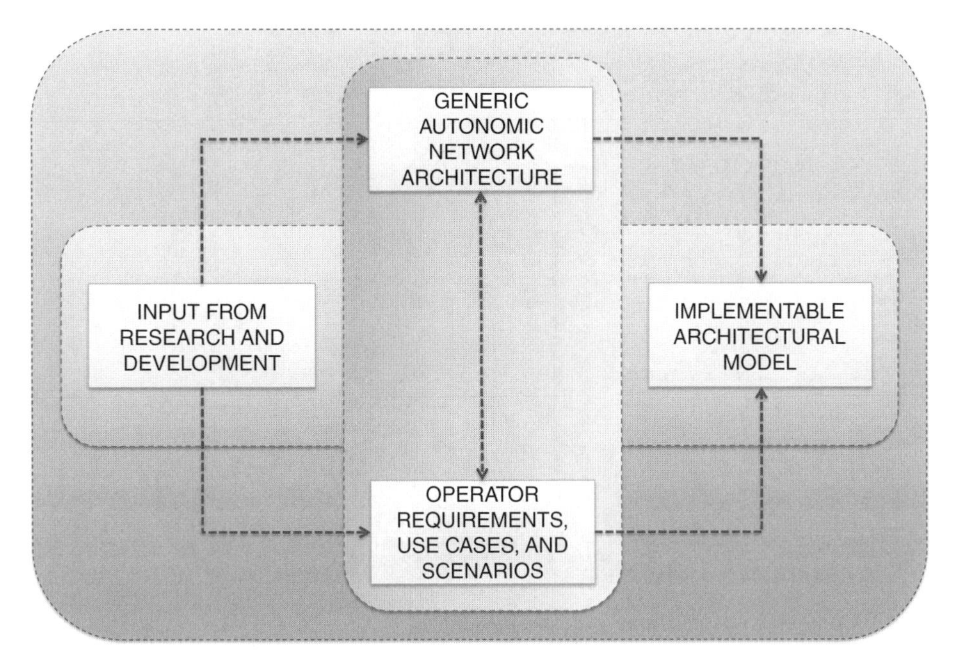

Figure 6.2 Approach to standardisation. Adapted from Wódczak et al. (2011).

that input from research and development projects would be invited not only to contribute to the refinement of the Generic Autonomic Network Architecture itself, but also to the update of the operator requirements, use cases, and scenarios. At the same time, a very tight interaction between the two WIs related to those branches was encouraged, so that the ultimate implementable architectural model could be devised as the major, and most comprehensive, outcome. Similarly, the inner process, as depicted in Figure 6.3, would be based on the implementing interaction between the two initial WIs, and, then, feeding the output directly to the third WI, yet, as the process would be iterative, feedback from the latter could be also conveyed to both the former WIs. Moreover, in addition to both the above-mentioned inner and outer processes, as depicted in Figure 6.4, thanks to specific liaisons outside ETSI, the ISG AFI was to become part of a wider ecosystem including organisations such as 3rd Generation Partnership Project (3GPP), Telemanagement Forum (TMF), International Telecommunication Union – Telecommunications (ITU-T), Internet Engineering Task Force (IETF), European Union (EU) Framework Programme (FP), Broadband Forum (BBF), and Next Generation Mobile Networks (NGMN) (Wódczak et al. 2011).

6.2.2 Staged Instantiation of Reference Model

Looking at the ecosystem and the profiles of the organisations identified therein, one may come to the conclusion that, comprehensive as it may be, when analysed from the conceptual design perspective, the rationale behind the intrinsic arrangement driving the workings of the Generic Autonomic Network Architecture appears not to have been devised accidentally, since it is ready to cover a vast range of potentially

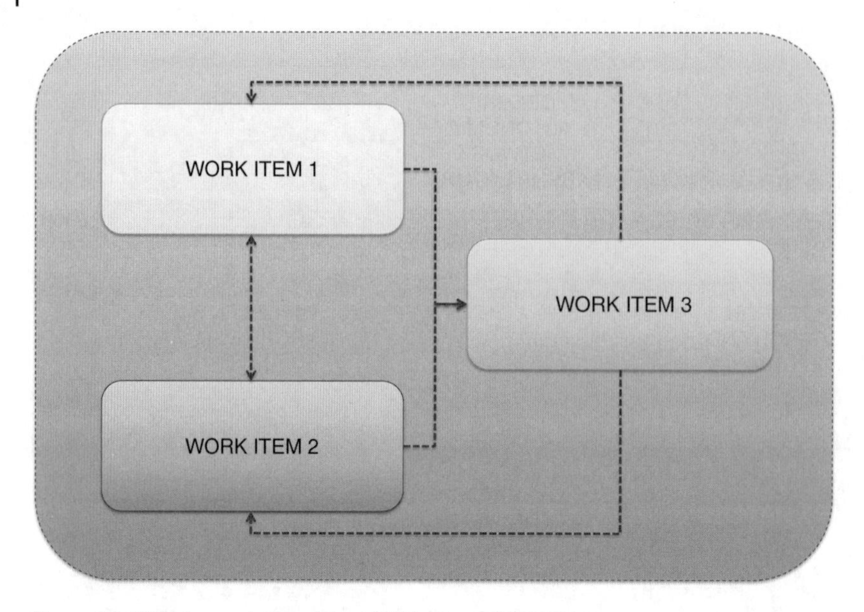

Figure 6.3 Work Items. Adapted from Wódczak et al. (2011).

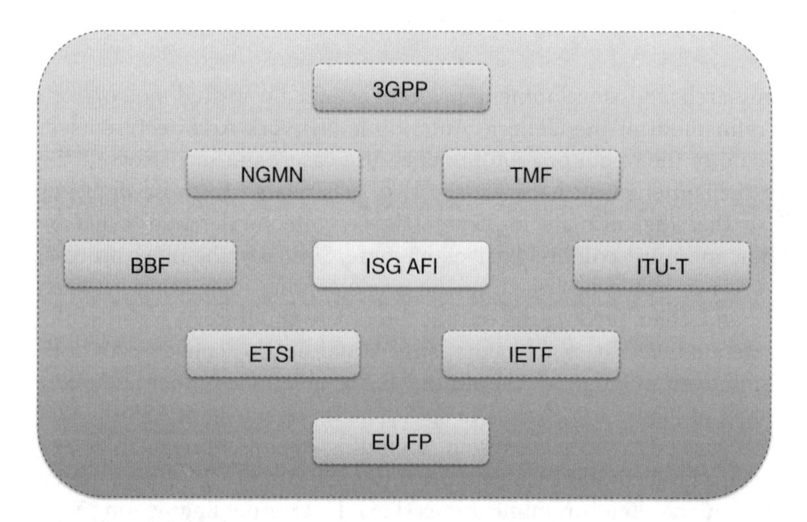

Figure 6.4 Standardisation ecosystem. Adapted from Wódczak et al. (2011).

compliant systems, requiring the incorporation of an autonomic overlay (AO) to enable virtually ubiquitous self-management. In particular, being generally targeted at the orchestration of vastly distributed networked systems, the Generic Autonomic Network Architecture is inherently based on the four conceptual levels of abstraction in the form of the protocol level, function level, node level, and network level, listed in the bottom-up order. A structure of this type, allowing to take direct control over any pertinent physical software or hardware components at the lowest level of the hierarchy only, while, in the majority of cases, making the entities located above responsible for the encapsulation of certain logic resulting in improved general flexibility in terms of the internal structure and the roles taken, is highly pertinent to the staged instantiation of the Reference Model behind the Generic Autonomic Network Architecture. Such an approach may be contemplated thanks to certain design assumptions, according to which the interrelated functional blocks (FBs) were introduced along with their respective reference points (RFPs). Interestingly, being arranged in a staged manner, this orchestration is supposed to proceed in a top–bottom order, as detailed below.

In general, the rationale behind the introduction of the Reference Model for the already outlined Generic Autonomic Network Architecture, as originally proposed by Chaparadza (2008) and, then, additionally advanced by Wódczak et al. (2011), is primarily related to the necessity of outlining the fully-fledged workings of the Autonomic Cooperative Networking Architectural Model, the ultimate version of which is to be further detailed in this chapter. One should note that, as such, the pertinent version of the existing GANA Reference Architecture would result not only from the European Union Seventh Framework Programme Integrated Project: Exposing the Features in IP version Six protocols that can be exploited/extended for the purposes of designing/building Autonomic Networks and Services (EFIPSANS), but especially from the follow-up standardisation effort carried out under the auspices of the European Telecommunications Standards Institute, and the Industry Specification Group on Autonomic network engineering for the self-managing Future Internet, in particular. To follow the original nomenclature introduced still in the past century and applied by Zimmermann (1980) to Open Systems Interconnection Reference Model, the notion of a Reference Architecture should be understood as a an incarnation, or maybe even an implementation of the said Reference Model. However, as the acronym of Generic Autonomic Network Architecture already contains the word 'Architecture', which makes the modifier of a Reference Architecture (RA) at least redundant, if not tautological, the author of this book would advise, for the sake of clarity, the usage either of the 'GANA Reference Architecture' or simply the 'Generic Autonomic Network Architecture'.

According to the ETSI-GS-AFI-001 (2011) and ETSI-GS-AFI-002 (2013) GSs, the instantiation of the Reference Model for the Generic Autonomic Network Architecture may be perceived from the perspective of introducing the interrelated functional blocks and their respective reference points, both for its own internal needs and in order to enable open interaction with architectures defined by other standards development organisations (SDOs), such as the 3GPP or BBF. Given the complexity of such a process, a staged introduction of the components was prescribed in a top-down manner, starting from the knowledge plane, where the network level decision element normally resides, as depicted in Figure 6.5, through the node level outlined in Figure 6.6 and the function level presented in Figure 6.7, down to the protocol level contained in Figure 6.8. Not only does such an approach facilitate the overall instantiation process, but it also allows for proper coordination of autonomic functions (AFs) for the sake of stability provisioning. This requires that the hierarchical autonomic control loops be designed to guarantee noncoupling and nonconflicting behaviour of the said autonomic functions understood in a generic way, and by no means limited by mechanisms of time scaling or decision ordering, as outlined in the ETSI-GS-AFI-002 (2013) GS. As much as the hierarchical autonomic control loops condition any interaction between such autonomic functions, it is necessary to note the predominant role of mutual relations

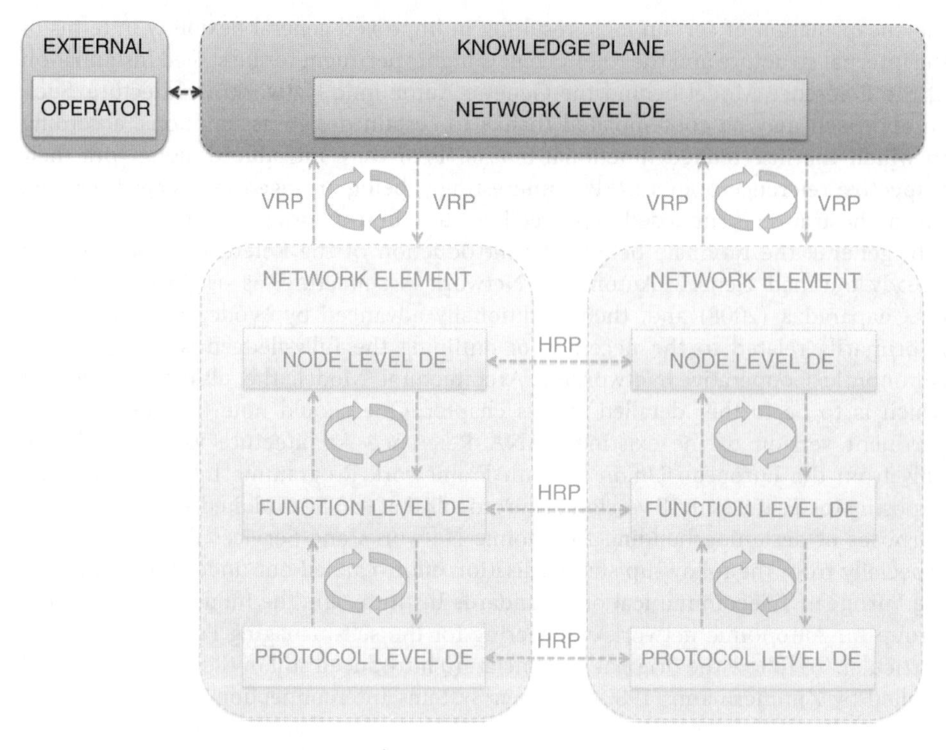

Figure 6.5 First stage of GANA instantiation. Adapted from ETSI-GS-AFI-002 (2013).

between decision elements, where a lower level decision element immediately becomes a managed entity.

Moving towards the internals of the Generic Autonomic Network Architecture, one should note that a generally holistic approach is advocated, where not only are distinct network nodes, in fact acting as autonomic nodes, supposed to express their respective, yet by default noncooperative, autonomic behaviour, but the overall distributed network is presumed to function in an entirely autonomic fashion. From such a perspective, it already appears clear that, even if cooperation were not precisely defined, it could undoubtedly be one of the most inherent features of such an autonomic networking system, where otherwise diminishable and simple information exchange might have a really constructive, and therefore cooperatively positive, influence on the overall system functioning. This might seem even more obvious when the assumption of the underlying biological connotation of self-management is kept in mind, according to which such an autonomic networking system would be expected to imitate the behaviour of a living organism, and particularly the behaviour of the human autonomic nervous system in terms of being continually driven by a substantial number of parallel processes, operating on their own yet still remaining in close correlation, making it possible to continually avoid any specific requirement for orchestration from a centrally positioned supervisory entity for the majority of the time of operation. To this end, it is claimed that such an autonomic networking system should monitor itself with the aid of hierarchical autonomic control loops running at different pace and located in the various levels of abstraction.

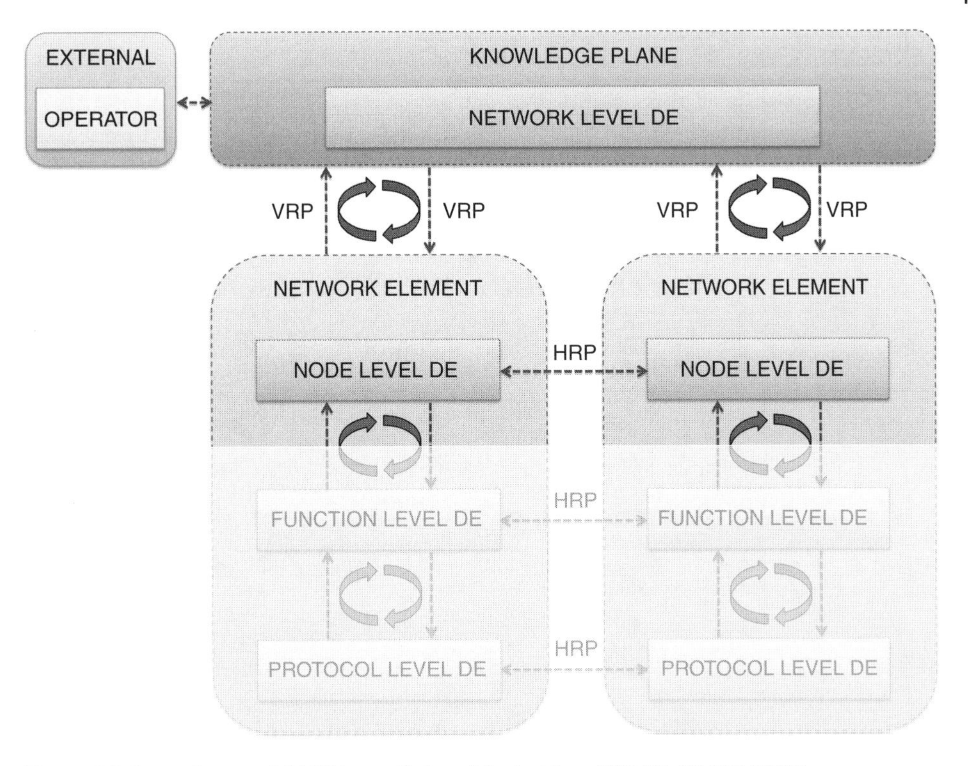

Figure 6.6 Second stage of GANA instantiation. Adapted from ETSI-GS-AFI-002 (2013).

6.2.3 Cross-Specification Extensions

Given the increasing complexity of the currently devised networked systems of the future, one may come across other standardisation activities, possibly being carried out in a separate or not explicitly integrated manner, which are similarly progressing with intentions highly synergetic with autonomics. One could then consider certain cross-specification-related investigations, keeping in mind the ultimate objective of releasing the human mind from the supervision of the related configuration and management processes to the highest possible extent, thereby also attempting to increase the overall stability and scalability of such advanced designs, as indicated by Wódczak (2014a). Being realistic, and taking into account the previously described role and pattern of a properly arranged standardisation cycle, one might become hesitant with regard to any possibility of a fully structured approach in this respect, until it becomes apparent that the institution of liaison could be highly useful in this respect. Regardless of any such uncertainties, the author would like to explore briefly a few areas of related standardisation efforts, where autonomics could benefit the other design, the other design could benefit autonomics, or, best of all, the two could capitalise on each other. As outlined in Figure 6.9, three different options are addressed below under the assumption that not all of them must be standardised by the same SDO. First comes software-defined networking (SDN), then machine-to-machine (M2M) communications, and finally intelligent transport systems (ITSs).

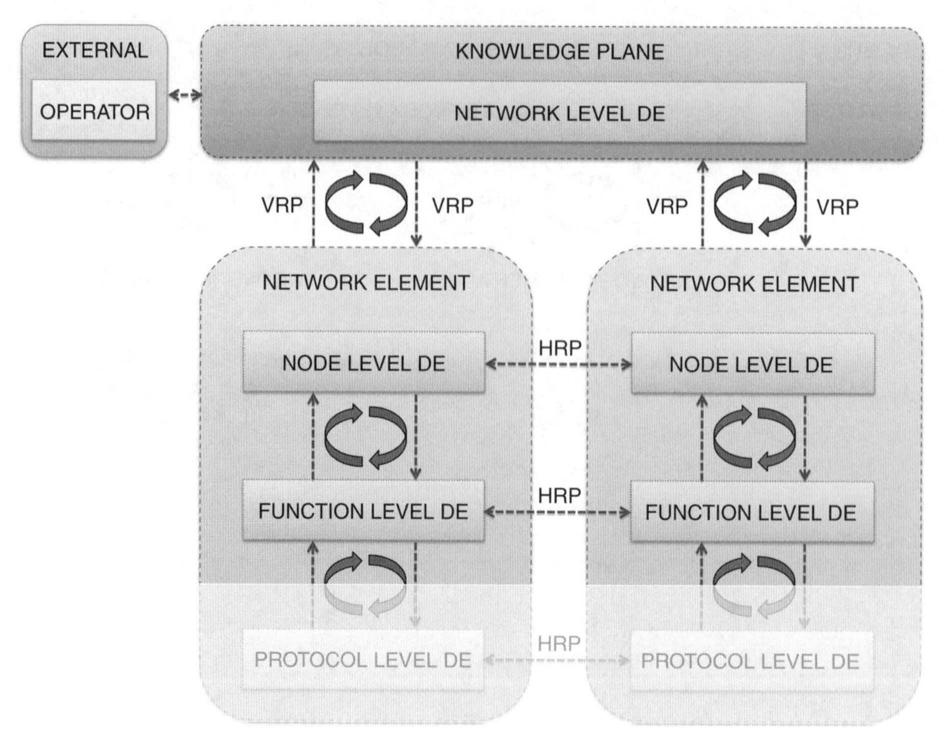

Figure 6.7 Third stage of GANA instantiation. Adapted from ETSI-GS-AFI-002 (2013).

In particular, by no accident is the rationale behind the programmability driven concept of software defined networking (SDN)[4] to be discussed prior to the other two approaches, as this one seems a little bit elevated making it to be fairly on par with the related routines of Generic Autonomic Network Architecture, just to consider Chaparadza et al. (2013). Even though the logic is different, the justification remains similar, as it stems from the fact that the distributed character of today's networked systems demands certain dose of facilitation in terms of their configurability and manageability. One should note, however, that, by comparison, the need for autonomics corresponds to the fact that even nowadays the more and more sophisticated policies and tasks might still demand implementation with the use of a legacy command line interface (CLI), which, in fact, allows for the use of a limited set of low-level device configuration directives only, as explained by Hyojoon and Feamster (2013). Apparently, still a while ago, such an issue did seem resolvable with the aid of the so-called dynamic scripting techniques, yet, as of now, even assuming that one could perform the entire configuration in such a way, the technological progress and the changing state of the continually evolving networked systems, would render a proper maintenance thereof virtually impossible (Wódczak, 2014a). This is why, among others, the concept of SDN assumes that the network behaviour be rather managed through

4 Usually, SDN goes hand in hand with network function virtualisation (NFV). However, given the specific character of this book, the latter rather remains beyond its scope. Should the reader be interested in the fusion of both approaches, they are referred, for instance, to Callegati et al. (2016).

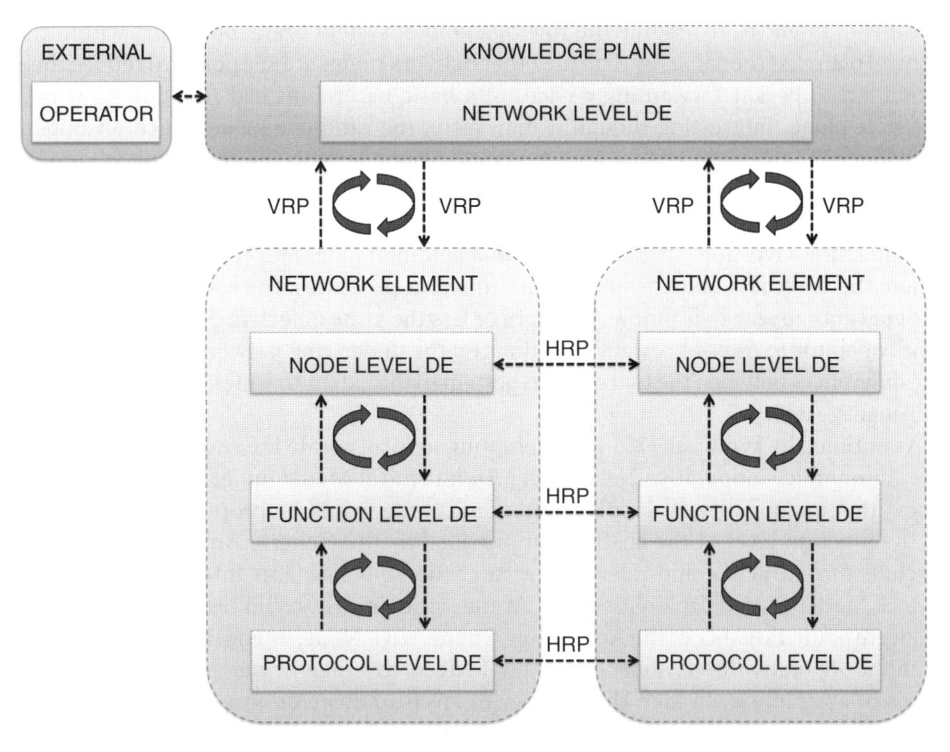

Figure 6.8 Fourth stage of GANA instantiation. Adapted from ETSI-GS-AFI-002 (2013).

Figure 6.9 Possible standardisation synergies.

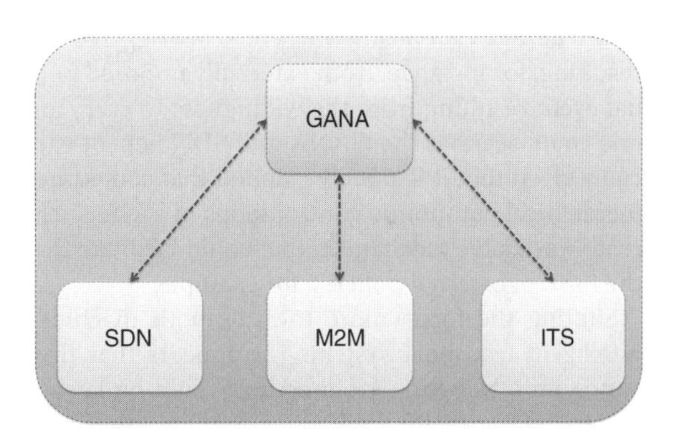

a set of specifically designed interfaces, as indicated by Haleplidis et al. (2015). While one's perception could be that this might remove a certain amount of leeway, in practice a great configuration relief should be expected instead.

There is no denying that the concept of SDN has been emerging for quite some time to finally gain an unprecedentedly rapid growth of interest (Wódczak, 2014a). Looking into the details, this approach assumes that the data plane and the control plane (CTP) be separated in the current understanding, as indicated by Hyojoon and Feamster (2013). In other words, the behaviour of the whole networked system is expected to be not only orchestrated but, in fact, dictated by a logically centralised yet physically distributed controller, intended to assume the role of a kind of system brain, operating within the control plane. At the same time, all the other network nodes are supposed to reduce their operation to packet forwarding devices of a basic nature, instead functioning within the data plane. Interesting though it may seem, the rationale behind such an understanding of SDN might at first appear not to be in line with the reasoning stemming from the Generic Autonomic Network Architecture. This perception may result from the somewhat inadequate understanding that while the Generic Autonomic Network Architecture advocates virtually ubiquitous automation, SDN promotes an approach where the control is taken from network inherent processes. In reality, however, both the concepts appear to lean towards more or less the same objective of allowing the network operator to impose certain directions on the underlying networked system, while the differences between the two seem to pertain to the extent to which they may exercise self-manageability.

As outlined by Wódczak (2014a), attempting to incorporate the routines of SDN into the Autonomic Cooperative Networking Architectural Model under discussion in this book, the functionality of the plain autonomic node would be proposed to be capped at the function level of the abstraction provided by the Generic Autonomic Network Architecture, which would relevantly restrict the Open Systems Interconnection Reference Model related capabilities, too. At the same time, it would be desirable that the respective functionality of the Autonomic Cooperative Node, intended to express Autonomic Cooperative Behaviour, offer fully-fledged coverage of the entire range of the levels of abstraction. In fact, the number of levels of abstraction would be subject to dynamic changes, since, once the promotion to the level of an Autonomic Cooperative Node took place, the functionality of autonomic networking would need to become elevated, as well as seamlessly reduced, should the conversion in the opposite direction hold true. Such a repositioning could be triggered by a dynamic network reconfiguration resulting, for instance, from externally imposed policies, or be conditioned by an internal event resulting from the willingness to carry and forward traffic. Given the above, one could see that the GANA and SDN benefit each other more or less equally, especially given that it is openly claimed that sometimes it could be necessary to open the hierarchical autonomic control loops of the former, while the latter already provides a readily available mechanism that could facilitate the injection of certain control data in the form of external policies, for example.

Shifting the focus more towards both machine-to-machine communications and intelligent transport systems,[5] it appears that the role and position of the Generic Autonomic Network Architecture would be fairly different in this respect, especially due to the fact that in both cases configurations resembling the mobile ad hoc network scenarios would be of interest. In other words, rather than serving as a mutually complementary concept, the rationale behind the Generic Autonomic Network Architecture could be applied to orchestrate each of the two, thereby reflecting the primary design objective of the same to act as a self-management overlay for networked systems. Looking into the definition provided by the ETSI-TS-102-689 (2010) Technical Specification, the machine-to-machine communications shall be established between two or more entities, and shall be assumed to be organised essentially without any

5 In this context, as advocated by Tonguz and Viriyasitavat (2016), one should note that there are already efforts undertaken to introduce self-organisation into the realm of ITSs.

specific necessity for direct involvement on the side of a human. What is more, as the related machine-to-machine services are to automate the relevant decision-making and communication processes, the overall concept thereof immediately appears to be highly synergetic with the already explained paradigm of self-manageability, so that an approach resembling the operation of the human autonomic nervous system, inherent in the Generic Autonomic Network Architecture, could be applicable. Such a characteristics are especially important in the case of highly complex networked systems, where the network nodes could equally well be instantiated by machine-to-machine devices, and where automation would be the only way forward in terms of sustainable and durable system operation, as outlined by Wódczak (2014a).

While the notion of a machine-to-machine system becomes more and more integrated with the idea of the Internet of Things (IoT),[6] as outlined in the ETSI-TR-103-375 (2016). Technical Report, despite the fact that the latter is still undergoing substantial efforts in terms of becoming ultimately defined, the initial understanding of the former called for its decomposition into two functional parts, the first covering both the machine-to-machine device and machine-to-machine gateway, while the second encompassing the machine-to-machine network (ETSI-TS-102-690, 2013). As much as typically such an M2M device would be expected to connect to the respective M2M network over an M2M gateway, a direct connection is not precluded, under the assumption that a proper capability is implemented. In the light of the above it appears that one could introduce a new notion of M2M cooperative gateways able to express not only Autonomic Cooperative Behaviour, but also their willingness to carry and forward traffic, as proposed by Wódczak (2014a). Such an approach would remain very much in line with the rationale behind the Autonomic Cooperative Networking Protocol discussed earlier in the book, especially that the said M2M cooperative gateways would be entitled to play the role of Autonomic Cooperative Nodes. Thus the M2M-driven design could be integrated into the Autonomic Cooperative Networking Architectural Model, where, depending on the nature of the M2M devices, certain elements of the discussed SDN extension could also apply.

Finally come the intelligent transport systems[7] which, given the global drive for worldwide deployment of efficient vehicular systems, are becoming a substantial element of the discussed cross-standardisation ecosystem: certain cooperative awareness related efforts are currently being undertaken, just to follow the ETSI-TS-102-868-2 (2017) Technical Specification.[8] Yet, going even further and assuming a broader perspective, one may notice that, on the one hand, certain advancement in regard to the theme of this book becomes conspicuous in the case of both the physical layer and link layer, where the emphasis is generally laid on the provision of wider bandwidth and lower transmission latency, as outlined by Li et al. (2012a). On the other hand, however, similarly following Li et al. (2012b), one cannot forget the related routines of the network layer, where the aspects of autonomic networking and cooperative routing seem to gain

6 For a more business-orientated approach to the growth of the IoT, the reader is referred to Ghanbari et al. (2017).

7 For a broader view on ITSs in terms of the future research perspectives, the reader is referred to Dressler et al. (2014).

8 In particular, one should take into account the growing inclination to refer to ITSs as 'Cooperative Intelligent Transportation Systems'.

more attention than one would expect these days.[9] Essentially, the very relevant question of self-management still appears to lack a proper answer, which would make it possible to understand how the vehicular network nodes could become elevated to Autonomic Cooperative Nodes and enabled to express Autonomic Cooperative Behaviour to be orchestrated by the Autonomic Cooperative Networking Protocol (Wódczak, 2011c). In fact, the incorporation of the relevant overlay routines may be facilitated through the integration with the Autonomic Cooperative Networking Architectural Model, which builds on top of the Generic Autonomic Network Architecture along with its inherent hierarchical autonomic control loops, as explained by Wódczak (2014a).

Looking into the details, as indicated by Wódczak (2014a), the basic approach could assume that the interaction with the decision elements of the Generic Autonomic Network Architecture is taking place at all the levels of abstraction, where the hierarchical autonomic control loops allow for the orchestration of their respective managed entities, implemented within the on-board unit (OBU). In fact, the OBU shall function as a networked device encompassing a whole composition of various protocols and manageable routines, altogether making it by no means perceivable as a single managed entity. What is more, similarly to network devices such as typical routers, the relevant portions of the Autonomic Cooperative Networking Architectural Model would need to be integrated into such an OBU in order to oversee its behaviour from the inside. In other words, moving across the levels of abstraction, the cars constituting a given vehicular ad hoc network (VANET) would act as Autonomic Cooperative Nodes and would be entitled to form autonomic cooperative sets (ACSs), thereby expressing Autonomic Cooperative Behaviour under the supervision of the Autonomic Cooperative Networking Protocol, in order to instantiate distributed space-time block coding enabled cooperative transmission over a virtual multiple-input multiple-output radio channel. All in all, as it is clearly visible, the order of presentation made it possible to commence the analysis with the highest-level SDN[10] so that, going through the M2M communications positioned somewhere in the middle, one could approach the ITSs, where certain concepts devised in this book would match directly.[11]

6.3 Cooperative Emergency Networking

6.3.1 Emergency System Requirements

An emergency communications network (ECN) is formed by the first responders (FRs), or, in fact, by the communication equipment they carry around in the area of an incident, where mostly mobile ad hoc network system configurations are possible, even though there are certain semi-fixed or rather movable components such as the mobile emergency operations centre (MEOC) of a completely different purpose. A configuration of this type appears to make the emergency communications network

9 For additional context, the reader is referred to Hsu et al. (2010) and Wódczak (2012c).

10 Interesting though it may seem, there exists a relation between Software Defined Networking (SDN) and Software Defined Radio (SDR). As much as it remains beyond the scope of this book, the reader is referred, for example, to Ramirez-Perez and Ramos (2016), for further information in this respect.

11 One should note that both notions of automation and cooperation appear to become more and more inherent in ITSs (Dias et al. 2015; Hobert et al. 2015).

a very relevant area for experimentation in terms of the application of the routines driving the Autonomic Cooperative Networking Architectural Model, as defined throughout this book. Essentially, as there is no denying that, simply because of the related highly dynamic topology changes creating an entirely mobile environment, where the network nodes embodied by the somewhat elevated chief first responders (CFRs) may be upgraded to offer the capability of fully-fledged Autonomic Cooperative Nodes, it should be possible to exercise Autonomic Cooperative Behaviour through the formation of ACSs. In other words, one may be entitled to employ the conceptual device of an equivalent distributed space-time block encoder (EDSTBE) between the preselected chief first responders, so that, being fed by a base station (BS) located on the roof of the vehicle acting as the mobile emergency operations centre, they could provide communication capability to the first responders in the field, as proposed by Wódczak (2012b). All the configurations considered in this respect are discussed in detail below, along with certain system requirements that need to be properly incorporated.

In general, the current advancements in the realm of novel infrastructure design for the emergency communications networks of the future appear to be critically demanding in terms of the technologies to be applied, as explained by Vassiliadis et al. (2010) and Calarco et al. (2010). This phenomenon is particularly visible in the mobile ad hoc network part of the system, where the devices carried by the first responders seek seamless and on-demand connectivity for radio transmissions requiring different quality of service (QoS) measures to be implemented (Wódczak, 2011b). Thus, the emergency communications network, formed mainly by the first responders operating in the area of an incident, seems to have become a very relevant field for the application of the concepts related to the Autonomic Cooperative Networking Architectural Model. This especially holds true for the clearly envisaged to be numerous, yet rather small in size, groups of first responders, as coordinated by their respective chief first responders (Wódczak, 2012b). According to the requirements resulting from the most typical mode of operation in such scenarios, these groups may be assumed to contain strictly from four to six first responders, which may immediately be translated into certain advantages and disadvantages when the network topology of preference is concerned, as indicated by Wódczak (2011d). In particular, this might equally well mean that multi-hop communications between the chief first responders and their first responders might not be preferred or even allowed, since the first responders would normally be expected to gather around their chief first responders, forming a star topology as presented in Figure 6.10.

Approaching the issue of the related requirements from a purely technical perspective, one may come to the conclusion that such a multi-hop[12] configuration cannot be entirely precluded, especially under the circumstances where a group might need to be spread much more apart, as outlined in Figure 6.11. What is more, the possibility of serving larger groups, or maybe even merging and splitting groups, of first responders might also need to be taken into consideration, at least theoretically, to understand and address the behaviour of the entire system should any of the above-mentioned options become a reality. Going even further, this might raise the very relevant question of how the transmission should be arranged so that the emergency communications network demonstrates increased resiliency to the dynamically fluctuating environment it

12 For additional analysis of multi-hop networks for disaster recovery, the reader is referred, for example, to Minh et al. (2014).

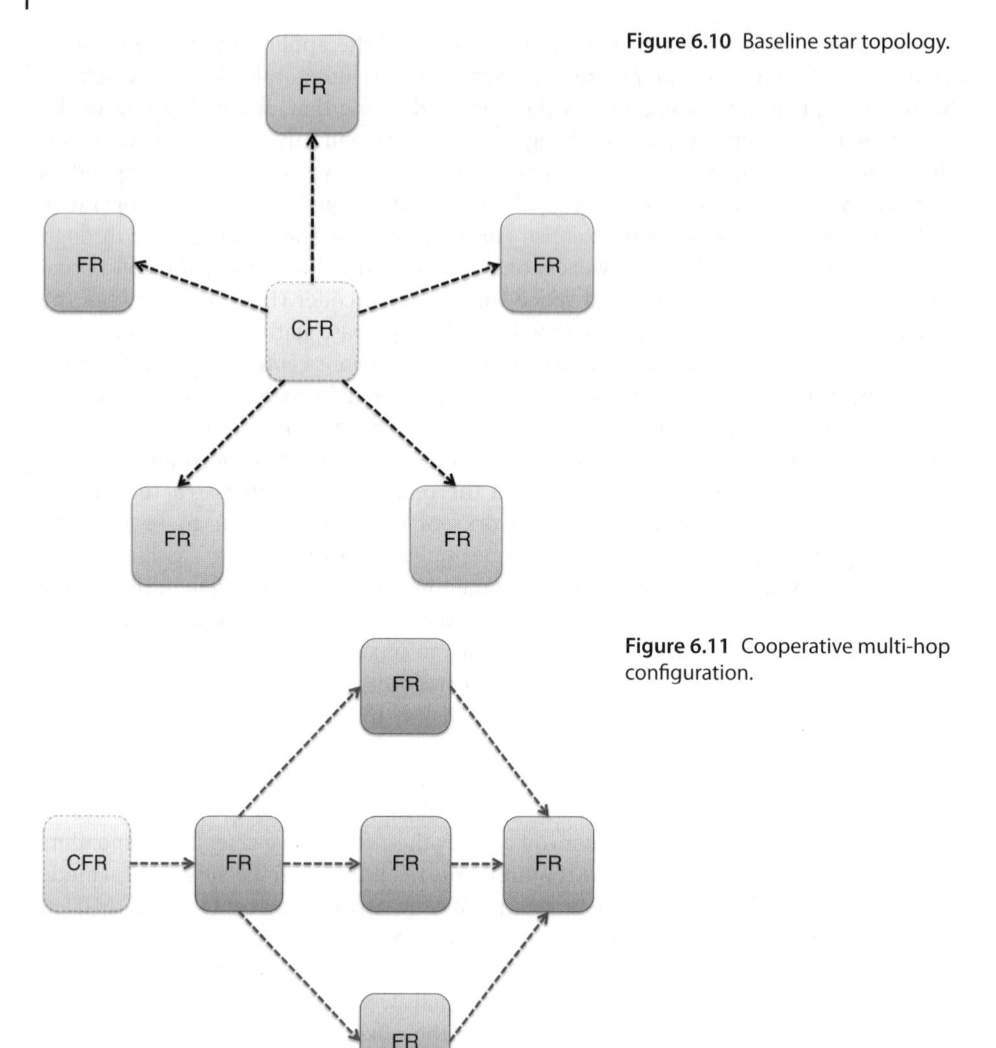

Figure 6.10 Baseline star topology.

Figure 6.11 Cooperative multi-hop configuration.

is required to operate in, as noted by Wódczak (2011e). In principle, the incorporation of the routines of the Autonomic Cooperative Networking Architectural Model, especially characterised by the inclusion of Autonomic Cooperative Behaviour, appears to constitute what could be referred to as a proper approach (Wódczak, 2011a). Its applicability is directly visible in Figure 6.11, which depicts a group of first responders where each of the members is perceived as an Autonomic Cooperative Node and could equally well be qualified to join an ACS, thereby becoming entitled to instantiate Autonomic Cooperative Behaviour, not only through autonomic cooperative transmission, but also autonomic cooperative re-routing (ACRR), as indicated by Wódczak (2014a).

As already signalled, there is yet another dimension to the investigated case, since, from the orchestration perspective of the emergency communications network, in addition to the chief first responders and first responders, the system also needs to feature

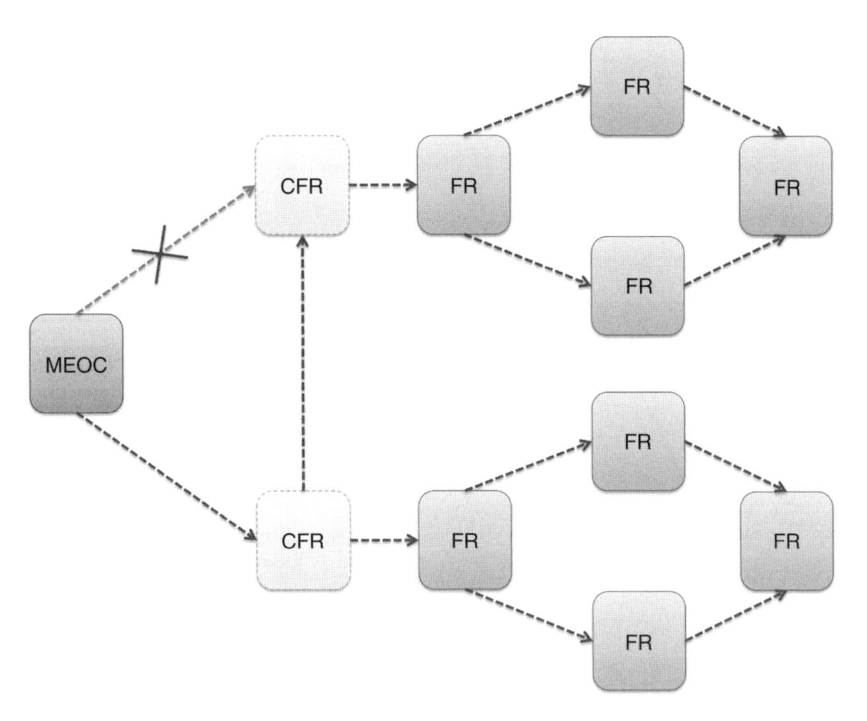

Figure 6.12 Supportive communication between CFRs.

both the emergency operations centre (EOC) and the mobile emergency operations centres. While the former is always located at a single fixed position, the latter typically coincide with fully mobile and specifically designed vehicles, always being relocated to the area of an incident. One may thus envisage two distinct types of Autonomic Cooperative Behaviour between the Autonomic Cooperative Nodes, instantiated by the respective chief first responders, to be taken into consideration (Wódczak, 2011d). In the first place, it is most commonly assumed that the process of communication involving two different groups of first responders, coordinated by two distinct chief first responders, will be assisted by those chief first responder only, who do not communicate directly, but rather over the mobile emergency operations centre, as outlined in Figure 6.12. Such an assumption would potentially need to be relaxed, however, in order to accommodate the possibility of a lack of communication between one of the chief first responder and the mobile emergency operations centre. One would also need to consider another situation, particularly in the initial response phase, when various groups of first responders might already be in a given location along with their respective chief first responders, while the mobile emergency operations centre could still be on its way.[13]

What is more, for legacy reasons stemming from a consolidated management approach imposed by the inherent human hierarchy, it could be required that only a given chief first responder route the data stream coming from the mobile emergency operations centre towards a given group of first responder, as no other chief first

13 For the context of emergency group communications under the umbrella of an LTE-driven design the reader is referred to Kim et al. (2016), while more information on the related topic of moving public safety networks (PSNs) may be found in Favraud et al. (2016).

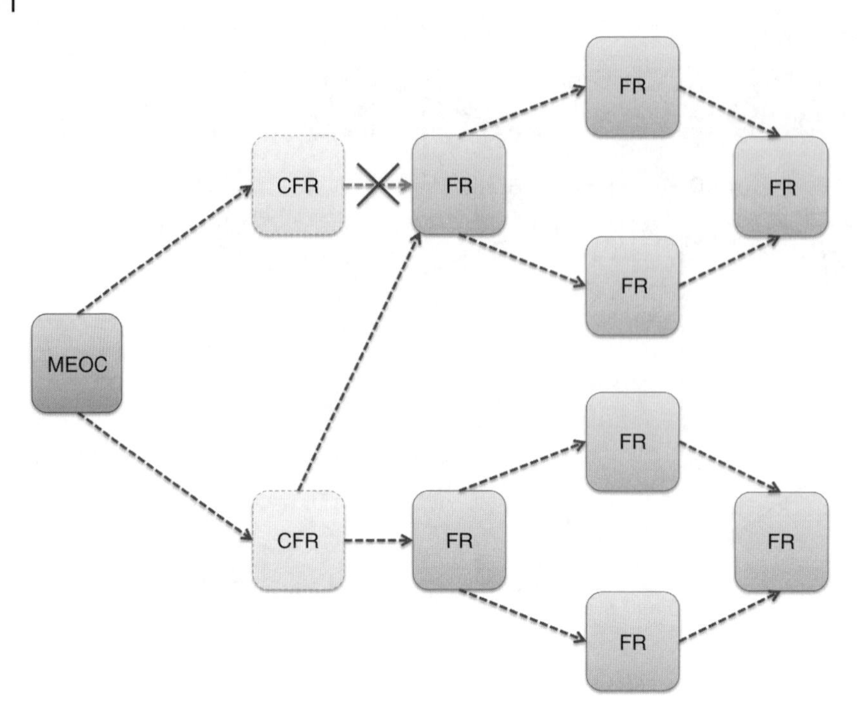

Figure 6.13 Supportive omission of affected CFR.

responder should be superior to the said group in terms of duty pertinent relations. As understandable as this is, there are no clear arguments from the system design perspective against the establishment of a logical link over such a chief first responder positioned externally to this group of first responders, as long as such operation remained transparent to the functioning of the emergency communications network so that the hierarchy could be perceived as if nothing had changed. Analysing such a case even further, as depicted in the example provided in Figure 6.13, for various reasons such as external obstacles to radio transmission or maximum time of service with a fully charged battery, one of the chief first responders may not be able to communicate with their first responders. As proposed by Wódczak (2012b), the system could address this situation through autonomic switching to a backup mode, where another chief first responder would act as an entity supporting the communication with the mobile emergency operations centre. Although it is technically feasible, such an approach could, in some sense, violate the said requirements in this respect.[14] In fact, there may be much more to it than the human hierarchy, as in emergency situations the responsibility must be clear and the persons involved should know exactly from where the orders are physically coming.

14 Such requirements come from the European Union Seventh Framework Programme Integrated Project: A Holistic Approach Towards the Development of the First Responder of the Future (E-SPONDER).

6.3.2 Autonomic Control Incorporation

The previously discussed issue related to the discrepancy between certain justified requirements and the technological approaches thereto may create a serious issue at the design stage. This is especially important when the related technologies are so advanced that they could easily solve any outstanding problem, yet, for reasons similar to the ones stated above, where being in command means the responsibility for direct physical coordination of all the related actions in the field, one cannot risk employing them. Clearly, it could be possible to contemplate certain complementary solutions, where schemes such as conventional relaying, understood in the sense depicted in Figure 4.2, could be deployed in order to allow for a given chief first responder to have the communication it originates supported fully consciously by another chief first responder of their choice. Yet, even if not susceptible to being charged with any violation in this respect, such an approach could still remain doubtful from the legal perspective. For such a reason, the cooperation-enabled approach advocated here focuses solely on the communication originating from the mobile emergency operations centre, as then none of the potential restrictions comes even close to being at risk of becoming classified as doubtful. Next, in the light of the above, first of all the rationale behind such a cooperative solution will be outlined in more detail, before it is possible to advance the analysis towards certain related interactions involving the mobile emergency operations centre and its chief first responders, so that, eventually, the focus may be shifted to the ACNAM-related considerations.

Moving into the details, an example approach where the communication between the mobile emergency operations centre and a given first responder is carried out with the assistance of an external chief first responder, in a way transparent to the system and not affecting the hierarchy, is presented in Figure 6.14. As outlined by Wódczak (2012b), taking into consideration that a given first responder might become exposed to fairly severe impairments induced by the radio channel, for example from obstacles to radio propagation or too big a distance towards their chief first responder, another chief first responder could step into such communication so that the diversity gain offered by the virtual multiple-input single-output radio channel could be efficiently exploited, as the target first responder would be served cooperatively by both the chief first responders in question. This would allow the application of DSTBC, since both the chief first responders would logically form a virtual antenna array and apply the related physical layer signal processing techniques in order to orthogonalise the wireless radio channel. Consequently, the data could be delivered in a more resilient manner, assuming that proper synchronisation can be guaranteed between the cooperating chief first responders. Essentially, shifting more towards the nomenclature inherent in the Autonomic Cooperative Networking Architectural Model, one could say that, as long as both the chief first responders are exposing the capabilities of and acting as Autonomic Cooperative Nodes, Autonomic Cooperative Behaviour would be instantiated in this respect.

To enable Autonomic Cooperative Behaviour one would need to employ the appropriate entities of the Autonomic Cooperative Networking Architectural Model, such as the Autonomic Cooperative Networking Protocol, particularly applicable in dense environments, where the neighbour discovery and link verification capabilities of the Optimised Link State Routing protocol would become highly useful. Thus, not only could the links

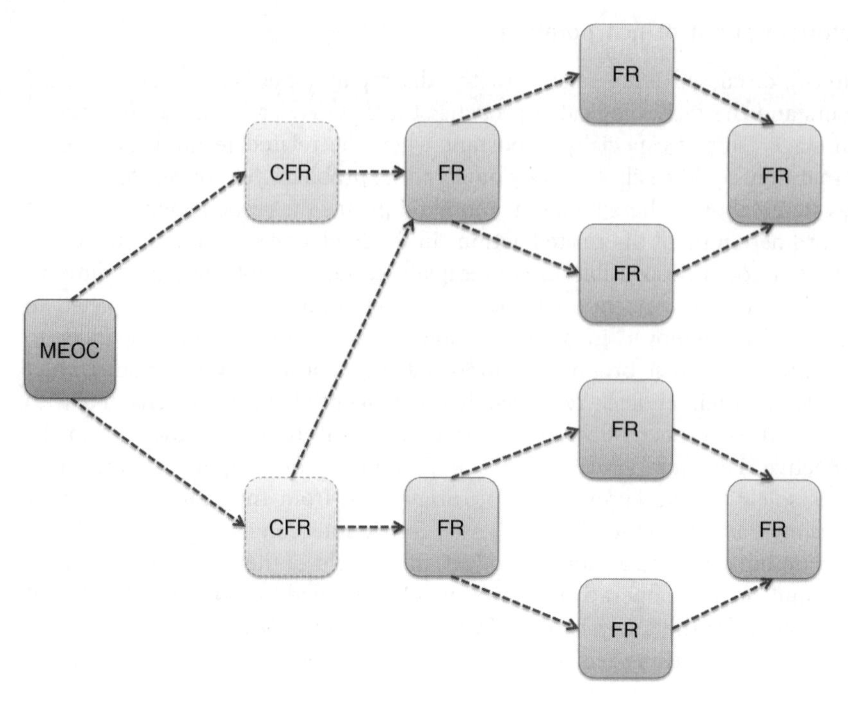

Figure 6.14 Cooperative mode of operation.

between chief first responders and their respective first responders be monitored, but the mobile emergency operations centre would also acquire sufficient data to identify when another chief first responder could potentially enter into cooperation to serve an first responders belonging to a different team. In general, better robustness could be offered thanks to the operations performed by all of the mobile emergency operations centre, chief first responders, and first responders, as outlined in Figure 6.15. Consequently, having the relevant global data describing the status of the emergency communications network, the mobile emergency operations centre would even have some leeway in terms of prearranging cooperation before link degradation occurs. In particular, should a transmission request from a chief first responder be rejected by a first responder due to a transmission problem, the mobile emergency operations centre would be entitled to issue cooperation requests sequentially to both chief first responders. If both reply in the affirmative, then a relevant cooperation indication would need to be issued by the mobile emergency operations centre before the chief first responders could enter the cooperation phase.

Essentially, the process leading from the evaluation of specific radio links towards the ultimate Autonomic Cooperative Behaviour could be perceived as being carried out in two complementary, though not completely disjoint, ways, as depicted in Figure 6.16. Similarly to what was originally discussed on the introduction of the classification governing the routing protocols for mobile ad hoc networks, such as the Optimised Link State Routing protocol, it would be possible to assume either a proactive or a reactive approach in this respect. The choice between the two would clearly be conditioned on the required level of resiliency, expected to differ at specific stages of

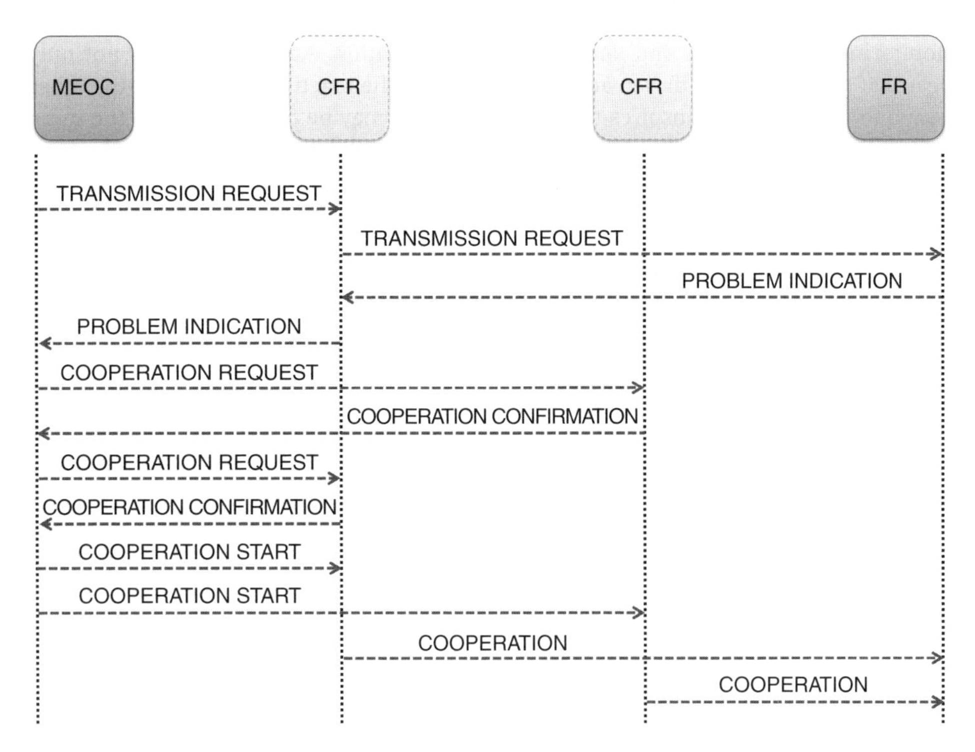

Figure 6.15 Instantiation of cooperation by MEOC.

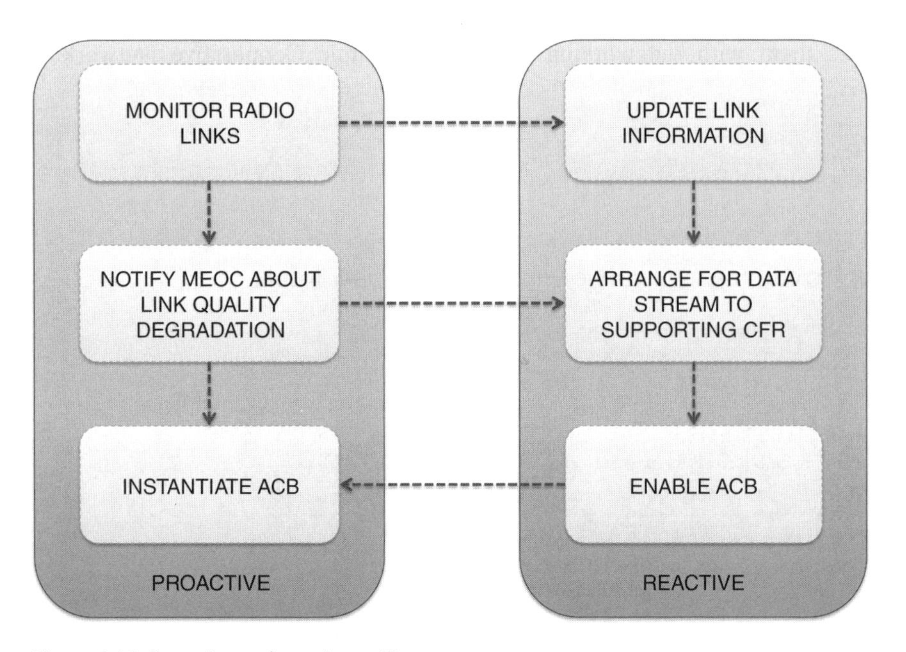

Figure 6.16 Proactive and reactive resiliency process.

operation. Scrutinising the process itself, the starting points are related both to the monitoring of radio links and updating their information. At this stage it is not only clear that the proactivity-driven action could trigger the reactive one and not the other way round, but it also transpires that both processes may be ongoing within the same or different Autonomic Cooperative Nodes. Assuming that the same protocol would be exploited by each Autonomic Cooperative Node, such as the Optimised Link State Routing protocol, the default proactivity could also take a much more reactive flavour, for example due to energy constraints resulting from battery drainage and leading to an intentional reduction in the level of activity. In this mood, the following steps would involve either the notification related to link quality degradation or the arrangement for directing data stream to the supporting chief first responder, so that eventually it would be possible to enable or instantiate the Autonomic Cooperative Behaviour, respectively.

Finally, analysing the integration aspects of the ECN with the Autonomic Cooperative Networking Architectural Model, as depicted in Figure 6.17, one may see that the pattern applicable in this respect would be not significantly different when compared to other networked configurations. In particular, one would need to accommodate both the perpendicularity or orthogonally interrelated, from the conceptual perspective, Open Systems Interconnection Reference Model and Generic Autonomic Network Architecture, which in the figure go side by side solely for graphical convenience reasons, as the major interfaces through which the Autonomic Cooperative Networking Architectural Model may orchestrate the behaviour of the ECN. Such an orchestration would, in the first step, be related to the elevation of the chief first responders to the functional role of Autonomic Cooperative Nodes in order to establish the link with the routines of the Autonomic Cooperative Networking Architectural Model. In particular, through the formation of ACSs, such Autonomic Cooperative Nodes would be able to express Autonomic Cooperative Behaviour. This way it would become possible to integrate them with the workings of the Autonomic Cooperative Networking

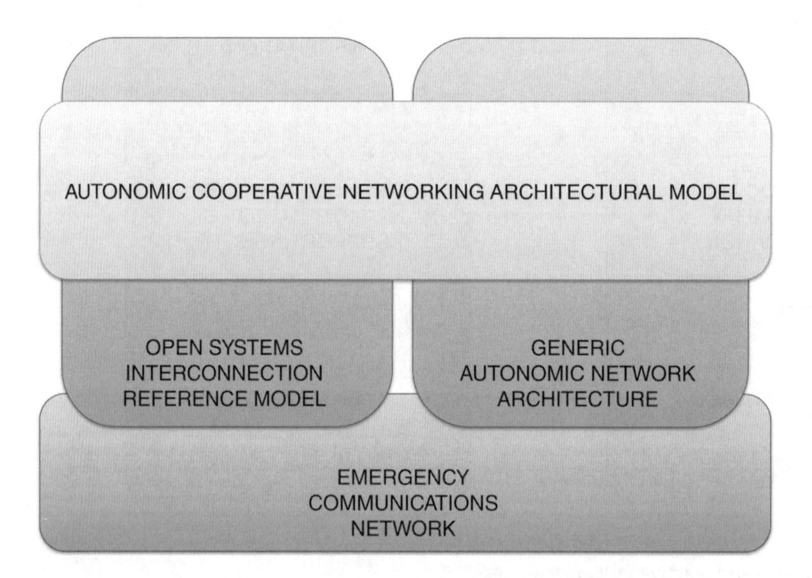

Figure 6.17 Integration into Autonomic Cooperative Networking Architectural Model.

Protocol as a kind of a wrapper for the OLSR-protocol-based extended routing information enhanced algorithm for cooperative transmission (EREACT). Consequently, DSTBC-driven cooperative transmission over the VMIMO radio channel could be instantiated between the related chief first responders.

6.3.3 Cooperative Enhancement Justification

In order to verify the potential gains that could result from the employment of the above-mentioned principles of governance, as imposed by the Autonomic Cooperative Networking Architectural Model, especially with respect to Autonomic Cooperative Behaviour, a performance investigation of the related cooperative transmission with several different snapshots of various configurations of chief first responders located at fixed positions was carried out.[15] To this end the baseline relay deployment indoor scenario was utilised, as originally depicted in Figure 4.20, along with the system parameters outlined in Table 4.5. Such an approach could be assumed since CFR deployments analysed at certain time stamps are equivalent to those featuring fixed relay nodes. The experiment was carried out in order to investigate whether an autonomic networking system in the form of an emergency communications network, with the capability of network monitoring and policy application, could benefit from cooperative deployment of chief first responders. One should note that normally the mobile emergency operations centre would be located outside such a building, yet, since it seems there are no standards defining what the relative positions of the two could be, it was assumed that the BS should be kept in the middle of the scenario, as in Figure 4.20. For each of the following figures, panel (b) shows the performance of the CFR2–CFR3 cooperative transmission, while panels (c) and (d) show the relative throughput attainable with single-path relaying via CFR2 and CFR3, respectively (Wódczak, 2012b).

First of all, it is assumed that the autonomic orchestration logic, to be upgraded later in this chapter to the notion of the cooperation orchestration decision element, would employ CFR3 at the same position in both the deployments presented in Figures 6.18 and 6.19. At the same time, the position of CFR2 would change, so that in the former case CFR2 would be placed closer to the BS thanks to which some improvement should be observable in comparison with the reference case detailed in Figure 4.22. In general, even though the throughput attainable thanks to CFR–CFR cooperation could be higher in certain regions, one might also consider employing the chief first responders disjointly in order to separately cover different regions of the area of interest. Consequently, as depicted in Figure 6.18(c), the case of single-path relaying using CFR2 should offer higher throughput in the rooms, and especially those immediately adjacent to the BS, and the corridor, while, according to Figure 6.18(d), CFR3 could better handle the remaining area. What is more, as becomes visible in the deployment shown in Figure 6.19, an attempt to compensate for the throughput degradation observable in the distant corners does not really provide satisfactory results, as the distance to CFR2 is too great. Thus, despite the reasonably satisfactory performance of CFR3, as confirmed by the results in Figure 6.19(d), overall CFR–CFR cooperation would be negatively affected and thus impractical, as shown in Figure 6.19(c) (Wódczak, 2012b).

---

15 Given that the cooperation orchestration decision element is yet to be introduced, the reader is referred to Wódczak (2012a, 2012e) for more advanced deployments to be analysed in the context thereof.

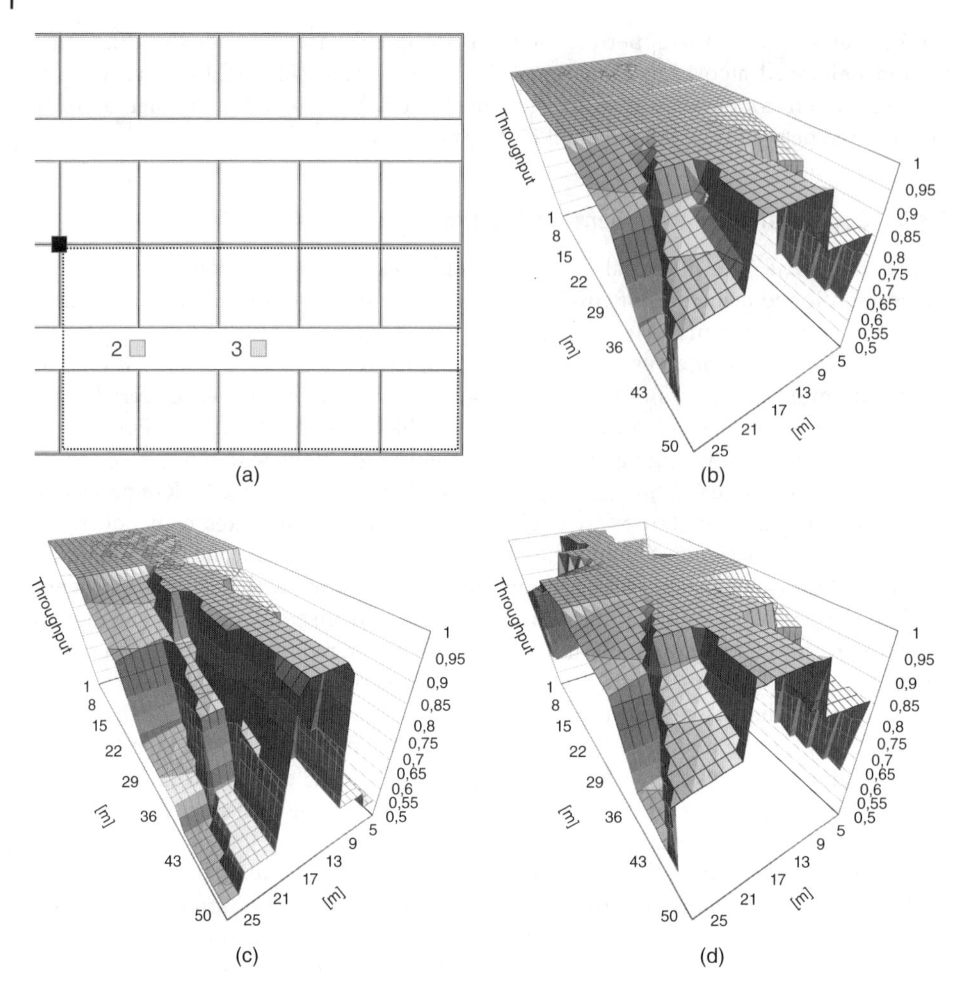

Figure 6.18 (a) Deployment of CFRs. (b) Relative throughput for CFR2–CFR3 cooperation. (c) Single-path relaying via CFR2. (d) Single-path relaying via CFR3.

Keeping in mind that in the case of the two above-mentioned scenarios, as well as the distance between them, as consequently the number of obstacles on the paths between the BS and each chief first responder can be different, two further possible deployments are evaluated as outlined in Figures 6.20 and 6.21. This time it is assumed that the autonomic orchestration logic ensures that chief first responders become selected for which the number of walls between them and the BS would be identical, while the distances were similar. Unfortunately, as much as the throughput achievable in the direct neighbourhood of the BS appears to be reasonably satisfactory, to follow Figure 6.20, the remaining area would not be covered sufficiently. All in all, it transpires that the issue of proper throughput provisioning in the case of the analysed indoor scenario is not a trivial question. On the one hand this might appear to be a serious drawback, yet, on the other hand, when the major theme of this book is taken into account, such a disadvantage should be perceived much more as a challenge. In fact, given the above context, as well as keeping in mind that the performance evaluation might be treated as providing at least

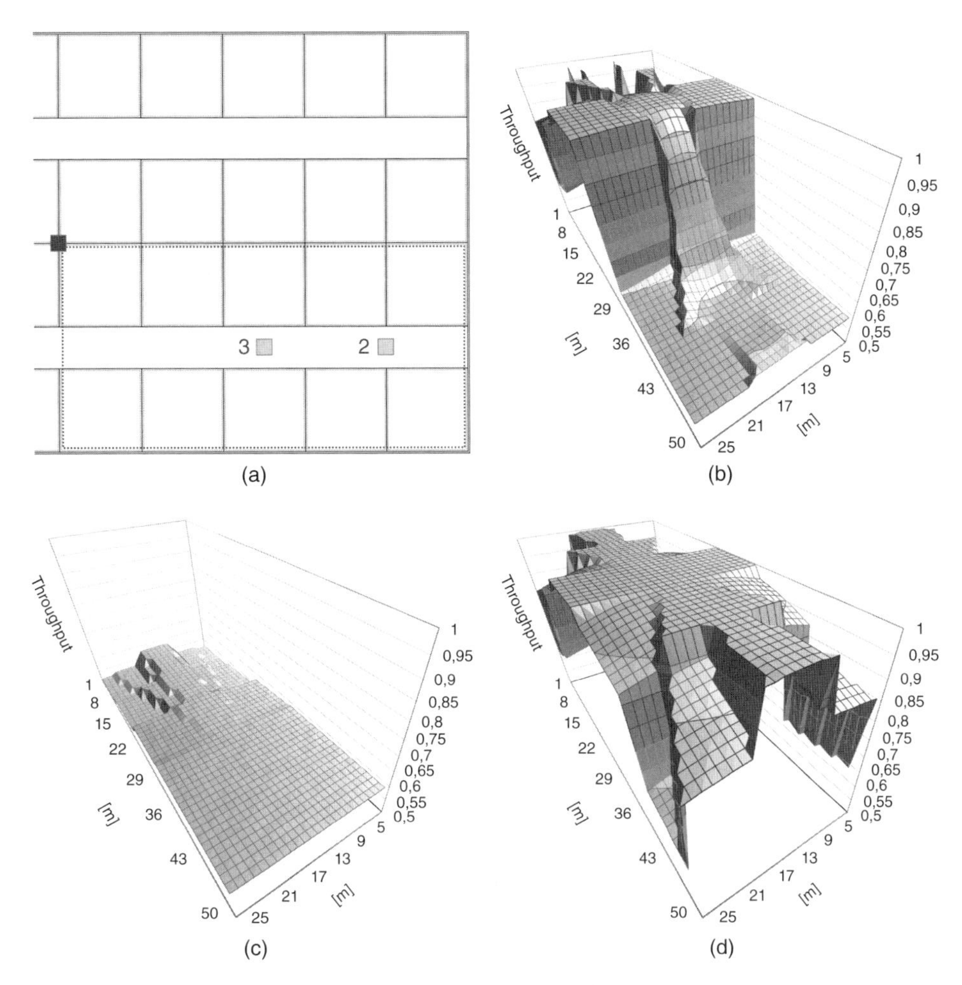

Figure 6.19 (a) Deployment of CFRs. (b) Relative throughput for CFR2–CFR3 cooperation. (c) Single-path relaying via CFR2. (d) Single-path relaying via CFR3.

a partial justification that cooperative enhancements may, in some cases, prove advantageous, the next step will be to extend the analysis to the aforementioned cooperation orchestration decision element of the network level. In particular, additional analyses of a variety of different deployment scenarios will be carried out, including modified positioning and denser setups, so that certain additional design directions can be provided.

6.4 Network Level Overlay Logic

6.4.1 Autonomic Cooperative Networking Architectural Model

Given that so far both the notions of the Autonomic Cooperative Networking Protocol and its related Autonomic Cooperative Behaviour have been scrutinised in utmost

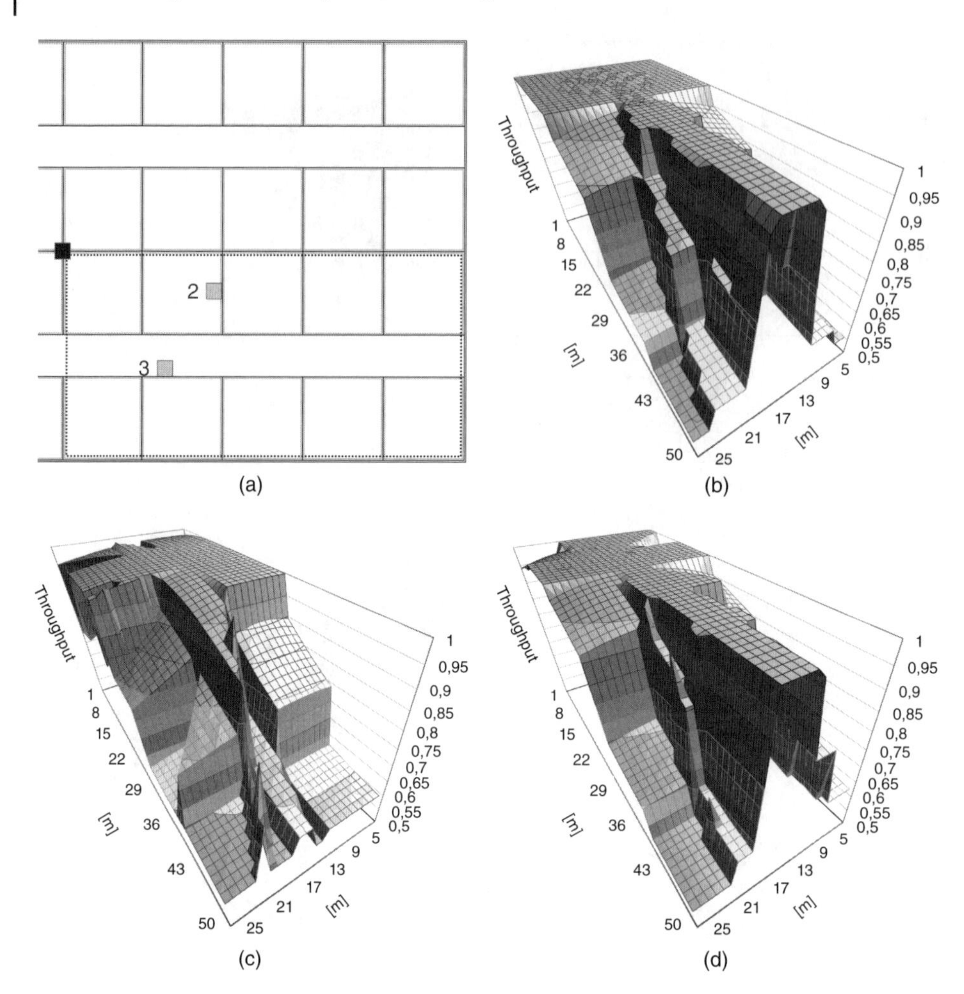

Figure 6.20 (a) Deployment of CFRs. (b) Relative throughput for CFR2–CFR3 cooperation. (c) Single-path relaying via CFR2. (d) Single-path relaying via CFR3.

detail, as well as each chapter has presented complementary considerations in the form of additional analyses related to the architectural integration aspects, the description of the concept of the Autonomic Cooperative Networking Architectural Model, being the ultimate enabler of the Autonomic Intelligence Evolved Cooperative Networking, could seem to have already been fully exhausted. Such a perception is natural in the light of the approach assumed throughout the book, where the decision to provide a bottom-up description resulted, to certain extent, in the necessity of outlining incrementally arranged and context-setting characteristics, as indicated in the opening chapter. One should note, however, that although being fairly ubiquitous across the entire book, the Autonomic Cooperative Networking Architectural Model is still attributed to the highest network level, making it highly abstract from the very outset, just like the complementary levels of abstraction. In other words, there are certain more or less explicitly voiced outstanding aspects to be addressed to make the overall picture complete. In particular, once a consolidated summary of the Autonomic Cooperative

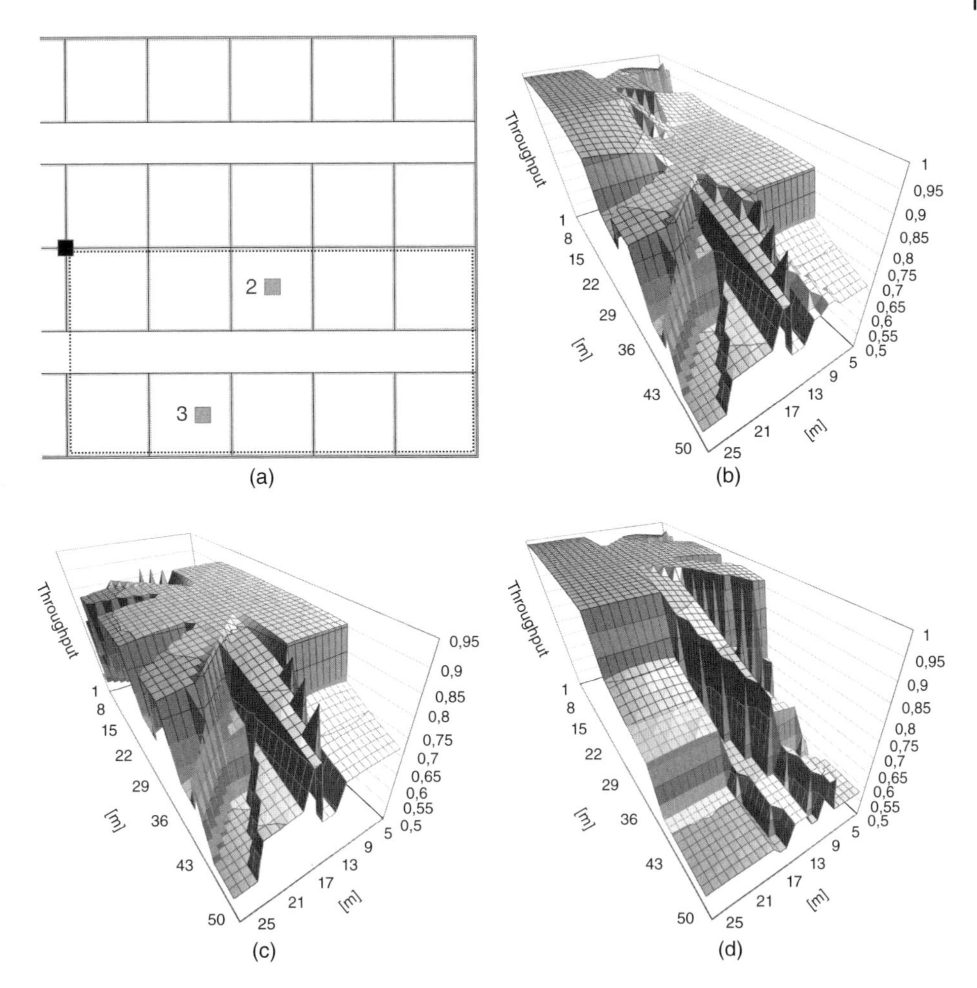

Figure 6.21 (a) Deployment of CFRs. (b) Relative throughput for CFR2–CFR3 cooperation. (c) Single-path relaying via CFR2. (d) Single-path relaying via CFR3.

Networking Architectural Model has been presented, the focus will initially shift to the question of the not yet fully definite relation between the Autonomic Cooperative Networking Protocol and the Autonomic Cooperative Behaviour, so that it is possible to advance the discussion and reflect upon the slightly dualistic role and place of the Autonomic Cooperative Networking Architectural Model.

In order to summarise the major conclusions arising from the incremental advancement of the workings of the said Autonomic Cooperative Networking Architectural Model, as well as provide sufficient background for the complementary analyses, a purposely elevated depiction of a consolidated version of the Autonomic Cooperative Networking Architectural Model is presented in Figure 6.22. In particular, special emphasis is laid on the representation of the frequently voiced perpendicular or even orthogonal positioning of the two major components of the Autonomic Cooperative Networking Architectural Model in the form of the specific developments attributable to each of the Open Systems Interconnection Reference Model and the Generic

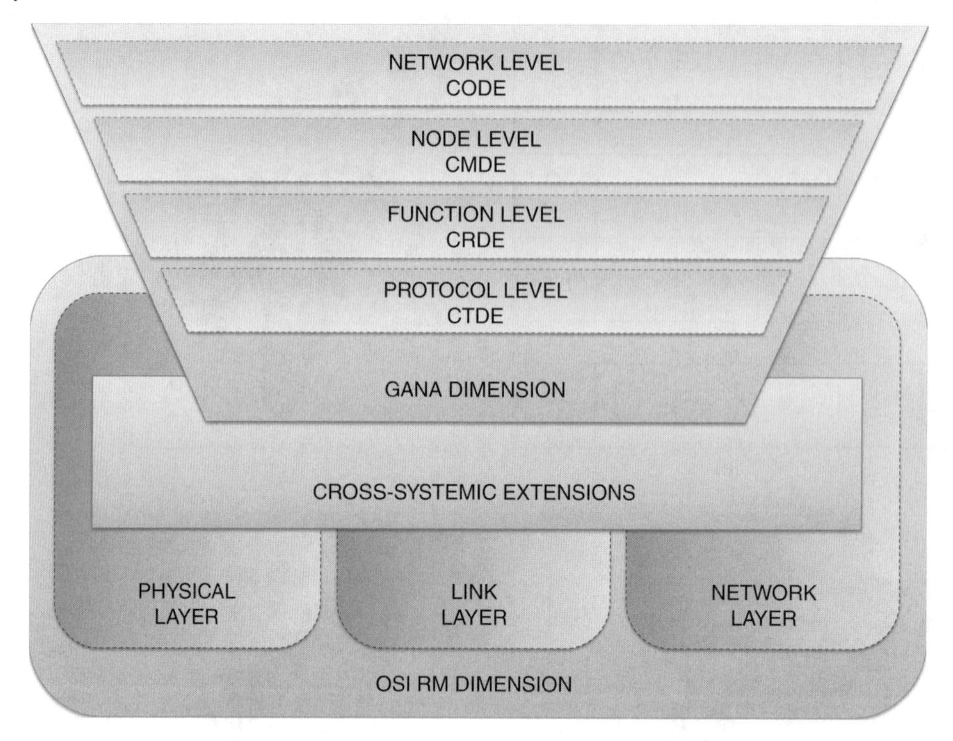

Figure 6.22 Consolidated view of the Autonomic Cooperative Networking Architectural Model.

Autonomic Network Architecture. While the notion of orthogonality may be a bit artificial in this respect, mostly because of the fact that there exist certain cross-systemic extensions, to be elaborated further below on the occasion of the comparison of the roles of the Autonomic Cooperative Networking Protocol and the Autonomic Cooperative Behaviour, it has been repeatedly stressed across the entire book to underline the fact that both the developments belong to different worlds. It is then clear that the ultimate role of the Autonomic Cooperative Networking Architectural Model is to function as a means for the fusion of selected aspects of the two, arranged in a highly synergetic approach. Such a perception may be clearer if the representation is viewed as if it had been drawn three-dimensionally. One should note, however, that only in going back to the two-dimensional perspective is it possible to identify the extent to which the two may realistically overlap.

Such an overlap, referred to as cross-systemic extensions, is where the two key ingredients of the Autonomic Cooperative Networking Architectural Model, the Autonomic Cooperative Networking Protocol and the Autonomic Cooperative Behaviour, coexist. What is of interest at this stage of the description is the outstanding analysis of their precedence, especially given that, as previously outlined in Figure 5.25, even though the Autonomic Cooperative Networking Protocol appears to contain or cap the Autonomic Cooperative Behaviour when perceived from the top of the network layer, it is the Autonomic Cooperative Behaviour[16] that interfaces with the cooperation orchestration

16 For additional information on Autonomic Cooperative Behaviour, the reader is referred to Wódczak et al. (2013) and Wódczak (2014b).

decision element (CODE) of the network level. This duality is apparently caused by the fact that the previous depiction is, once again, two-dimensional, making it difficult to maintain the perspective of perpendicularity or orthogonality. In other words, as underlined more conspicuously in Figure 6.23, by belonging slightly more to the dimension of the Open Systems Interconnection Reference Model or that of the Generic Autonomic Network Architecture, the two notions need to be perceived separately. Attempting to understand such a division one would need to explore the roots of both Autonomic Cooperative Networking Protocol and Autonomic Cooperative Behaviour. In fact, as the former builds on top of the Optimised Link State Routing protocol, the latter stems from the notion of autonomic behaviour. While such origins could justify the reason for varying precedence, one should note that there is an undeniable common point between the two in the form of the multi-point relay (MPR) station selection heuristics.

Moving slightly away from abstract discussion and attempting to understand the practical side of the issue in question, one may ask whether it would be either the Autonomic Cooperative Networking Protocol or the Autonomic Cooperative Behaviour to take precedence in reality. Such a question appears especially valid given the architectural dependencies, as initially outlined in Figure 2.19, where it seemed to be the former that orchestrates the latter, thereby creating a possible contradiction with what was later concluded in Figure 5.25. In particular, assuming the baseline case of a mobile ad hoc network comprising a single domain where all the network nodes display equivalent capabilities, one could assume that the depiction presented

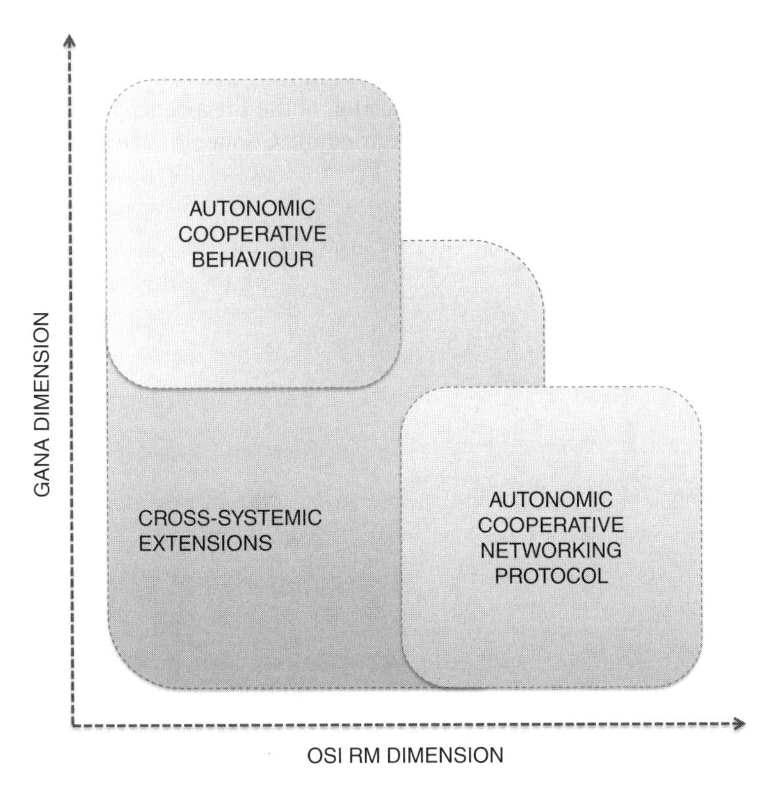

Figure 6.23 Autonomic Cooperative Networking Protocol and Autonomic Cooperative Behaviour.

in Figure 2.19 would generally hold, especially should all the said network nodes be elevated to function as Autonomic Cooperative Nodes. This would be so since, in such a case, the Autonomic Cooperative Networking Protocol could be perceived as the entity orchestrating the Autonomic Cooperative Behaviour instantiated by such Autonomic Cooperative Nodes. At this point one could refer directly to the Open Systems Interconnection Reference Model in order to elicit what the notion of orchestration may mean in this respect. This is necessary because orchestration tends to be understood as being performed in a centralised manner, where a specifically elevated entity is responsible for the entire process. After some consideration, however, one may conclude that the same task could be accomplished in a distributed manner if the presence of such a centralised entity were not expected or even allowable by default.

What the above explanation is intended to suggest is that an entity of considerable scope, possibly playing an inherently pivotal role in the whole design, may be perceived in a not entirely exhaustive way, just as if it had lost its formally elevated status, while in reality its privileged role still prevails. Referring more directly to the investigated case, especially keeping in mind that the Autonomic Cooperative Networking Protocol stems from a modified version of the Optimised Link State Routing protocol, which encompasses all the upgrades introduced with EREACT, one realises that the role of an otherwise network layer protocol is to be implemented within network nodes and function as a kind of interface between or among the same, over which they may establish a networked setup, so that neighbour nodes could exercise the tasks of communication and data transfer. A more concrete incarnation of such an understanding is outlined in Figure 6.24, where the Autonomic Cooperative Networking Protocol may suddenly be perceived as if it were reduced in terms of its role, while, in fact, this is by no means so. Given the above context, it transpires that the question of the precedence between Autonomic Cooperative Networking Protocol and Autonomic Cooperative Behaviour

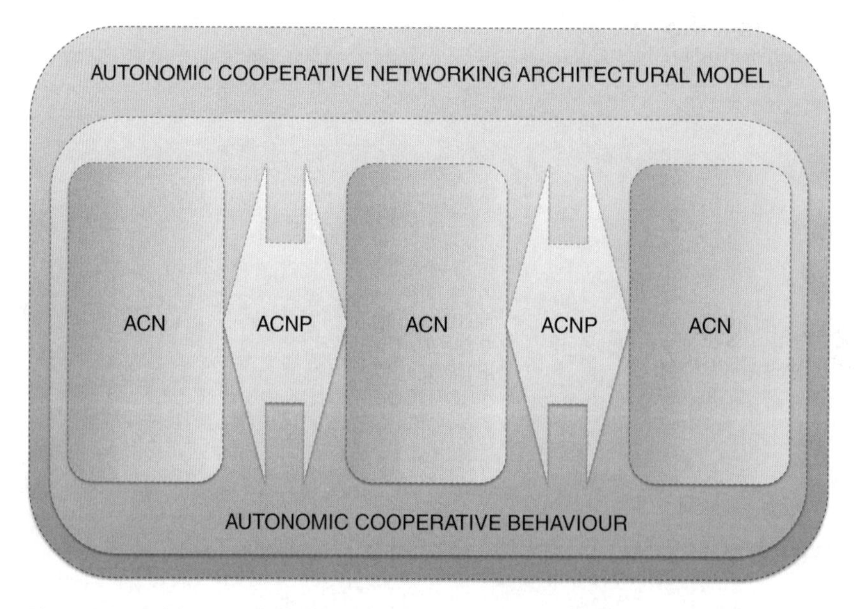

Figure 6.24 Relative position of Autonomic Cooperative Networking Protocol.

narrows down to identifying whether it is a high-level design that is scrutinised or rather a more specific instantiation thereof. In the former case it is much more reasonable to present the Autonomic Cooperative Networking Protocol on top, while in the latter the Autonomic Cooperative Behaviour takes the formal, yet not entirely factual, precedence.

Last but not least, there appears the question of the actual place and scope of the Autonomic Cooperative Networking Architectural Model, especially when the major theme of Autonomic Intelligence Evolved Cooperative Networking is concerned. In fact, the answer might, once again, not be too immediate or straightforward, since the perception and context of discussion may play a significant role in this respect. As far as the dimension of the Autonomic Cooperative Networking Architectural Model is concerned, where it builds on top of the Generic Autonomic Network Architecture, commonly known as an autonomic overlay, the same perception of an elevated entity may become the best representation thereof, especially since, in such a case, it also builds on top of the Open Systems Interconnection Reference Model related developments. Yet, quite contradictory to the Generic Autonomic Network Architecture itself, which remains quite detached from the said Open Systems Interconnection Reference Model, the Autonomic Cooperative Networking Architectural Model incorporates certain modified workings of the same, for example the Autonomic Cooperative Networking Protocol and its constituents. Consequently, a more complex depiction appears to present itself where the Autonomic Cooperative Networking Architectural Model is viewed not only as being an overlay, especially in more generic considerations, but also as encompassing all the discussed aspects, calling for a more detailed insight into it to be exercised. All in all, it appears that the major potentially outstanding issues have been concluded, and the focus can shift to the cooperation orchestration decision element.

6.4.2 Cooperation Orchestration Decision Element

Approaching the topmost cooperation orchestration decision element it is necessary to note that, as already explained, being located at the network level it should be perceived as if it were more of a virtual inclination. What is more, fairly alike it was the case for the protocol level of the Generic Autonomic Network Architecture abstraction structure, the nomenclature might pose certain dose of ambiguity, potentially inducing a similar level of uncertainty, since the notion of the network level should not be confused with the network layer of the Open Systems Interconnection Reference Model. In the case of the Generic Autonomic Network Architecture, the network level is usually referred to as being of a more intangible nature since it is only formed by selected or elevated network nodes. Most importantly, however, while, the network layer of the Open Systems Interconnection Reference Model generally pertains to the relevant networking protocols, along with their inherent addressing and routing schemes, the network level in question seems, to the contrary, to be a volatile incarnation of a particular networked setup as a whole. What is more, keeping such a context in mind, the cooperation orchestration decision element additionally appears to be an excellent illustration of the most recent considerations related to the precedence of the Autonomic Cooperative Networking Protocol and Autonomic Cooperative Behaviour. Thus, the following description will commence with an account of the prioritised role of Autonomic Cooperative Behaviour, so that it is possible to follow with specifically chosen scenarios and the assumed algorithmic description, along with the evaluation results.

Algorithm 6.1 Logic of CODE.

1: **if** ($\mathbf{BER}(n_1, n^2) \geq \theta$ **and** $\mathbf{BER}(n_2, n^2) \geq \theta$) **then**
2: $\quad$ $\mathbf{ACB}(ACS(n_1, n_2))$
3: **end if**

The question of priority between the Autonomic Cooperative Networking Protocol and Autonomic Cooperative Behaviour results, most of all, from the nature of the assumed set of evaluation scenarios, which not only stem from the earlier discussed relay-enhanced cell, but additionally follow the hierarchical requirements imposed by the recently scrutinised specifics of emergency communications networks. In essence, it is assumed that a carefully selected configuration of a mesh-like grid of Autonomic Cooperative Nodes, by all means entitled to express Autonomic Cooperative Behaviour, is elevated above the operation of the Autonomic Cooperative Networking Protocol for reasons in line with the specifics of the highly illustrative case of chief first responders. Such an approach is possible because of a conceptual experiment where the Autonomic Cooperative Nodes are assumed to be fed over wired connections, rather than the wireless ones as it has been prescribed thus far, so that there would be no penalty for having a BS located centrally indoors or mounted externally on a MEOC. This way, the parameters described in Table 4.5 still hold with the exception of the first hop links, which in this context may be considered lossless, making it possible to apply the concept of the equivalent distributed space-time block encoder in its most canonical form. In other words, while the ACNP-driven communication of a mobile ad hoc network type would be not precluded, the Autonomic Cooperative Nodes would be entitled to express the Autonomic Cooperative Behaviour only.

The precedence of Autonomic Cooperative Behaviour clearly results from the nature of the scenario in question, and it is further highlighted by the structure of Algorithm 6.1, where the major routine is to be executed should the bit error rate threshold criterion be met for both network nodes under inspection. In fact, as indicated previously, such network nodes would be elevated to act as Autonomic Cooperative Nodes, and by forming an ACS they would instantiate the Autonomic Cooperative Behaviour. In this context, the results pertaining to the layouts of Figures 6.25, 6.26, and 6.27 are presented in Figures 6.28, 6.29, and 6.30, respectively (Wódczak, 2014a).[17] Most importantly, the inner configurations denoted by 1, 2, 3, and 4 are arranged in an increasing order such that the Autonomic Cooperative Nodes belonging to each of the groups they represent are located farther and farther from the centre of the relay-enhanced cell.[18] In fact, none of the above inner configurations may cover the entire area, thereby making the instantiation of the related Autonomic Cooperative Behaviour virtually indispensable.[19] This is shown by the cumulative distribution functions (CDFs) indicating that layout B, and particularly the inner configurations labelled 2 and 3, appear to perform best. Although for layout A, with a more circular nature, the similarly almost

17 For a more direct visualisation of the relative throughput pertaining to each of the inner configurations, the reader may also refer to Figures A.26, A.27, A.28, and A.29 covering layout A, Figures A.30, A.31, A.32, and A.33 covering layout B, and Figures A.34, A.35, A.36, and A.37 covering layout C.
18 For additional context the reader may also refer to Wódczak (2012d, 2013).
19 For this reason, Figures A.38–A.49 show relative comparisons between the inner configurations denoted by the same numbers and belonging to layouts A and B, layouts B and C, as well as layouts C and A, respectively.

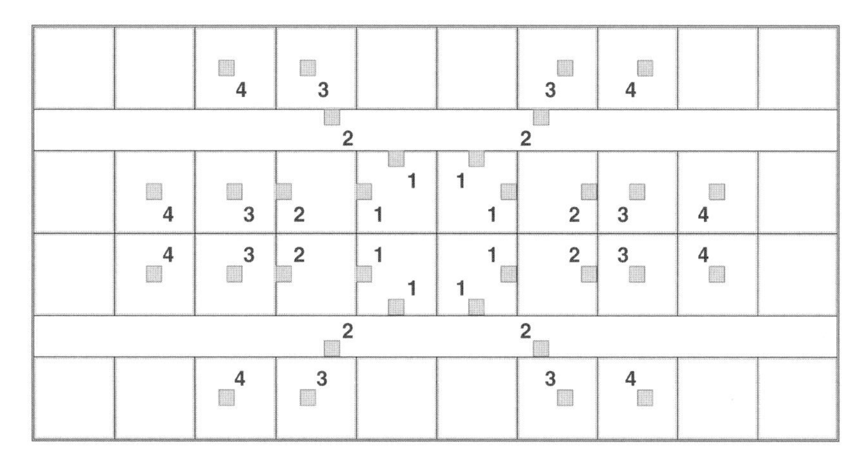

Figure 6.25 Layout A.

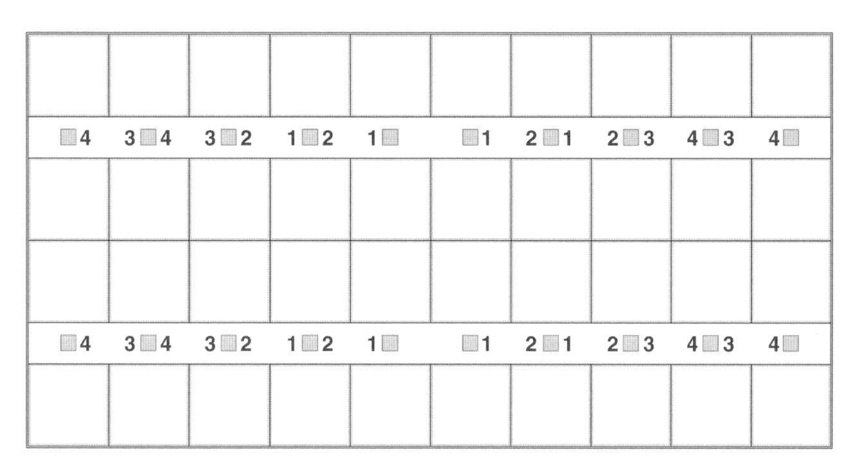

Figure 6.26 Layout B.

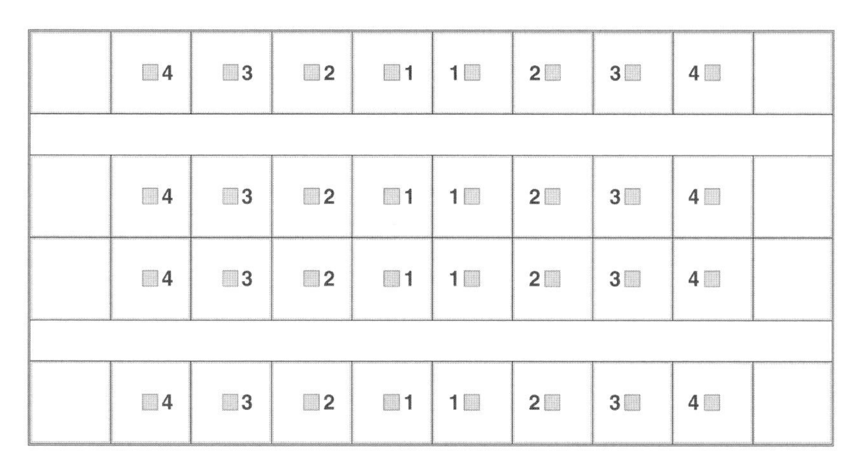

Figure 6.27 Layout C.

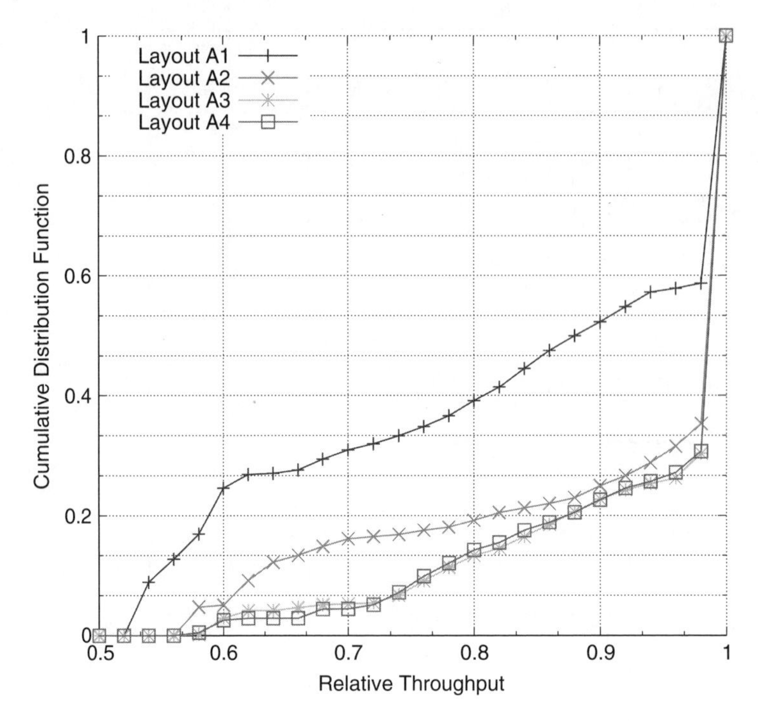

Figure 6.28 Cumulative distribution functions for layouts A1–A4.

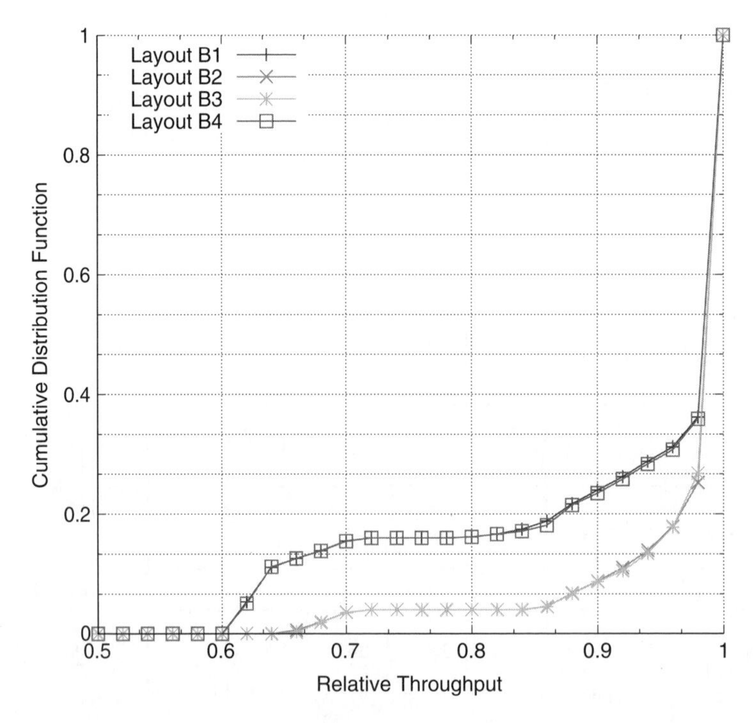

Figure 6.29 Cumulative distribution functions for layouts B1–B4.

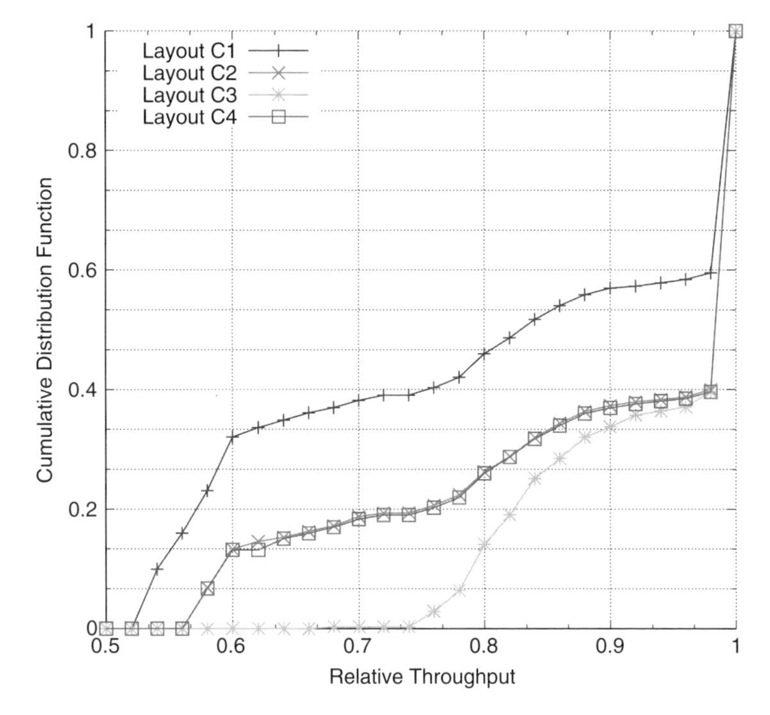

Figure 6.30 Cumulative distribution functions for layouts C1–C4.

identical inner configurations labelled 3 and 4 are most the advantageous,[20] such an arrangement pretends to be less efficient in comparison with the previous one. Last, but not least, layout C is also less efficient from this perspective, even though this time the inner configuration labelled 3 operates most effectively.

6.4.3 Architectural Integration Aspects

The time has come for some closing remarks related to the reoccurring architectural integration aspects. Given the fact that the outstanding issues pertaining to the Autonomic Cooperative Networking Architectural Model have already been discussed, the last topic of relevance relates to the mutual dependencies and synergies between all the decision elements introduced in this book, keeping in mind their residence at distinct levels of abstraction. Such a need becomes more evident the higher one moves in these levels of abstraction, being most clearly visible in the case of the cooperation orchestration decision element. Given the fact that the reliance one upon one another so evident among the decision elements is clearly directed top-down, a bottom-up approach to the analysis appears to be of most relevance, thanks to which the dependencies may be most exhaustively addressed. To this end, based on the algorithmic descriptions of the decision elements introduced thus far, illustrative structural depictions thereof will be presented, starting from the cooperative

20 As such, layout A was chosen to exemplify a combined illustration, where the relative throughput maps are cast on the pertinent grid of Autonomic Cooperative Nodes, as depicted in Figures A.50, A.51, A.52, and A.53.

transmission decision element (CTDE), through both the cooperative re-routing decision element (CRDE) and cooperation management decision element (CMDE), up to the cooperation orchestration decision element (CODE). The reason for such an approach is related to the existence of certain functional blocks within the decision elements responsible for interfacing between themselves to make the entire Autonomic Cooperative Networking Architectural Model fully operational. In other words, an incremental analysis is assumed, intended to result in a comprehensive depiction of the above-mentioned dependencies and synergies.

Looking at the structure of the bottommost cooperative transmission decision element depicted in Figure 6.31, one may notice that, given the previously defined roles of the decision elements residing at certain levels of abstraction, such a decision element clearly belonging to the protocol level is bound to interface on the lower side with external entities only, at least when the top-down direction is assumed, being, in fact, equivalent to the vertical one given the introduced perpendicularity or orthogonality of the Open Systems Interconnection Reference Model and the Generic Autonomic Network Architecture. What is important from the perspective of this incremental description, however, is related to the fact that based on such an external interfacing the cooperative transmission decision element becomes entitled to offer certain services to some other decision element or decision elements located either directly at the level of abstraction immediately above or indirectly at even higher levels of abstraction. In the case of the cooperative transmission decision element such a service is related to the provision of increased robustness through the instantiation of DSTBC using specifically chosen code matrices. As a result of this approach a given decision element entitled to use the service of the cooperative transmission decision

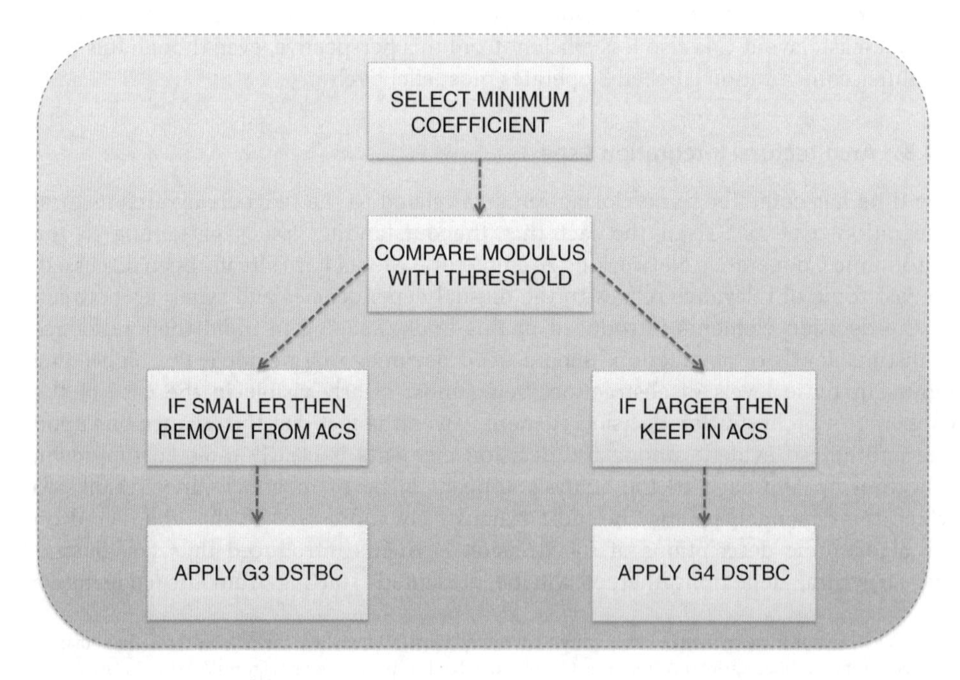

Figure 6.31 Structure of the CTDE.

element, in the direct or indirect manner, becomes separated from having any influence on the choice made at the protocol level with regard to the code matrix to be applied, other than being able to impose certain requirements to induce specific behaviour thereof should it be physically possible to have it accommodated without penalty.

This is, in fact, what happens on the side of the cooperative re-routing decision element, located at the function level of the Generic Autonomic Network Architecture, which is responsible for the instantiation of either the Autonomic Cooperative Behaviour or fast re-routing. Looking at Figure 6.32, one may discern that, once again, there is a certain decision-making process ongoing, according to which it becomes necessary to identify what will happen when the quality of specific radio links falls below the expected threshold. However, even if the outcome of such a decision process is positive, meaning that Autonomic Cooperative Behaviour should be instantiated, it may only happen assuming that the ACS is not empty. In other words, in the narrow version referred to above, assuming the nomenclature behind the EDSTBE, the cooperative transmission decision element allows solely for switching between the E_G^3 and E_G^4 mode, yet, in general, it could also indicate that the operation of DSTBC may not be possible at all, thereby indirectly making the cooperative re-routing decision element switch to the alternative operation mode offered by fast re-routing. While this is what could happen given the local scope of the analysis, one should also take into consideration that, when perceived from the perspective of higher levels of abstraction, there could equally well appear a decision to reschedule some other transmissions in order to make room for imposing the demanded Autonomic Cooperative Behaviour, should it be really required to take place. This is, however, where the cooperation management decision element comes into the global picture.

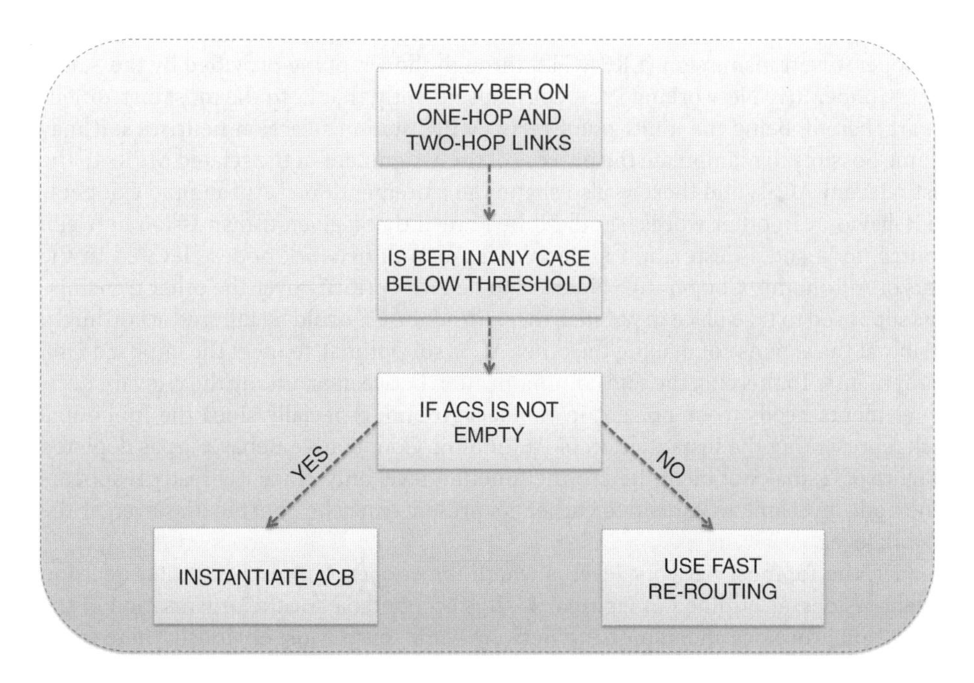

Figure 6.32 Structure of the CRDE.

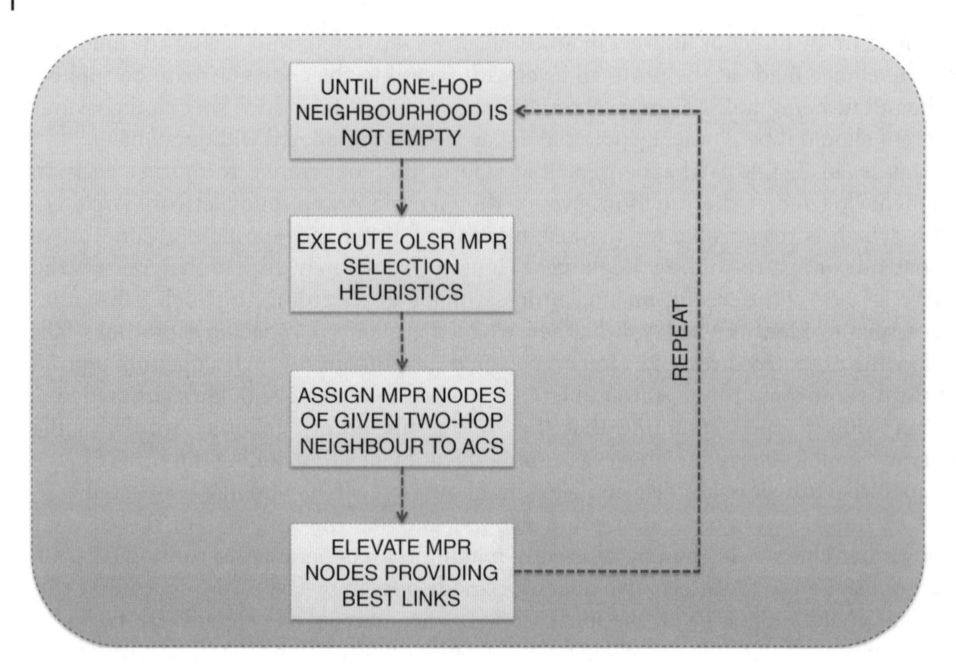

Figure 6.33 Structure of the CMDE.

As it may transpire from Figure 6.33, the cooperation management decision element has a much better overview of the mobile ad hoc network in question, as it interacts directly with the routines of the Optimised Link State Routing protocol, or rather its upgraded version in the form of the extended routing information enhanced algorithm for cooperative transmission (EREACT), through the wrapping provided by the Autonomic Cooperative Networking Protocol. Consequently, thanks to the most instrumental part thereof, being the multi-point relay (MPR) station selection heuristics, it may become possible to manipulate the process of the assignment of the related MPRs to the most relevant ACSs, and thereby also shape the aforementioned Autonomic Cooperative Behaviour. In other words, should it be known that a given transmission between a source node and a destination node over a group of network nodes elevated to the status of Autonomic Cooperative Nodes is to be given priority over the other transmissions supposed to take place in parallel, the pertinent ACS could be tailored accordingly, possibly at the expense of making the other ACSs suboptimal, to meet the imposed QoS requirements. To this end, the said communication between or among the relevant decision elements needs to be present in the background, especially since the functional block representing the instantiation of Autonomic Cooperative Behaviour, as depicted in Figure 6.32, may not be limited to the function level only, since it already reappears at the node level and, as explained earlier, its presence may be conspicuous even at the network level.

In fact, the topmost network level is where the cooperation orchestration decision element resides, as outlined in Figure 6.34. It is accountable for the organisation of the cooperation processes understood in a network-wide sense. Most obviously, there could be a single decision element of this type employed when a single-domain mobile ad hoc network is considered. However, one should not exclude multi-domain configurations,

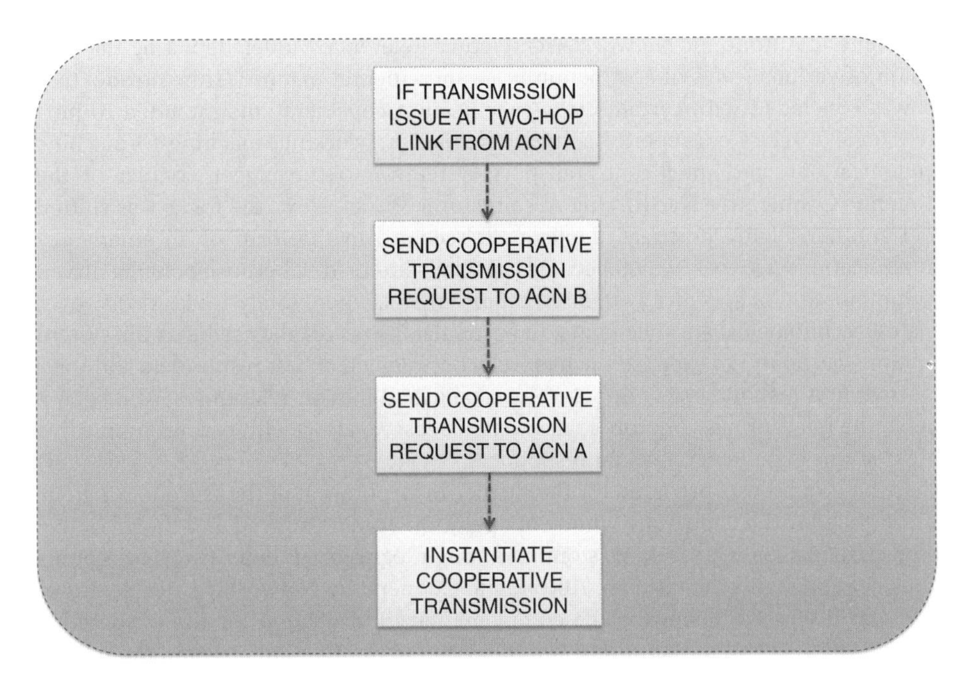

Figure 6.34 Structure of the CODE.

where the functionality of the cooperation orchestration decision element would need to be proliferated in a distributed manner, so that a logically centralised yet physically distributed entity would be obtained. Looking into its logic, one may discern that to some extent it operates slightly aside from the mainstream decision elements belonging to the levels of abstraction below the network level. It is so because of the elevated or even somewhat detached nature of the cooperation orchestration decision element, which operates in such a part of the overall abstraction that instantiates a nonexistent network device, being rather a composition of purposely promoted Autonomic Cooperative Nodes. Consequently, it no longer adds to the composition of Autonomic Cooperative Behaviour in general, but rather orchestrates the same, yet in a different way than is done for mobile ad hoc network internal transmission, taken care of by the cooperation management decision element. In other words, it is most applicable when the complexity of such a mobile ad hoc network becomes substantial and a hierarchical structure needs to be imposed, where cross-domain Autonomic Cooperative Behaviour may be established as in the case of the emergency communications network.

6.5 Conclusion

In this chapter, first of all, the standardisation-orientated design was introduced, where the research and investment driven perspective was assumed in order to explain where the Autonomic Cooperative Networking Architectural Model stems from, touching upon issues related to standardisation of the Open Systems Interconnection Reference Model and emphasising the role of prestandardisation related to the Generic Autonomic Network Architecture. Then followed a description of the staged instantiation of the Generic Autonomic Network Architecture Reference Model, depicting the progression of various levels of abstraction in an incremental manner. This introductory part was concluded with certain cross-specification considerations intended to purposely involve select concepts of software-defined networking, machine-to-machine communications, and intelligent transport systems into the bigger context of the Autonomic Cooperative Networking Architectural Model. Next, the focus was shifted to yet another, highly practical, deployment scenario in the form of an emergency communications network, which became especially interesting because of being driven by a combination of specifically tailored requirements, where safety took priority over the latest technological advancements. In particular, it was emphasised that the system operation was bound to exercise the hierarchy between chief first responders and their respective first responders, as imposed by established human relations. Given such a context, the relevant network topologies were discussed, along with certain supportive configurations of pairs of chief first responders.

Thus the ground was prepared for the further incorporation of the Autonomic Routines, since, after the cooperative mode of operation was introduced and the proactive and reactive resiliency process was outlined, the integration of emergency communications networks into the ultimate Autonomic Cooperative Networking Architectural Model was discussed. Following the complementary justification for the cooperative enhancement in question, supported with performance evaluation analysis, the related network level overlay logic was brought into the overall picture to encompass any still outstanding workings of the Autonomic Cooperative Networking Architectural Model. Thus the mutual relation between the Autonomic Cooperative Networking Protocol and Autonomic Cooperative Behaviour was presented from the perspective of the priority between the two, on the grounds of their being inherent in the respective dimensions of the Open Systems Interconnection Reference Model and the Generic Autonomic Network Architecture. Based on this, the notion of the cooperation orchestration decision element was introduced to emphasise more tangibly when the Autonomic Cooperative Behaviour may be prioritised over the Autonomic Cooperative Networking Protocol. In particular, the relay-enhanced cell scenario was revisited under certain additional assumptions allowing for a more accurate evaluation of the second hop. Finally, the architectural integration aspects were raised to address the mutual operation of all the discussed decision elements to introduce additional synergy to the already exhaustive depiction of the Autonomic Cooperative Networking Architectural Model.

References

Calarco G, Casoni M, Paganelli A, Vassiliadis D, and Wódczak M 2010 A satellite-based system for managing crisis scenarios: The E-SPONDER perspective. *Fifth Advanced Satellite Multimedia Systems Conference*, Cagliari, Italy.

Callegati F, Cerroni W, Contoli C, Cardone R, Nocentini M, and Manzalini A 2016 SDN for dynamic NFV deployment. *IEEE Communications Magazine* **54**(10), 89–95.

Chaparadza R 2008 Requirements for a Generic Autonomic Network Architecture (GANA), suitable for standardizable autonomic behaviour specifications of decision-making elements (DMEs) for diverse networking environments. *International Engineering Consortium (IEC) Annual Review of Communications.*

Chaparadza R, Ciavaglia L, Wódczak M, Chen CC, Lee B, Liakopoulos A, Zafeiropoulos A, Mancini E, Mulligan U, Davy A, Quinn K, Radier B, Alonistioti N, Kousaridas A, Demestichas P, Tsagkaris K, Vigoureux M, Vreck L, Wilson M, and Ladid L 2009a ETSI Industry Specification Group on Autonomic network engineering for self-managing Future Internet (ETSI ISG AFI). *10th International Conference on Web Information Systems Engineering*, Poznań, Poland.

Chaparadza R, Meriem TB, Radier B, Szott S, Wódczak M, Prakash A, and Ding J 2013 SDN enablers in the ETSI AFI GANA reference model for autonomic management and control (emerging standard), and virtualization impact. *Fifth IEEE MENS at GLOBECOM 2013*, 9–13 December, Atlanta, USA.

Chaparadza R, Papavassiliou S, Kastrinogiannis T, Vigoureux M, Dotaro E, Davy A, Quinn K, Wódczak M, and Toth A 2009b Creating a viable evolution path towards self-managing future internet via a standardizable reference model for autonomic network engineering. In *Towards the Future Internet – A European Research Perspective*, eds. Tselentis G, Domingue J, Galis A, Gavras A, Hausheer D, Krco S, Lotz V, and Zahariadis T. IOS Press.

Dias J, Rodrigues J, Kumar N, and Saleem K 2015 Cooperation strategies for vehicular delay-tolerant networks. *IEEE Communications Magazine* **53**(12), 88–94.

Dressler F, Hartenstein H, Altintas O, and Tonguz O 2014 Inter-vehicle communication: Quo vadis. *IEEE Communications Magazine* **52**(6), 170–177.

ETSI-GS-AFI-001 2011 *Autonomic network engineering for the self-managing Future Internet (AFI); Scenarios, Use Cases and Requirements for Autonomic/Self-Managing Future Internet*. ETSI Group Specification.

ETSI-GS-AFI-002 2013 *Autonomic network engineering for the self-managing Future Internet (AFI); Generic Autonomic Network Architecture (An Architectural Reference Model for Autonomic Networking, Cognitive Networking and Self-Management)*. ETSI Group Specification.

ETSI-TR-103-375 2016 *SmartM2M; IoT Standards Landscape and Future Evolutions*. ETSI Technical Report.

ETSI-TS-102-689 2010 *Machine-to-Machine communications (M2M); M2M service requirements*. ETSI Technical Specification.

ETSI-TS-102-690 2013 *Machine-to-Machine communications (M2M); Functional architecture*. ETSI Technical Specification.

ETSI-TS-102-868-2 2017 *Intelligent Transport Systems (ITS); Testing; Conformance test specifications for Cooperative Awareness Basic Service (CA); Part 2: Test Suite Structure and Test Purposes (TSS & TP)*. ETSI Technical Specification.

Favraud R, Apostolaras A, Nikaein N, and Korakis T 2016 Toward moving public safety networks. *IEEE Communications Magazine* **54**(3), 14–20.

Gelonch-Bosch A, Marojevic V, and Gomez I 2017 Teaching telecommunication standards: Bridging the gap between theory and practice. *IEEE Communications Magazine* **55**(5), 145–153.

Ghanbari A, Laya A, Alonso-Zarate J, and Markendahl J 2017 Business development in the Internet of Things: A matter of vertical cooperation. *IEEE Communications Magazine* **55**(2), 135–141.

Haleplidis E, Pentikousis K, Denazis S, Hadi Salim J, Meyer D, and Koufopavlou O 2015 *Software-Defined Networking (SDN): Layers and Architecture Terminology*. RFC 7426.

Hobert L, Festag A, Llatser I, Altomare L, Visintainer F, and Kovacs A 2015 Enhancements of V2X communication in support of cooperative autonomous driving. *IEEE Communications Magazine* **53**(12), 64–70.

Hsu IY, Wódczak M, White R, Zhang T, and Hsing T 2010 Challenges, approaches, and solutions in intelligent transportation systems. *Second International Conference on Ubiquitous and Future Networks*, Jeju Island, Korea.

Hyojoon K and Feamster N 2013 Improving network management with software defined networking. *IEEE Communications Magazine* **51**(2), 114–119.

Kim J, Choi S, Shin WY, Song YS, and Kim YK 2016 Group communication over LTE: A radio access perspective. *IEEE Communications Magazine* **54**(4), 16–23.

Li J, Wódczak M, Wu X, and Hsing T 2012a Vehicular networks and applications: challenges, requirements and service opportunities. *Journal of Communications (JCM)* **7**(5), 365–373.

Li J, Wódczak M, Wu X, and Hsing T 2012b Vehicular networks and applications: Challenges, requirements and service opportunities. *International Conference on Computing, Networking and Communications (ICNC)*, Maui, Hawaii, USA.

Minh Q, Nguyen K, Borcea C, and Yamada S 2014 On-the-fly establishment of multihop wireless access networks for disaster recovery. *IEEE Communications Magazine* **52**(10), 60–66.

Ramirez-Perez C and Ramos V 2016 SDN Meets SDR in self-organizing networks: Fitting the pieces of network management. *IEEE Communications Magazine* **54**(1), 48–57.

Tanenbaum A and Wetherall D 2011 *Computer Networks*. Prentice Hall.

Tonguz O and Viriyasitavat W 2016 A self-organizing network approach to priority management at intersections. *IEEE Communications Magazine* **54**(6), 119–127.

Vassiliadis D, Garbi A, Calarco G, Casoni M, Paganelli A, Morera R, Chen CM, and Wódczak M 2010 Wireless networks at the service of effective first response work: The E-SPONDER vision. *IEEE International Symposium on Wireless Pervasive Computing*, Modena, Italy.

Wódczak M 2011a Aspects of cross-layer design in autonomic cooperative networking. *IEEE Third International Workshop on Cross Layer Design*, Rennes, France.

Wódczak M 2011b Autonomic cooperation in ad hoc environments. *Fifth International Workshop on Localised Algorithms and Protocols for Wireless Sensor Networks (LOCALGOS) in conjunction with IEEE International Conference on Distributed Computing in Sensor Systems (DCOSS)*, Barcelona, Spain.

Wódczak M 2011c Autonomic cooperative networking for wireless green sensor systems. *International Journal of Sensor Networks (IJSNet)* **10**(1–2), 83–93.

Wódczak M 2011d Deployment aspects of autonomic cooperative communications in emergency networks. *Third International Congress on Ultra Modern Telecommunications and Control Systems, IEEE ICUMT*, Budapest, Hungary.

Wódczak M 2011e Resilience aspects of autonomic cooperative communications in the context of cloud networking. *IEEE First Symposium on Network Cloud Computing and Applications*, Toulouse, France.

Wódczak M 2012a Autonomic cooperative communications for emergency networks. *Fourth IEEE MENS at GLOBECOM 2012*, 3–7 December, Anaheim, California, USA.

Wódczak M 2012b *Autonomic Cooperative Networking*. Springer.

Wódczak M 2012c Autonomic cooperative vehicular communications. *Proc. 11th International Conference, ADHOC-NOW 2012*, Belgrade, Serbia.

Wódczak M 2012d Evaluation of dense cooperative relay deployments for autonomic emergency communications. *Fourth International Conference, MONAMI*, Hamburg, Germany.

Wódczak M 2012e Towards autonomic emergency communications: Evolution of cooperative networking. *Eighth IEEE, IET International Symposium on Communication Systems, Networks and Digital Signal Processing (CSNDSP 2012)*, Poznań, Poland.

Wódczak M 2013 Evaluation of dense cooperative relay deployments for autonomic emergency communications. In *Mobile Networks and Management (Lecture Notes of the Institute for Computer Sciences, Social Informatics and Telecommunications Engineering)*, eds. Timm-Giel A, Strassner J, Agüero R, Sargento S, and Pentikousis K. Springer.

Wódczak M 2014a *Autonomic Computing Enabled Cooperative Networked Design*. Springer.

Wódczak M 2014b Autonomic Cooperative Behaviour enabled emergency system design. *The Mediterranean Journal of Computers and Networks*, **10**(1).

Wódczak M, Meriem TB, Radier B, Chaparadza R, Quinn K, Kielthy J, Lee B, Ciavaglia L, Tsagkaris K, Szott S, Zafeiropoulos A, Liakopoulos A, Kousaridas A, and Duault M 2011 Standardizing a reference model and autonomic network architectures for the self-managing future internet. *IEEE Network* **25**(6), 50–56.

Wódczak M, Szott S, and Chaparadza R 2013 Autonomic Cooperative Behaviour in ETSI AFI scenario for autonomicity enabled ad hoc and mesh network architecture. *Fifth IEEE MENS at GLOBECOM 2013*, 9–13 December, Atlanta, Georgia, USA.

Zimmermann H 1980 OSI Reference Model – The ISO model of architecture for open systems interconnection. *IEEE Transactions on Communications* **28**(4), 425–432.

7

Conclusion

The book was opened with a comprehensive depiction of the concept of autonomic computing from the perspective of a modern networked system design, where the notion of the human autonomic nervous system was elevated with emphasis on the key dimensions to self-management in the form of self-configuration, self-optimisation, self-healing, and self-protection, yet keeping in mind the context of agent systems. In order to prepare the ground for the target Autonomic Cooperative Networking Architectural Model, the Generic Autonomic Network Architecture was characterised through analysis of role of decision elements and hierarchical autonomic control loops, as well as their respective levels of abstraction, arranged in an incremental order. In this respect, the Vertical Technological Pillars and the Horizontal Architectural Extensions were introduced, given the perpendicularity of the layers of the Open Systems Interconnection Reference Model and the levels of the Generic Autonomic Network Architecture, as were the architectural constituents in the form of the Autonomic Cooperative Node, the Autonomic Cooperative Behaviour, and the Autonomic Cooperative Networking Protocol. The protocol level cooperative transmission decision element was presented, with its responsibility for cooperative relaying. Next, the function level cooperative re-routing decision element was deployed as a trigger for resiliency driven cooperative re-routing. Moving forward, the node level cooperation management decision element was introduced to integrate routing mechanisms. Finally, the network level cooperation orchestration decision element was outlined to comprehensively oversee the overall system.

The focus then shifted to spatio-temporal processing, with the initial emphasis on the multiple-input multiple-output radio channel. Then, the diversity related aspects were discussed to justify the role of and the necessity for the later deployment of spatio-temporal processing, as well as the singular-value decomposition theorem was explained to introduce the equivalent virtual multiple-input multiple-output radio channel along with its capacity proved to scale linearly with the number of generic transmitters or generic receivers. Moreover, an external model for radio channel coefficient calculation was described, as well as the difference between coding gain and diversity gain was addressed. Going further, special attention was paid to the technology of space-time block coding, where the question of its being perceived more as a modulation than a coding technique was visited, the derivation process of the decoding metrics for a selected set of code matrices was outlined with the aim of clarifying certain inconsistencies existing in the source materials, and the extension towards space-time trellis coding was presented to account for additional coding gain.

Autonomic Intelligence Evolved Cooperative Networking, First Edition. Michał Wódczak.

Moving to the architectural integration aspects, the notion and internal structure of the Autonomic Cooperative Node was introduced, where the discussion of the relation between autonomics and cooperation was triggered. Finally, the cooperative transmission decision element of the protocol level was analysed along with specifically tailored adaptive logic allowing the switching of relevant code matrices on the basis of the radio channel parameters, and the pertinent architectural integration aspects were addressed to pave the way for further extensions.

Moving forward, both conventional relaying and cooperative relaying were addressed from the classificatory perspective, as well as supportive protocols and collaborative protocols were introduced as specifically interrelated subcategories of the generic cooperative protocol. Going further, the concept of multi-tier virtual antenna arrays was outlined, where the special operation mode, in the form of distributed space-time block coding, was discussed. Next, the attention was shifted towards a fixed deployment concept of a grid-based Manhattan scenario characterised by the fact that although the pattern formed by the buildings could become critically important for the suppression of interference among the fixed relay nodes, making it impossible to exercise any cooperative relaying based on virtual antenna arrays among the same, certain adaptation with regard to the framing structure and the buffer memory could still prove advantageous. Similarly, a cooperation enabled relay-enhanced cell indoor scenario was analysed, where the major emphasis was put on the aspects related to the link layer. Then the focus shifted towards the function level overlay logic to encompass the roots of Autonomic Cooperative Behaviour, including its enablers and the role of the equivalent distributed space-time block encoder in particular. Consequently, the cooperative re-routing decision element was presented, including not only its transition from the node level to the function level, but also the logic behind cooperative re-routing. Finally, the architectural integration aspects were discussed and complemented with the introduction of an extended version of the Autonomic Cooperative Node.

Going further, the workings of the Optimised Link State Routing protocol were discussed along with its multi-point relay station selection heuristics and information storage repositories, to provide the required context for further developments and to arrange for the later introduction of the virtual antenna array selector set and its related virtual antenna array selector tuples. Then came an analysis of the algorithmic description of the related routing information enhanced algorithm for cooperative transmission as the method for applying additional network layer information to enable and orchestrate cooperative transmission at the link layer, subsequently upgraded to the extended routing information enhanced algorithm for cooperative transmission along with its evolved messaging structure in order to lay the groundwork for the target Autonomic Cooperative Networking Protocol. In this respect, both address auto-configuration and duplicate address detection were discussed before the focus shifted towards the umbrella formed by the function level overlay logic, where the workings of the Autonomic Cooperative Networking Protocol were outlined to justify the place of the evolved messaging structure in the process of Autonomic Cooperative Node preselection, together with the layout of the routing table. Next, the extended algorithmic description of the cooperation management decision element was scrutinised not only to evaluate its advantages, but also to address the protocol overhead aspects. Finally, once again, the question of architectural integration was visited, involving the conceptual transitions and the dependencies among the related entities.

Eventually, the standardisation-orientated design was outlined, where the research and investment perspective accounted for the place of the Autonomic Cooperative Networking Architectural Model, not only by referring to the Open Systems Interconnection Reference Model and emphasising the instantiation of the Generic Autonomic Network Architecture along with its levels of abstraction, but also introducing certain cross-specification considerations related to software-defined networking, machine-to-machine communications, and intelligent transport systems. Next, the attention was shifted to emergency communications network, characterised by specifically tailored requirements making the system operation dependent on the hierarchy between chief first responders and their respective first responders, where the cooperative mode of operation, supported with performance evaluation analysis, was introduced. The proactive and reactive resiliency process was also outlined, to allow proper integration into the Autonomic Cooperative Networking Architectural Model. Then, the related network level overlay logic was brought up to explain the mutual relation between the Autonomic Cooperative Networking Protocol and Autonomic Cooperative Behaviour, as well as the cooperation orchestration decision element was introduced from the viewpoint of providing prioritisation between the two. Last, but not least, the relay-enhanced cell scenario was revisited for a more accurate evaluation of the second hop, and the concluding architectural integration aspects were raised to ensure the overall consistency of the proposed depiction of the Autonomic Cooperative Networking Architectural Model.

A

Appendix

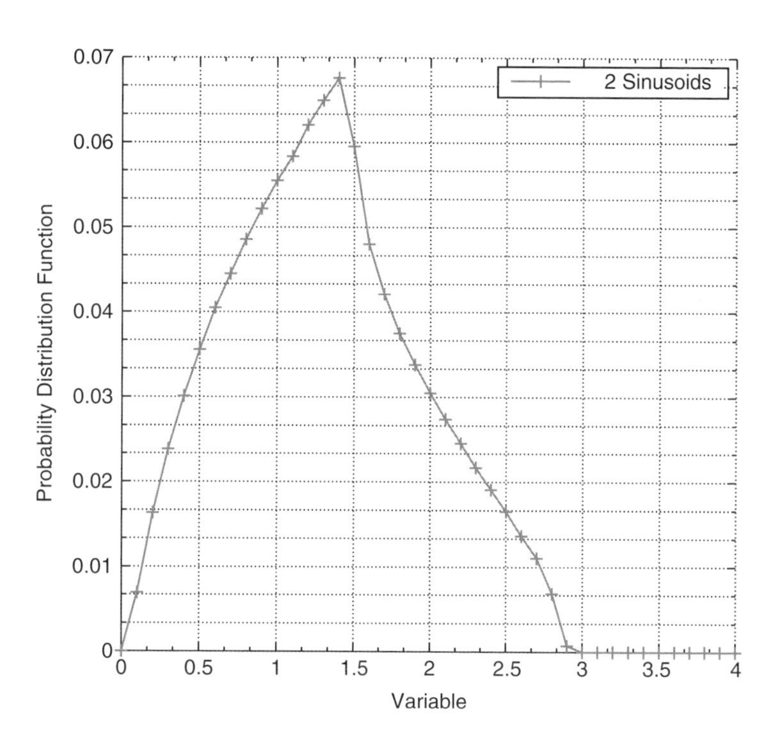

Figure A.1 Probability density function for two sinusoids.

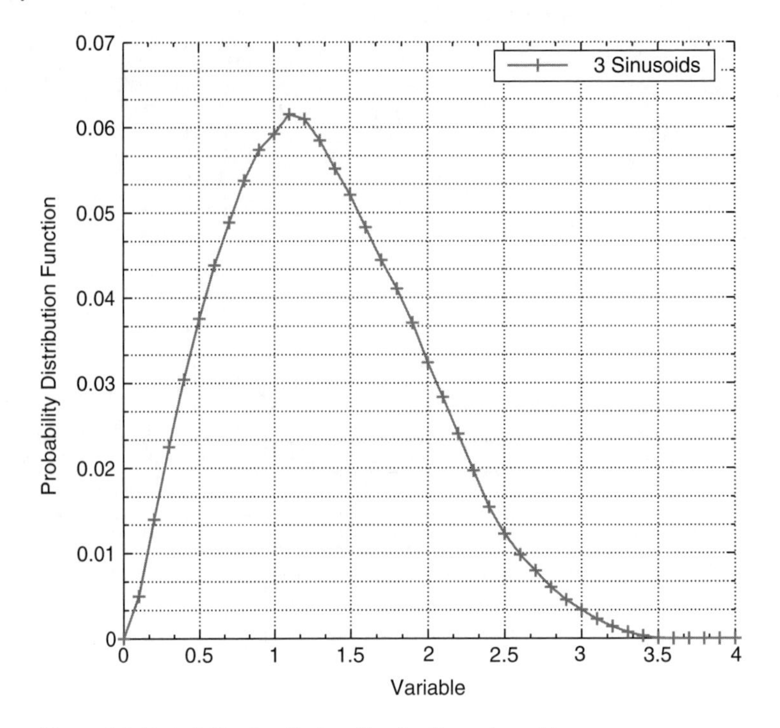

Figure A.2 Probability density function for three sinusoids.

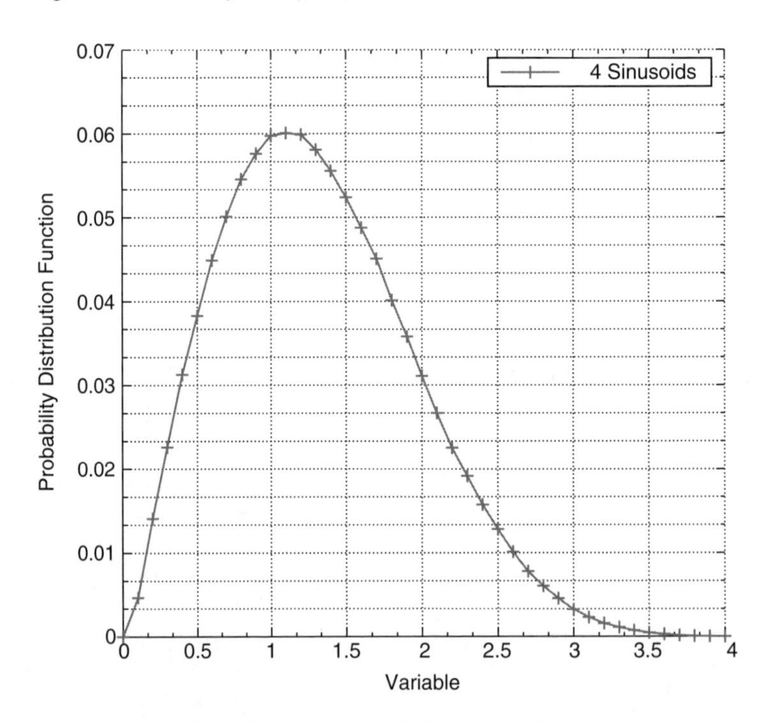

Figure A.3 Probability density function for four sinusoids.

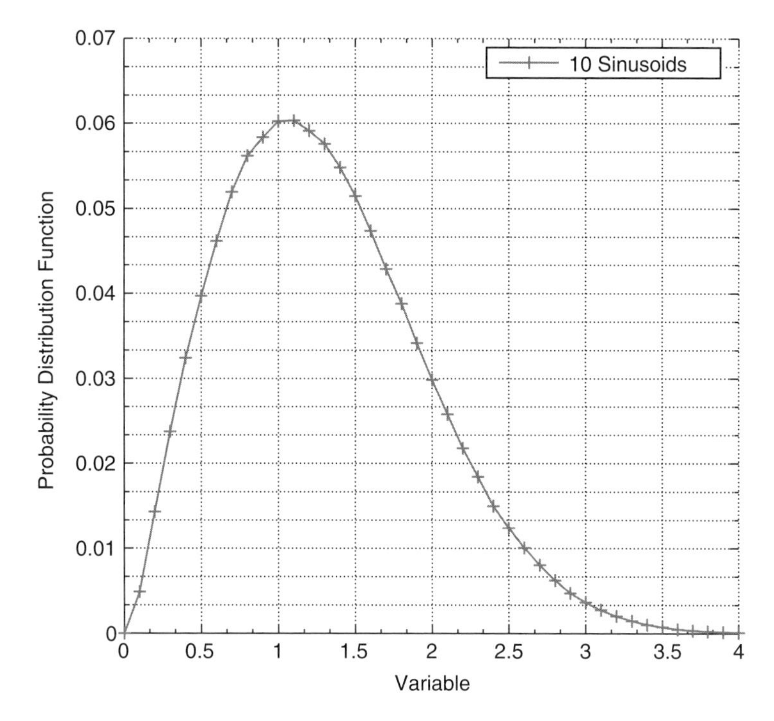

Figure A.4 Probability density function for ten sinusoids.

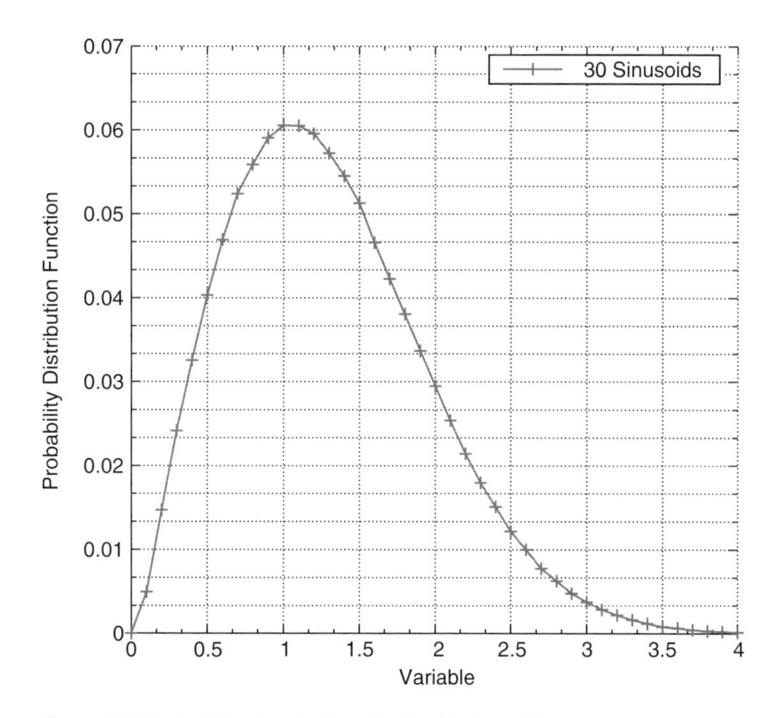

Figure A.5 Probability density function for 30 sinusoids.

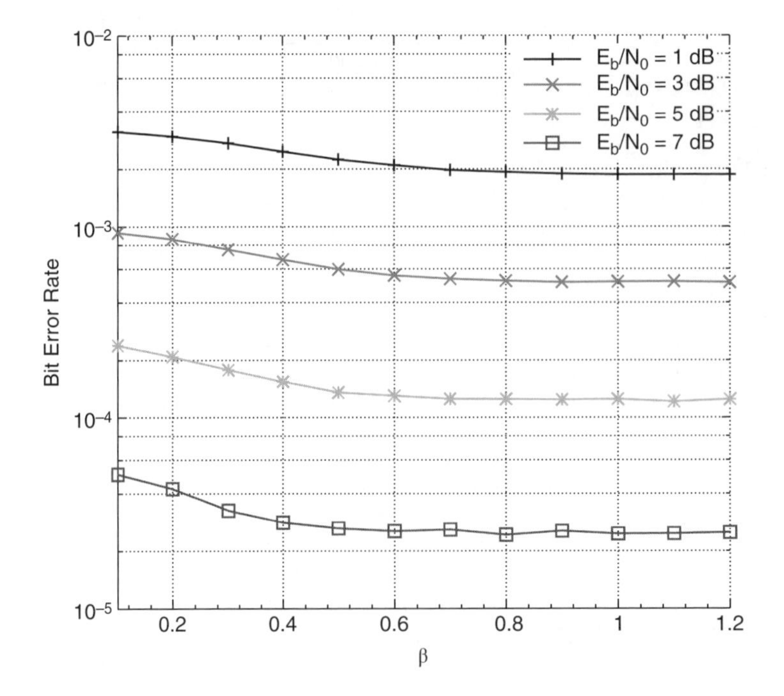

Figure A.6 CTDE in relation to additional β and E_b/N_0 values.

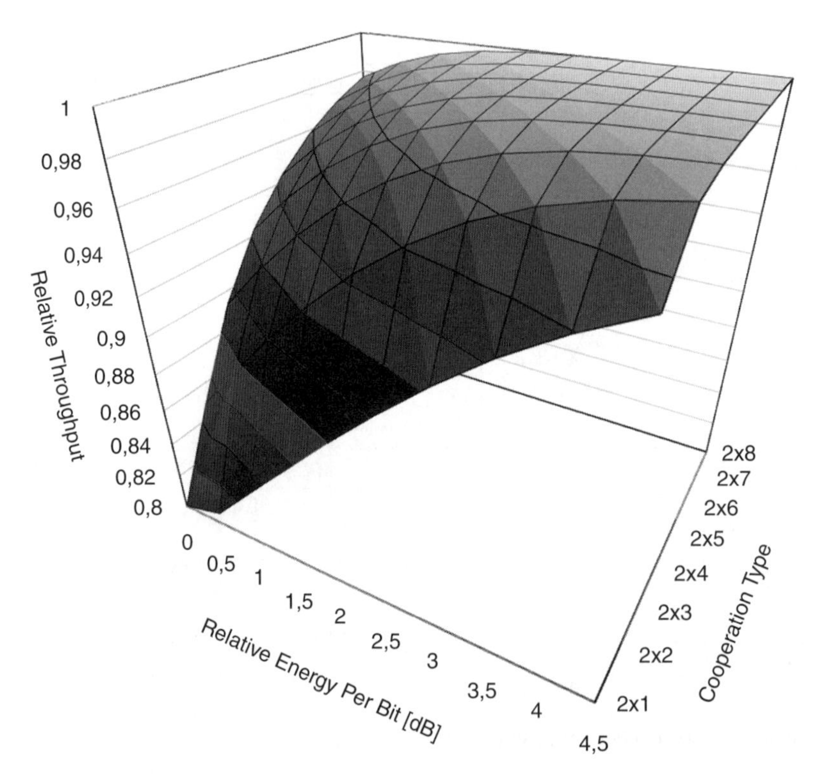

Figure A.7 Relative throughput for E_2^G code based Autonomic Cooperative Behaviour in AWGN Channel.

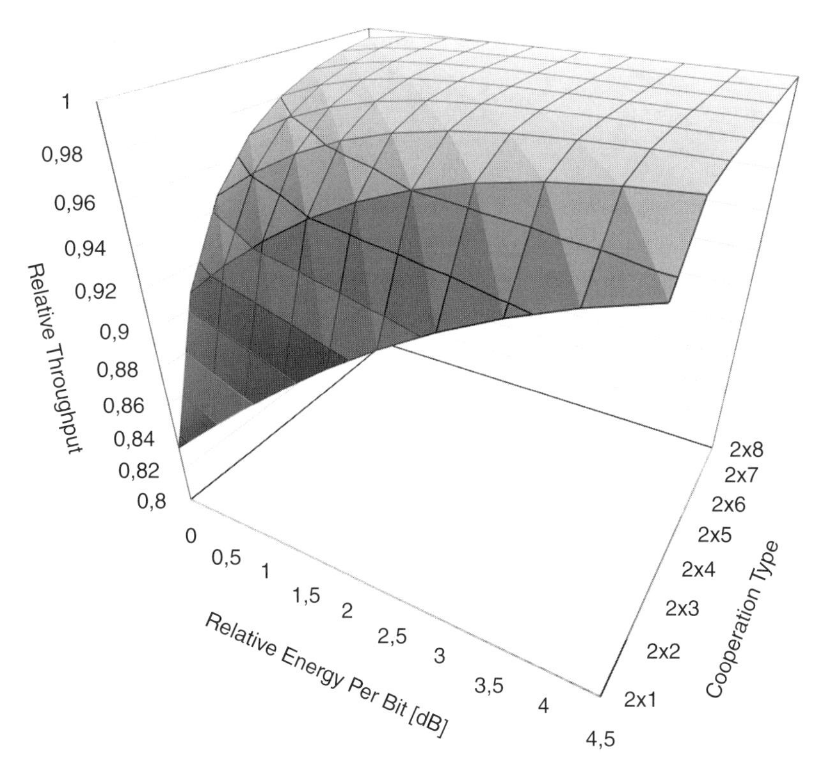

Figure A.8 Relative throughput for E_2^G code based Autonomic Cooperative Behaviour in Rayleigh channel.

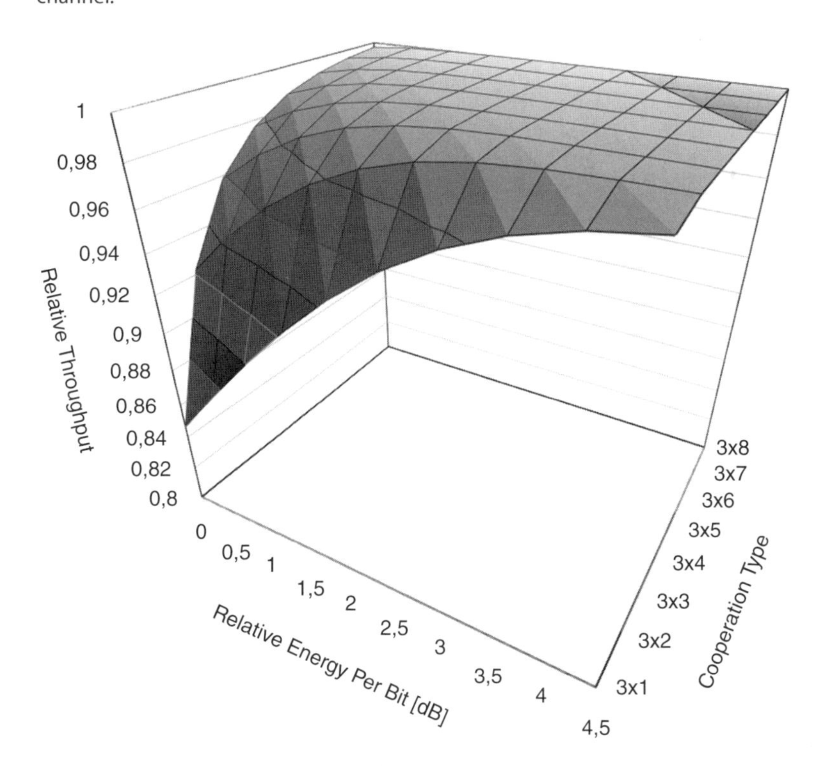

Figure A.9 Relative throughput for E_3^G code based ACB in AWGN channel.

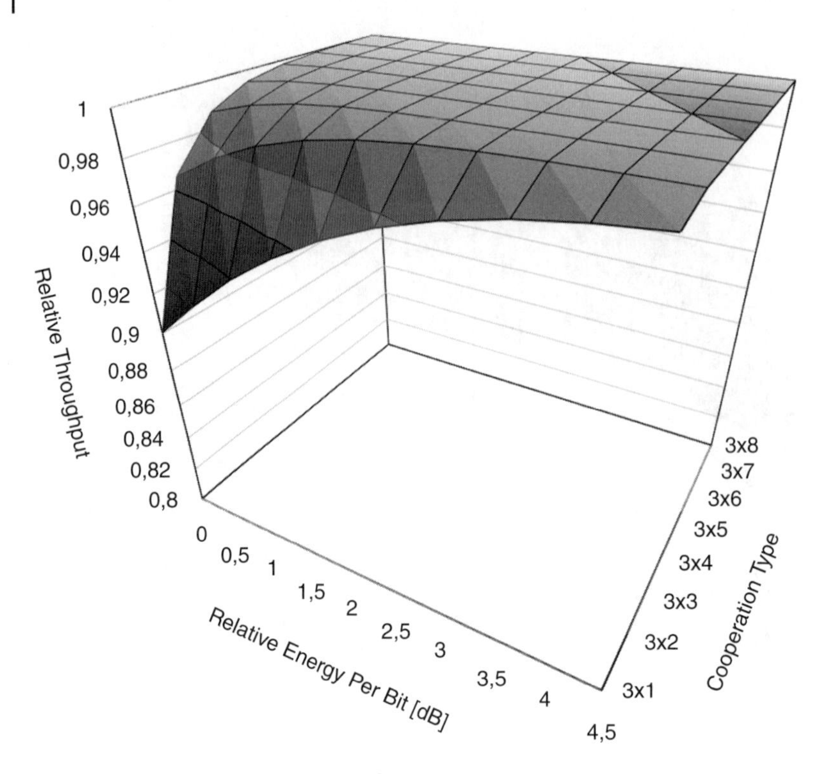

Figure A.10 Relative throughput for E_3^G code based Autonomic Cooperative Behaviour in Rayleigh channel.

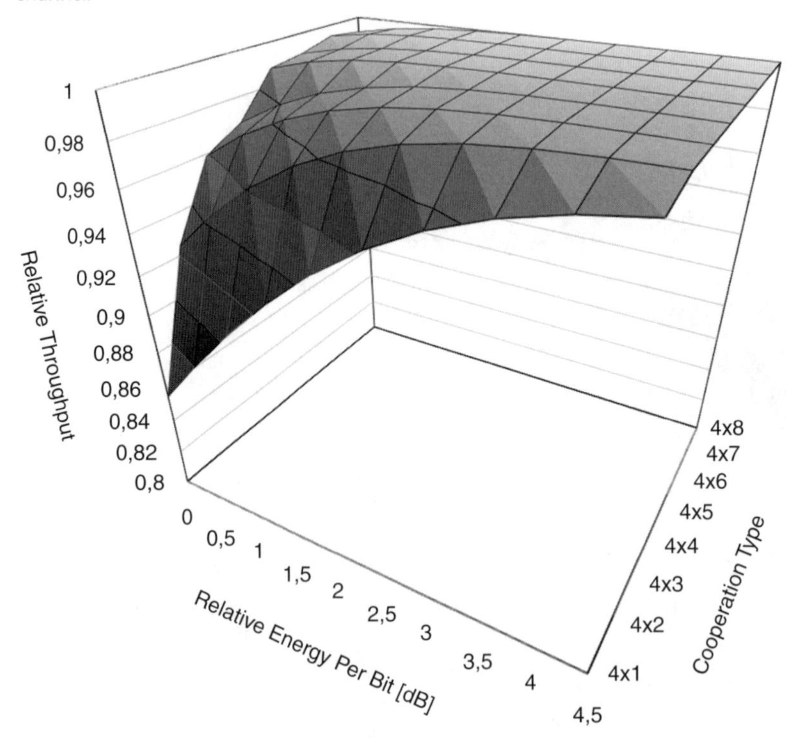

Figure A.11 Relative throughput for E_4^G code based Autonomic Cooperative Behaviour in AWGN channel.

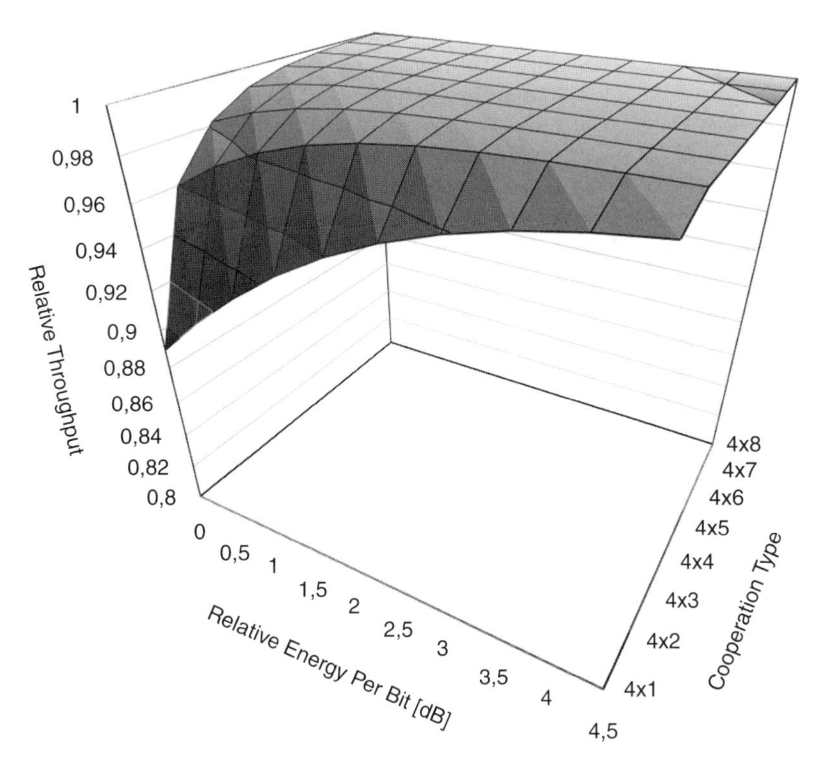

Figure A.12 Relative throughput for E_4^G code based Autonomic Cooperative Behaviour in Rayleigh channel.

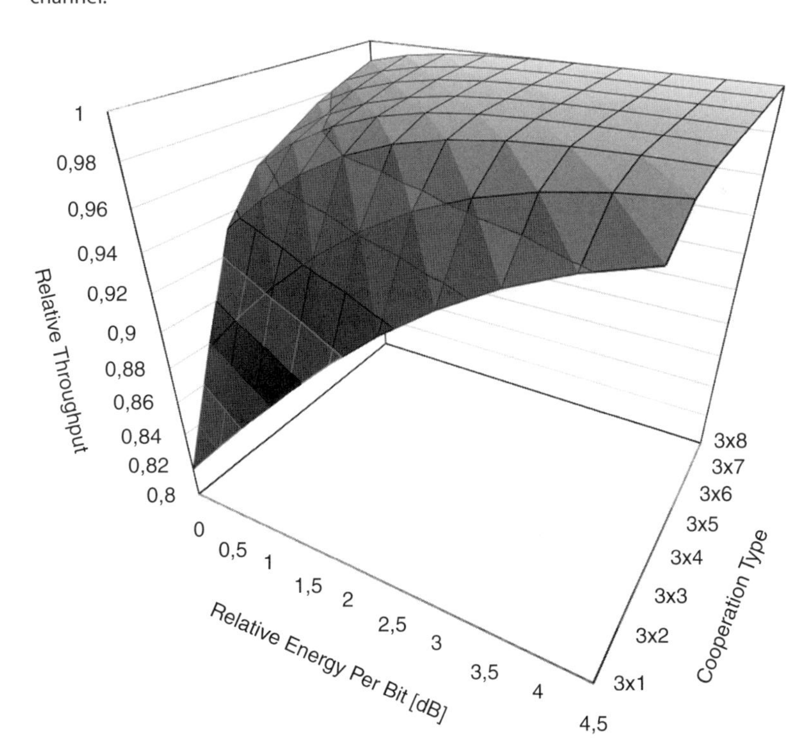

Figure A.13 Relative throughput for E_3^H code based Autonomic Cooperative Behaviour in AWGN channel.

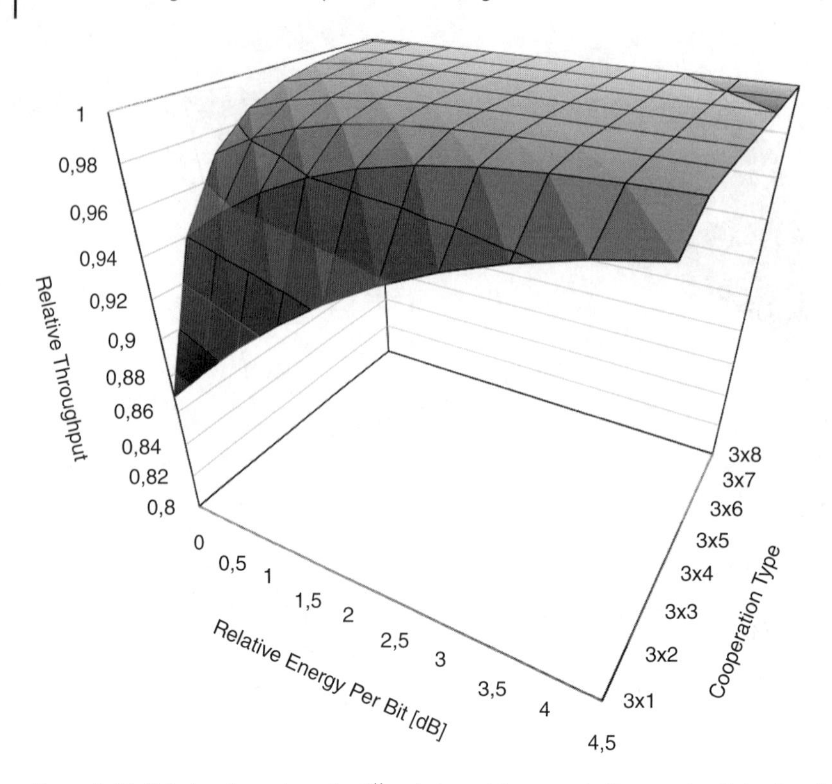

Figure A.14 Relative throughput for E_3^H code based Autonomic Cooperative Behaviour in Rayleigh channel.

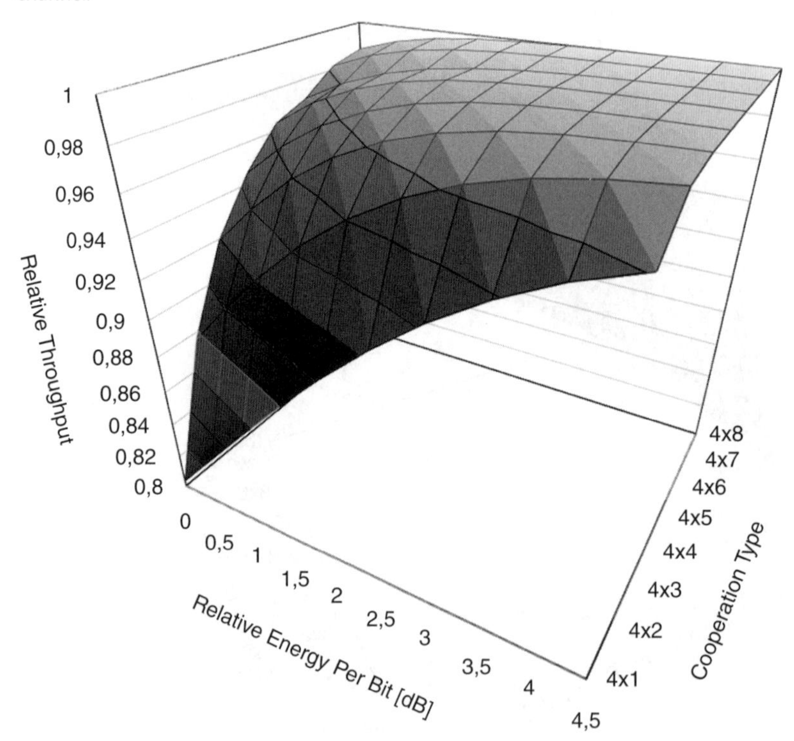

Figure A.15 Relative throughput for E_4^H code based Autonomic Cooperative Behaviour in AWGN channel.

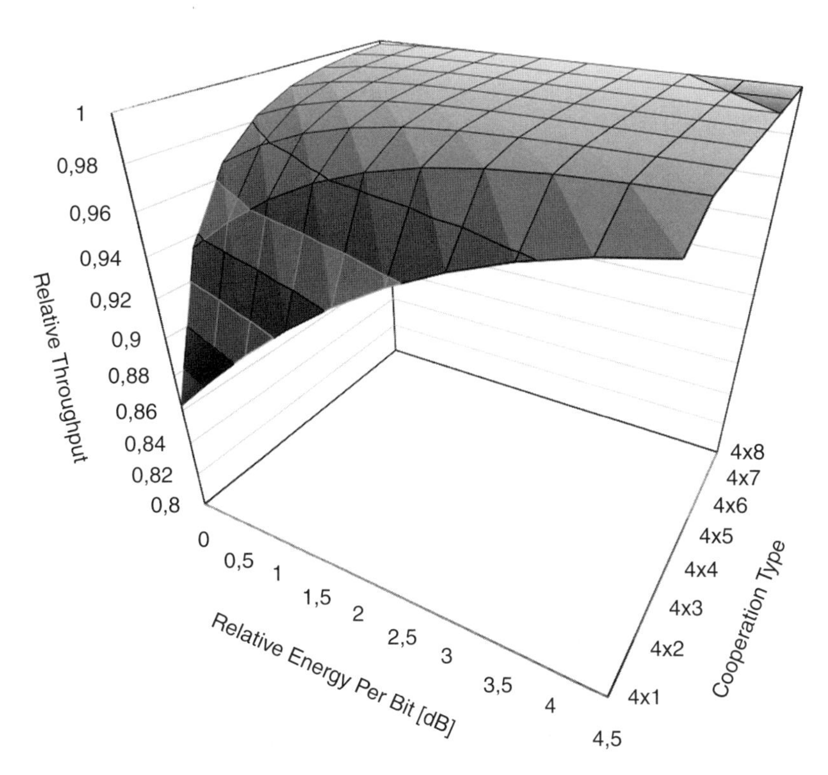

Figure A.16 Relative throughput for E_4^H code based Autonomic Cooperative Behaviour in Rayleigh channel.

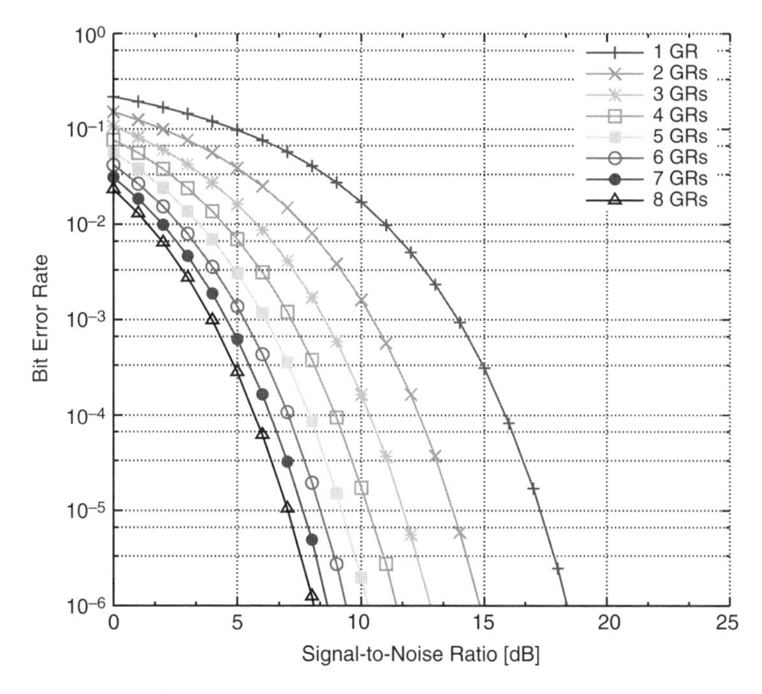

Figure A.17 E_2^G equivalent distributed space-time block encoder in AWGN channel.

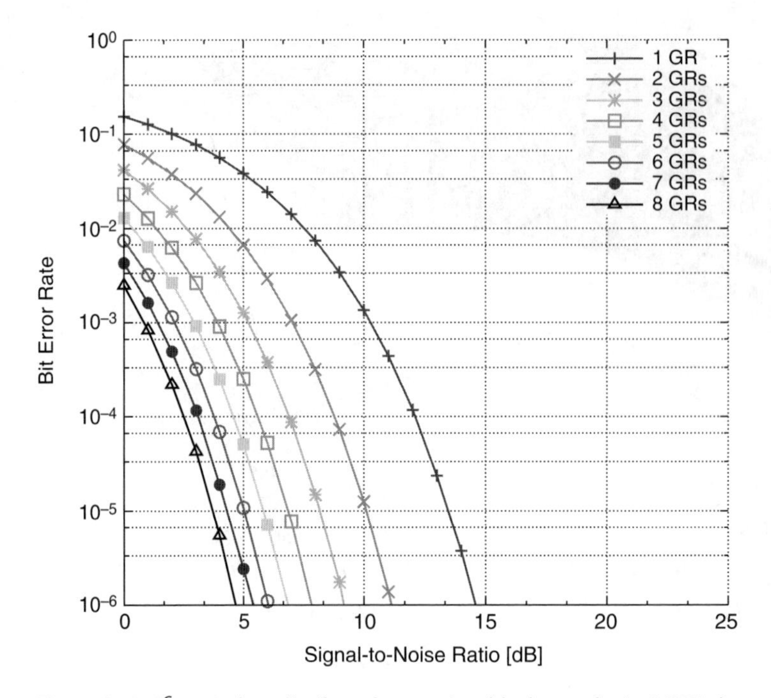

Figure A.18 E_3^G equivalent distributed space-time block encoder in AWGN channel.

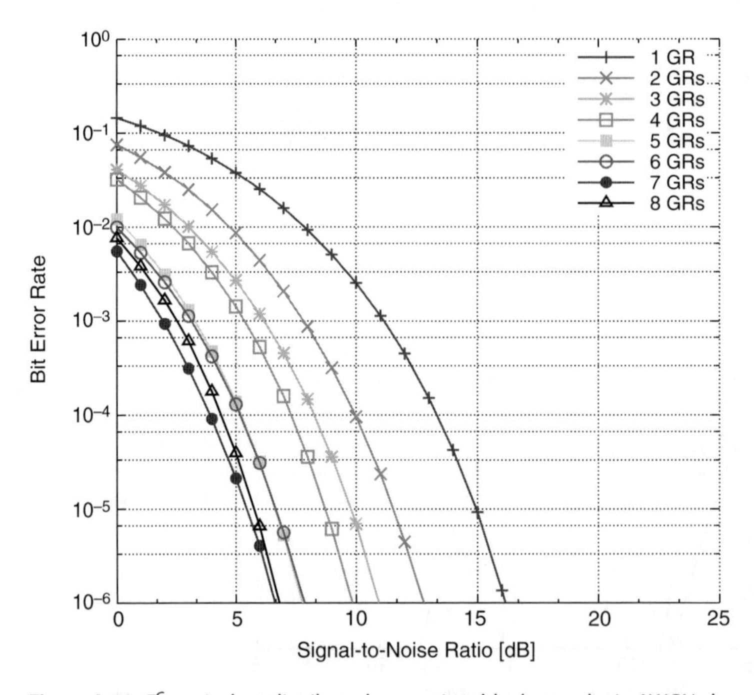

Figure A.19 E_4^G equivalent distributed space-time block encoder in AWGN channel.

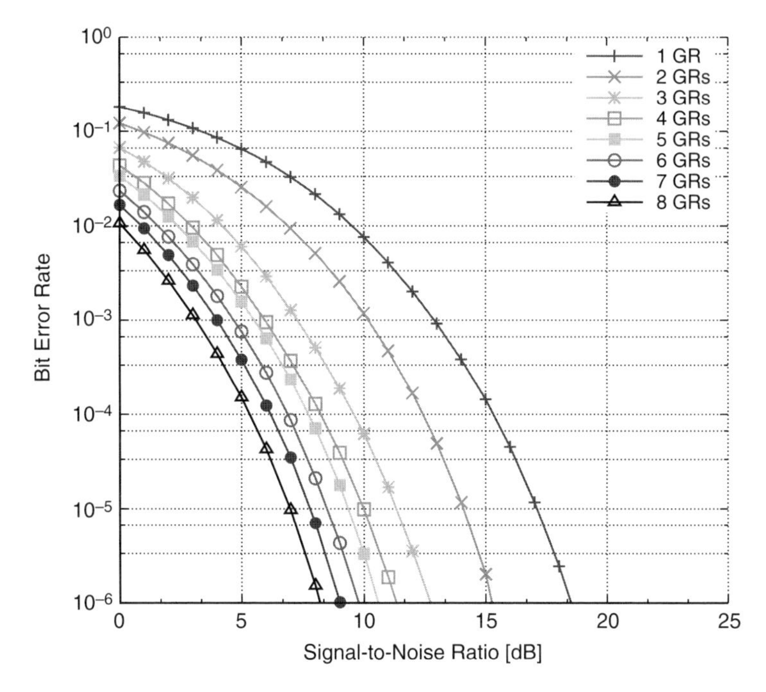

Figure A.20 E_3^H equivalent distributed space-time block encoder in AWGN channel.

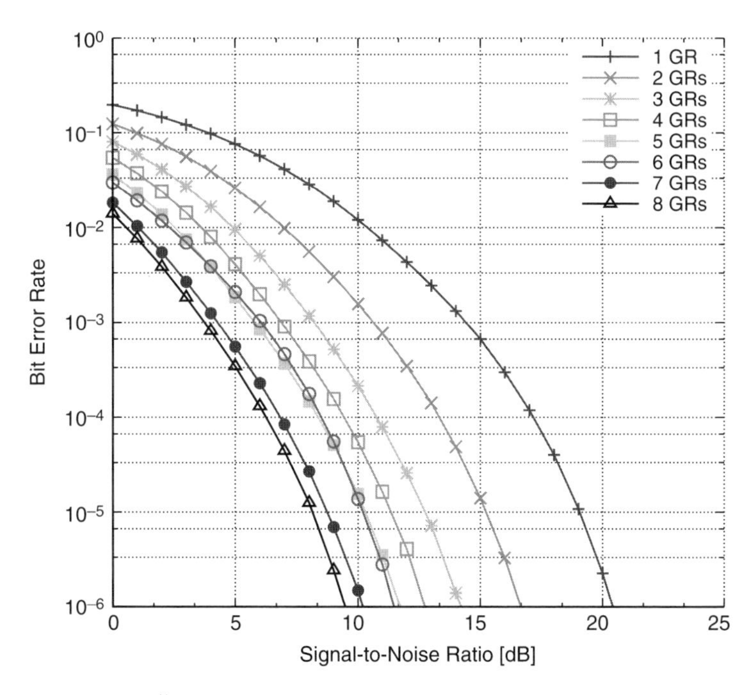

Figure A.21 E_4^H equivalent distributed space-time block encoder in AWGN channel.

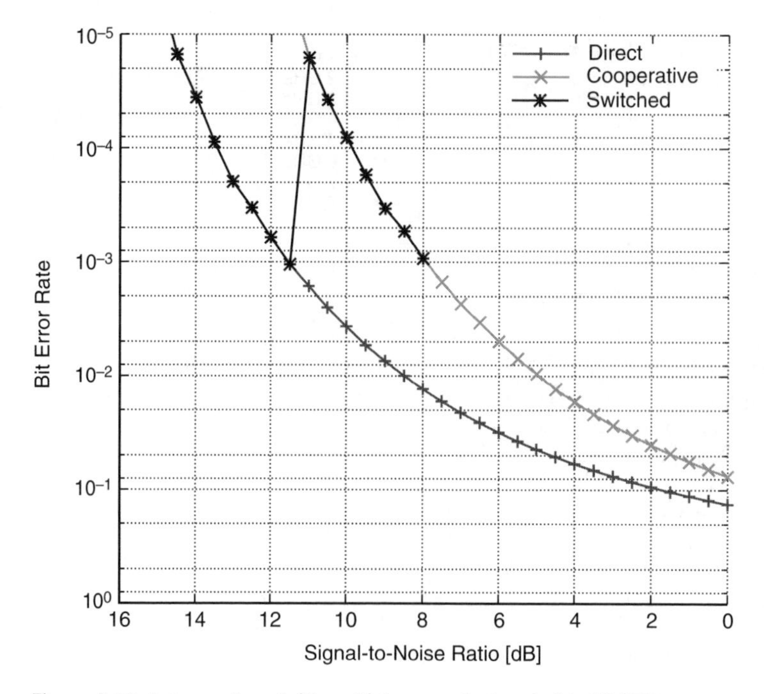

Figure A.22 Autonomic switching at bit error rate threshold of 0.0010.

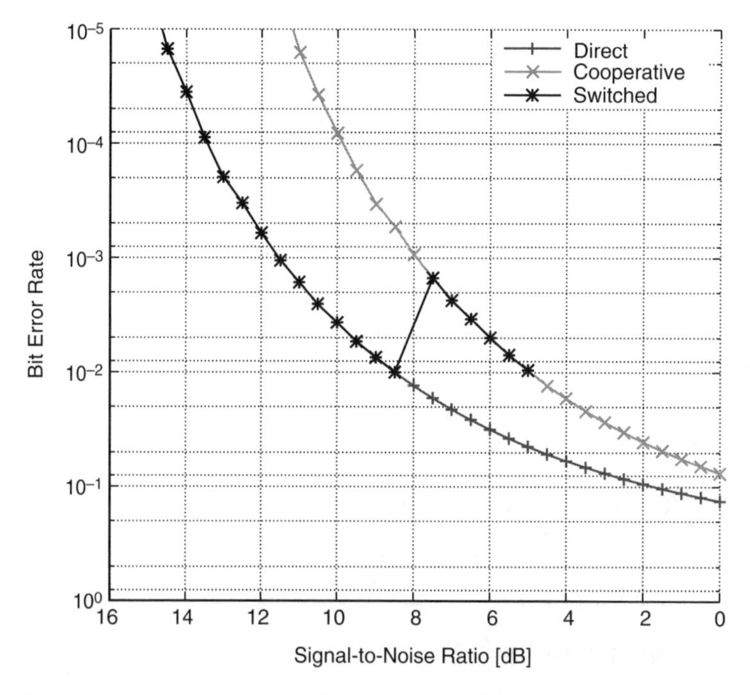

Figure A.23 Autonomic switching at bit error rate threshold of 0.0100.

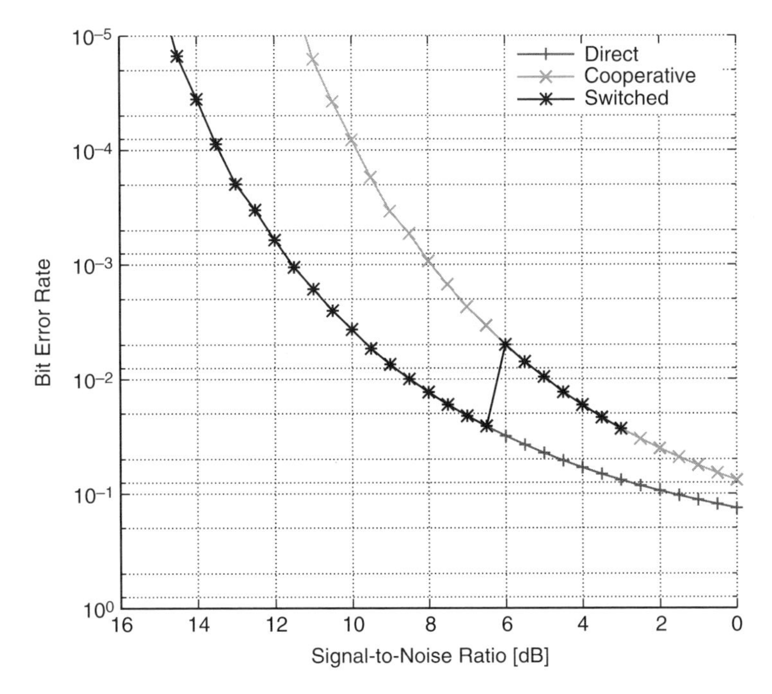

Figure A.24 Autonomic switching at bit error rate threshold of 0.0250.

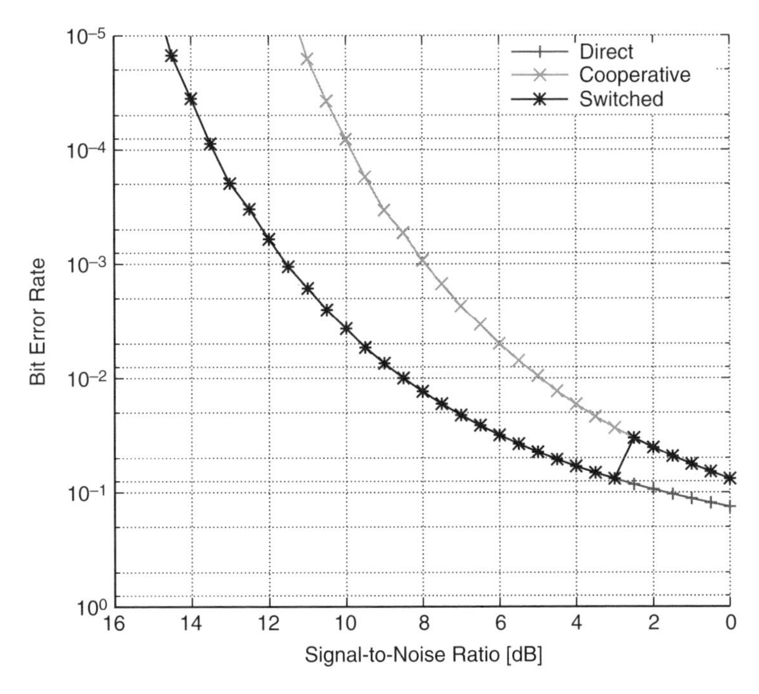

Figure A.25 Autonomic switching at bit error rate threshold of 0.0750.

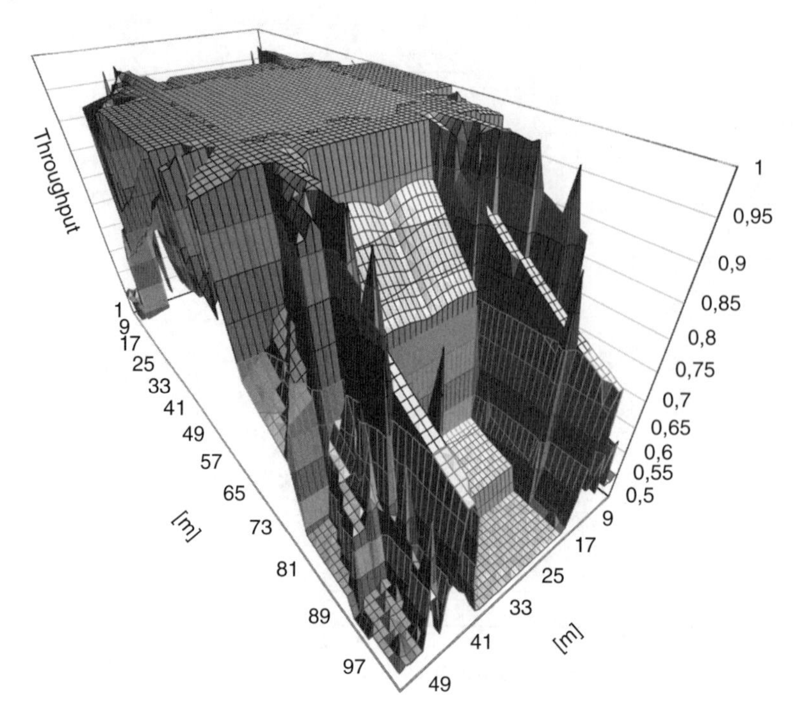

Figure A.26 Relative throughput for layout A1.

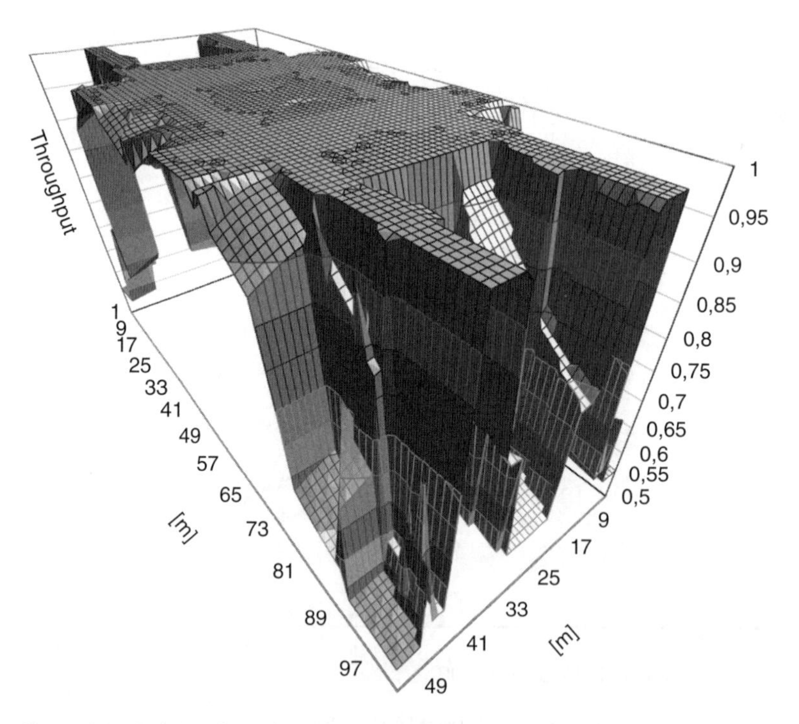

Figure A.27 Relative throughput for layout A2.

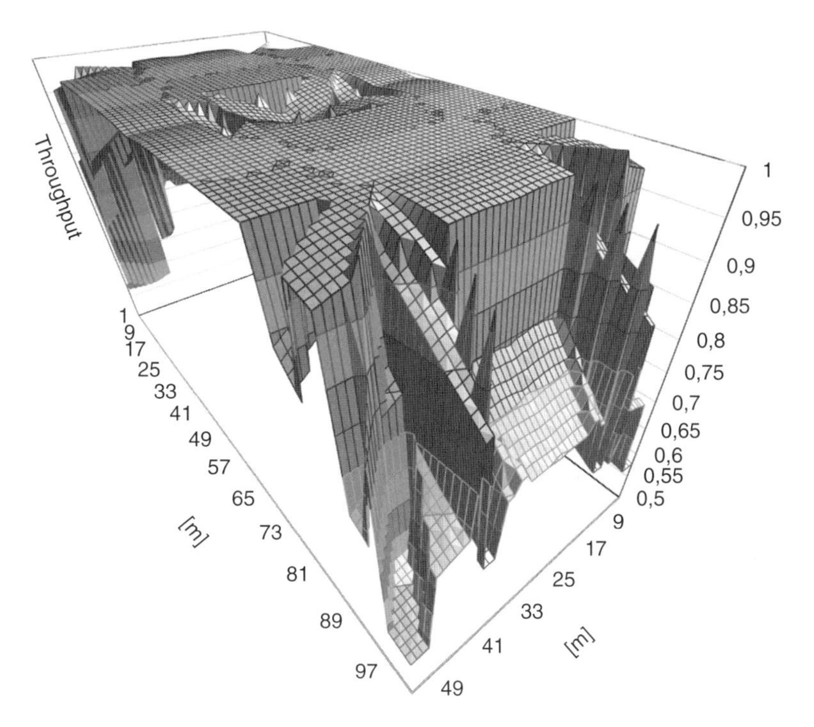

Figure A.28 Relative throughput for layout A3.

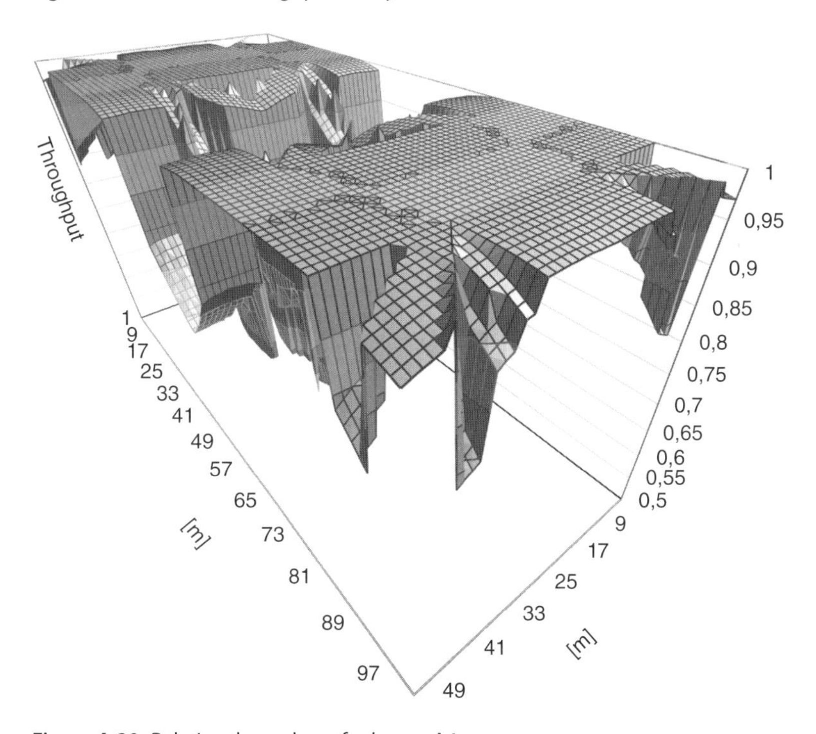

Figure A.29 Relative throughput for layout A4.

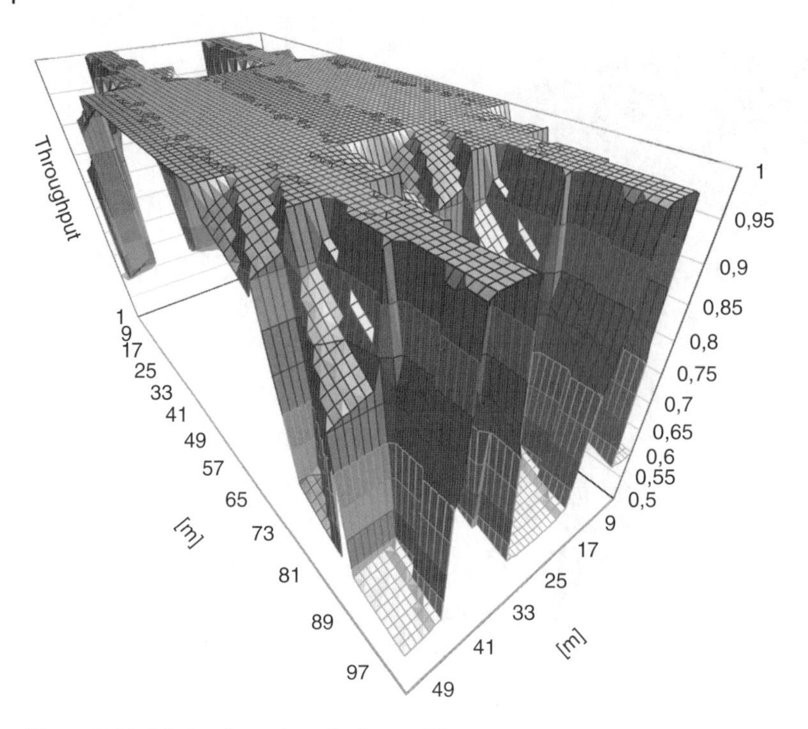

Figure A.30 Relative throughput for layout B1.

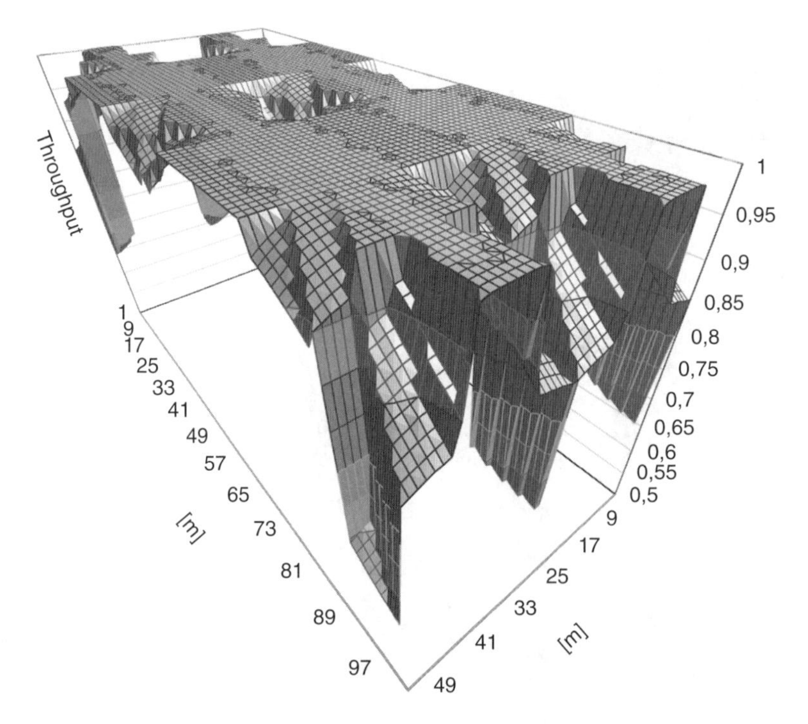

Figure A.31 Relative throughput for layout B2.

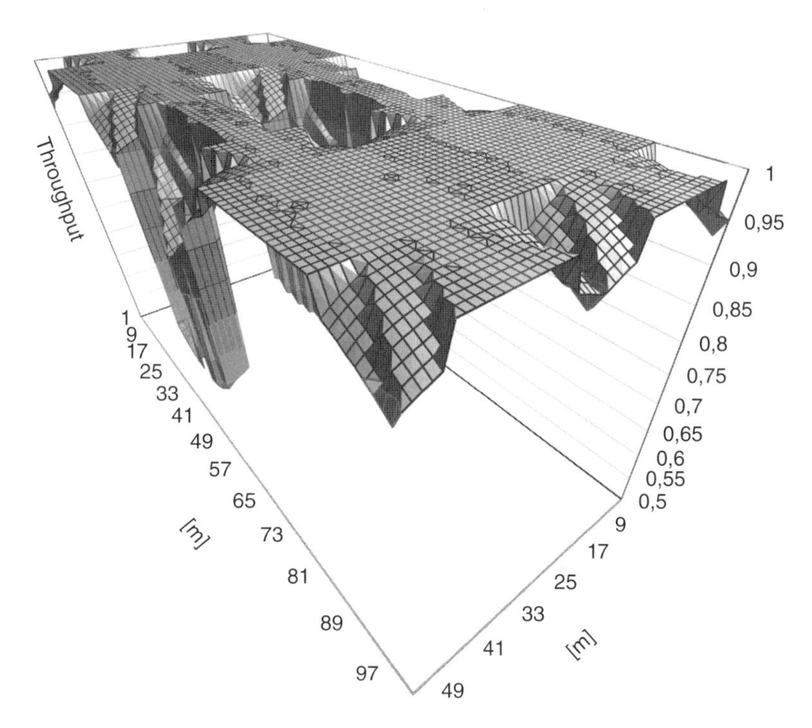

Figure A.32 Relative throughput for layout B3.

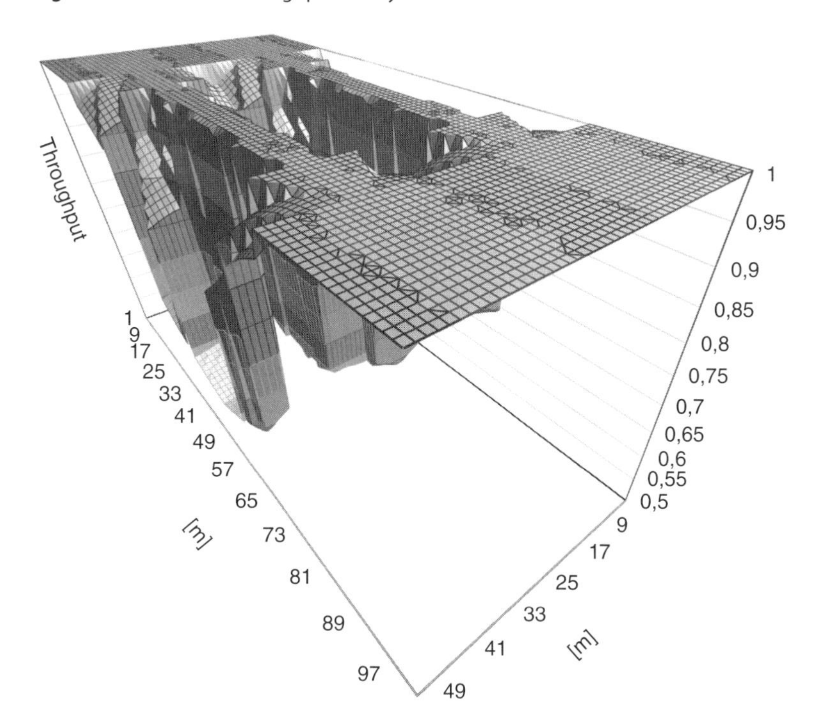

Figure A.33 Relative throughput for layout B4.

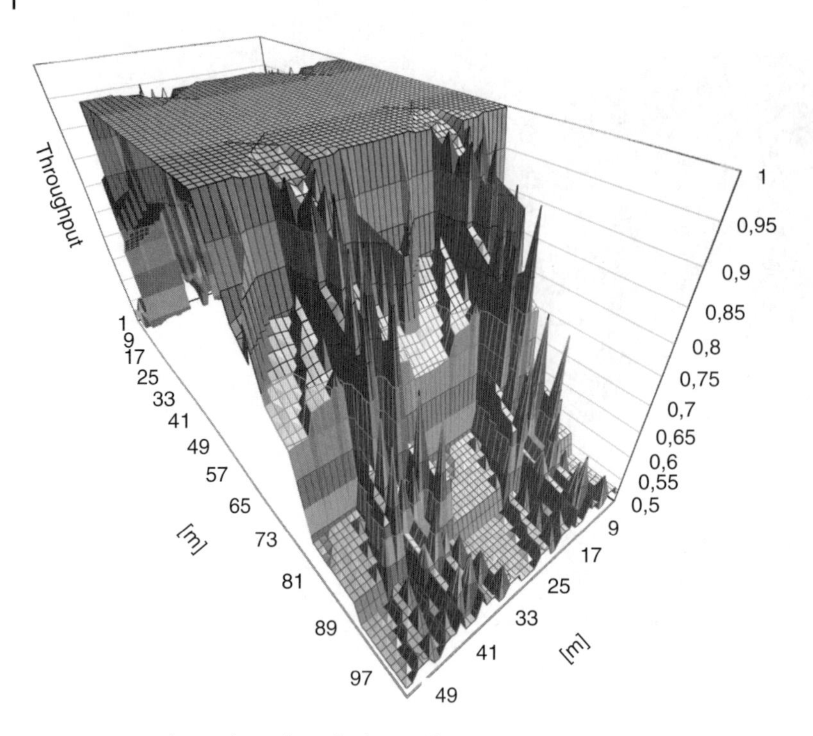

Figure A.34 Relative throughput for layout C1.

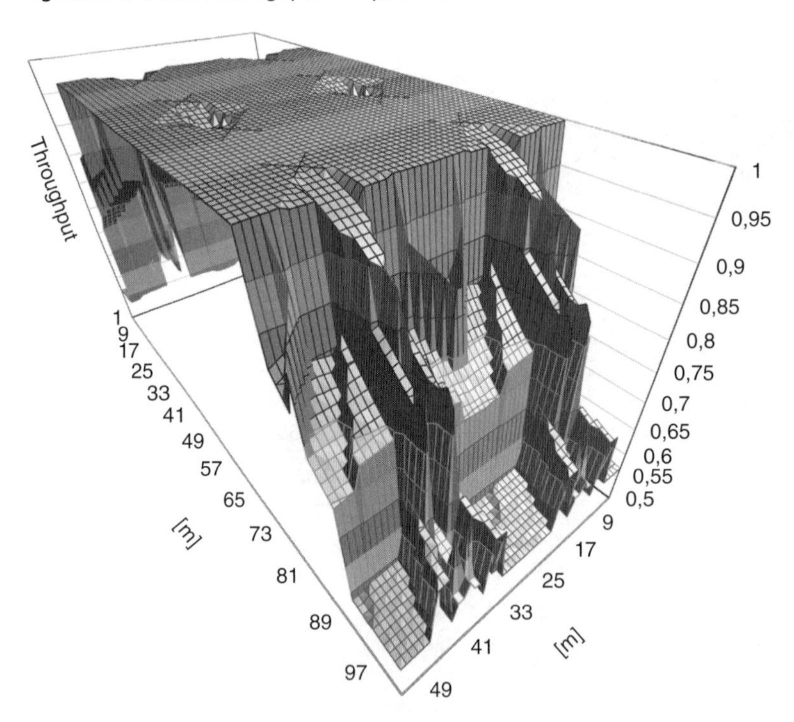

Figure A.35 Relative throughput for layout C2.

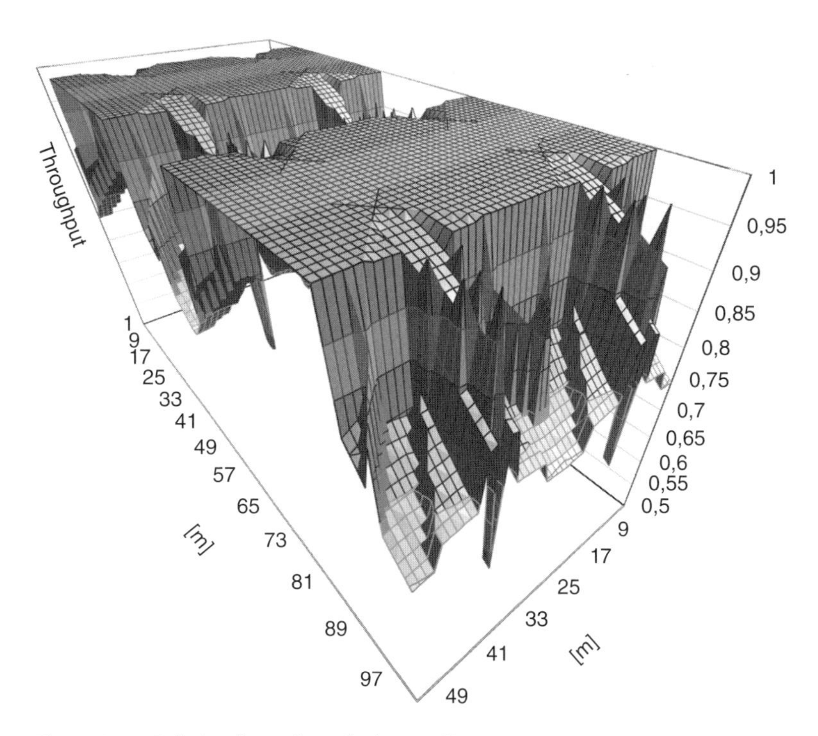

Figure A.36 Relative throughput for layout C3.

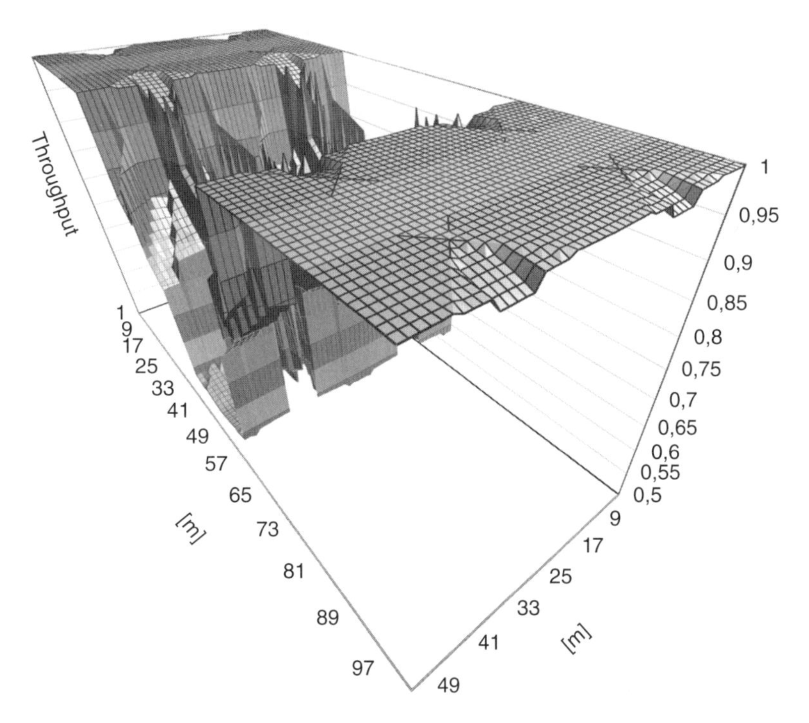

Figure A.37 Relative throughput for layout C4.

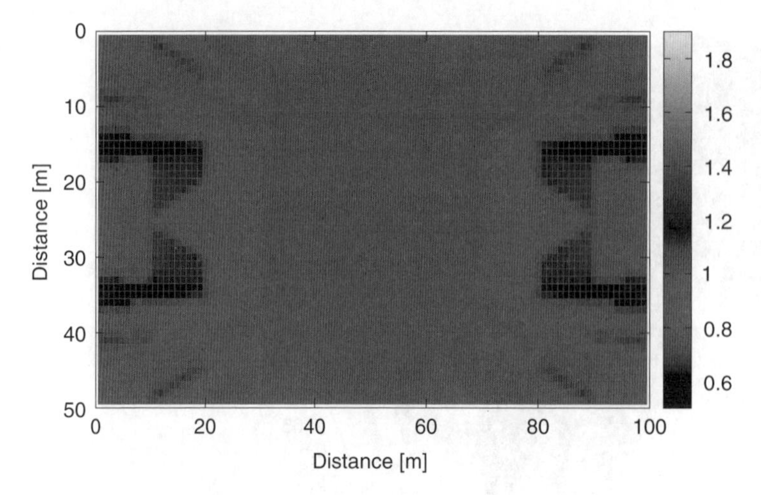

Figure A.38 Layout A1 in relation to layout B1.

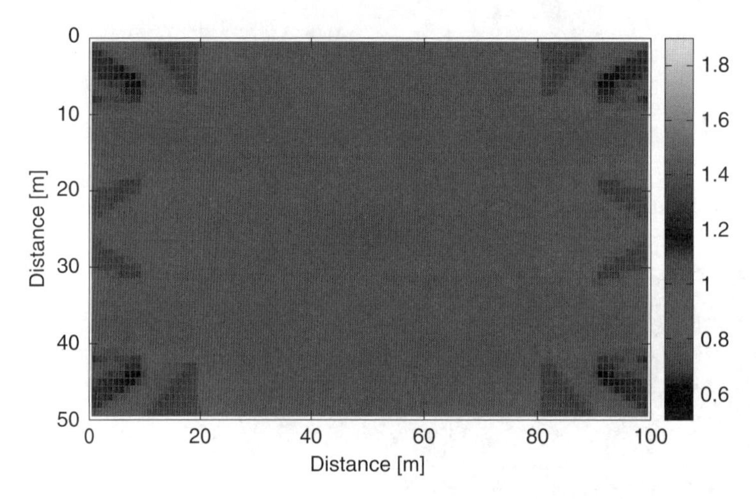

Figure A.39 Layout A2 in relation to layout B2.

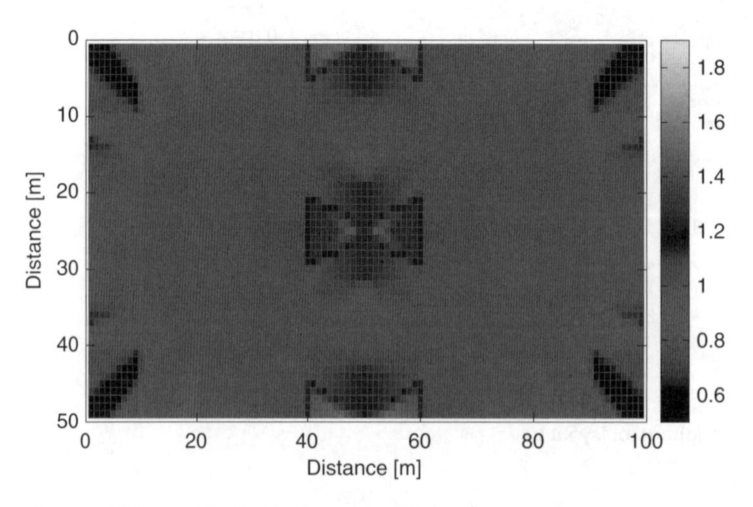

Figure A.40 Layout A3 in relation to layout B3.

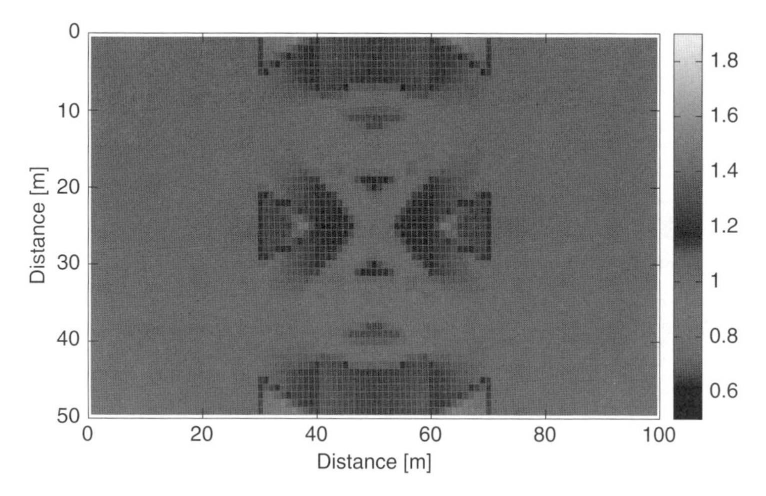

Figure A.41 Layout A4 in relation to layout B4.

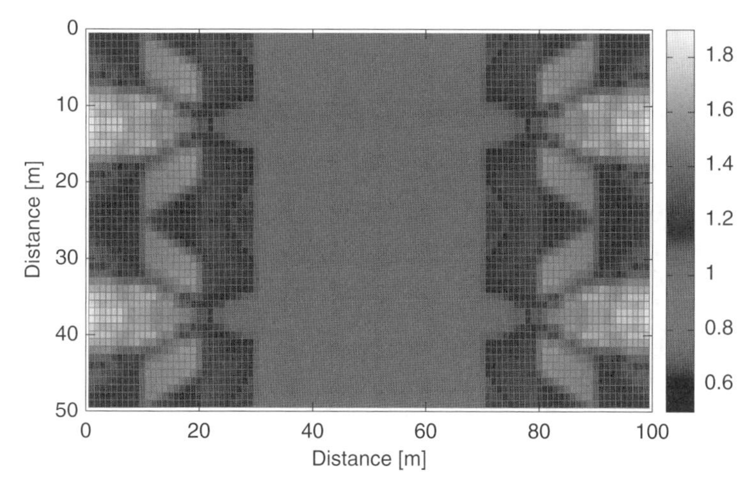

Figure A.42 Layout B1 in relation to layout C1.

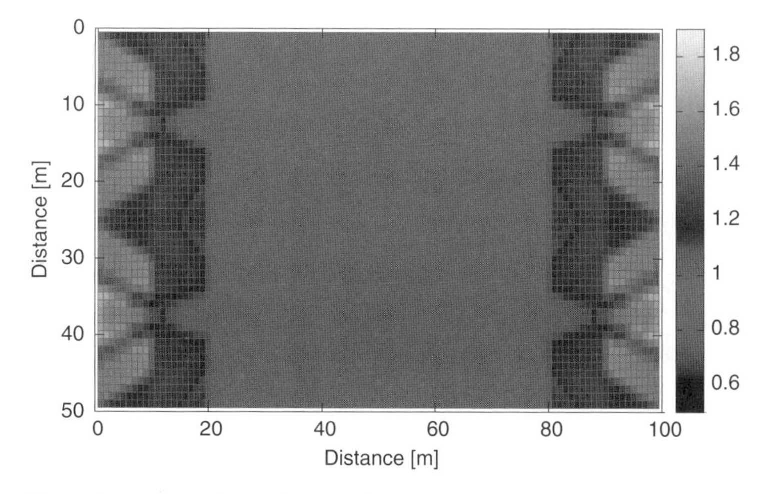

Figure A.43 Layout B2 in relation to layout C2.

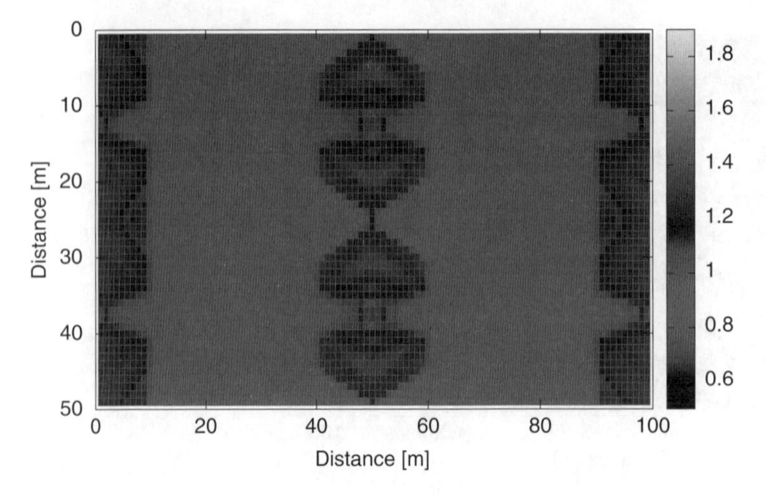

Figure A.44 Layout B3 in relation to layout C3.

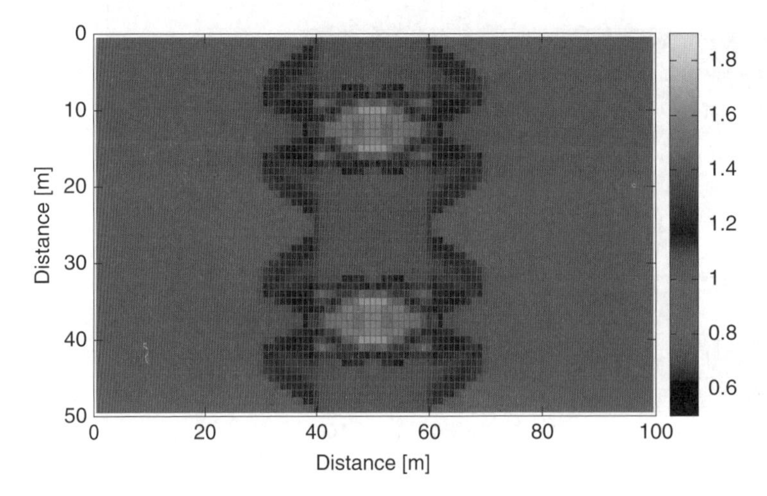

Figure A.45 Layout B4 in relation to layout C4.

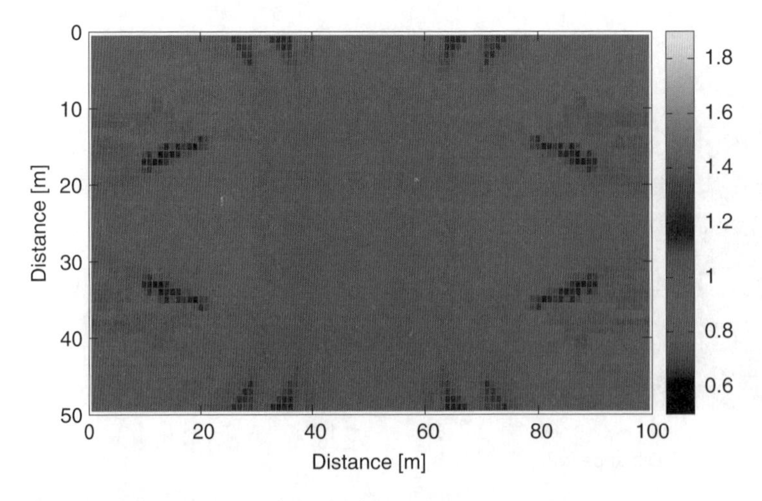

Figure A.46 Layout C1 in relation to layout A1.

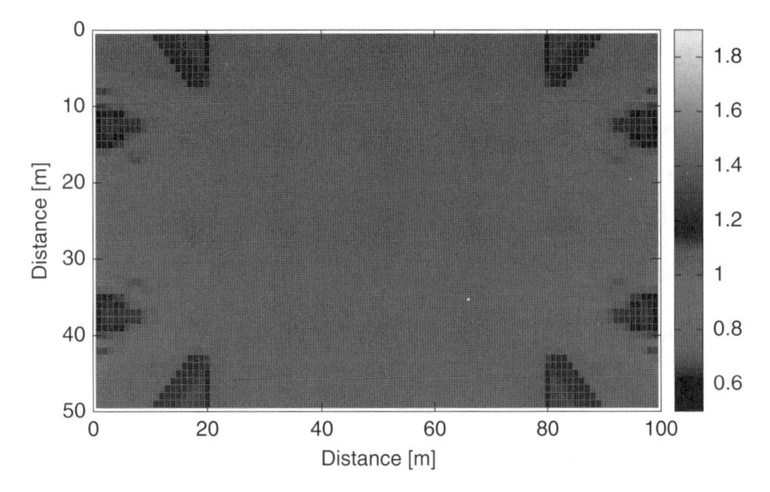

Figure A.47 Layout C2 in relation to layout A2.

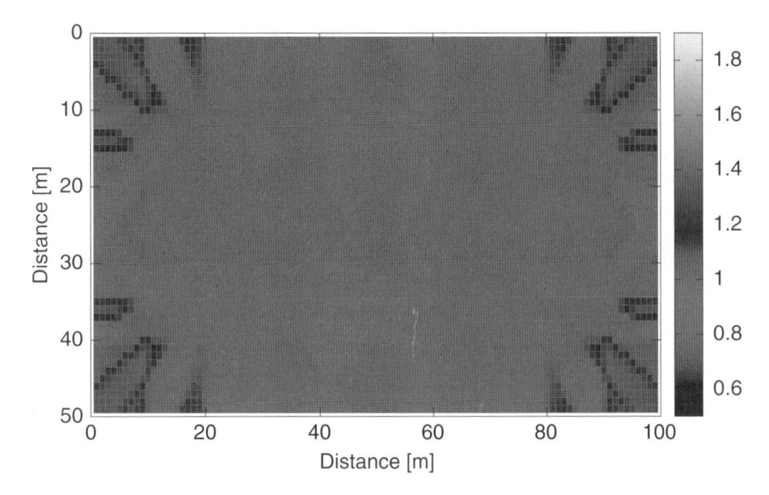

Figure A.48 Layout C3 in relation to layout A3.

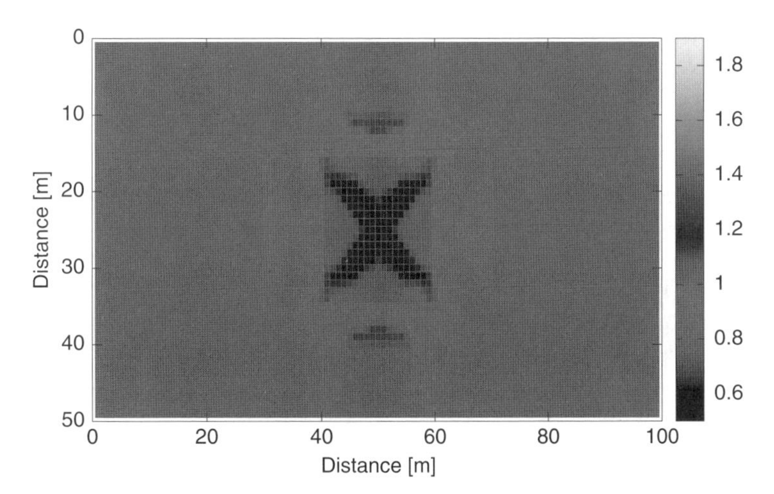

Figure A.49 Layout C4 in relation to layout A4.

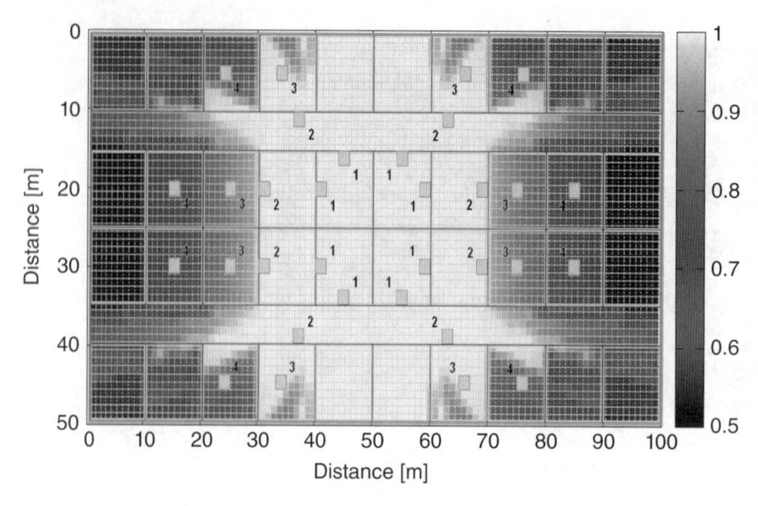

Figure A.50 Layout A1 combined with relative throughput.

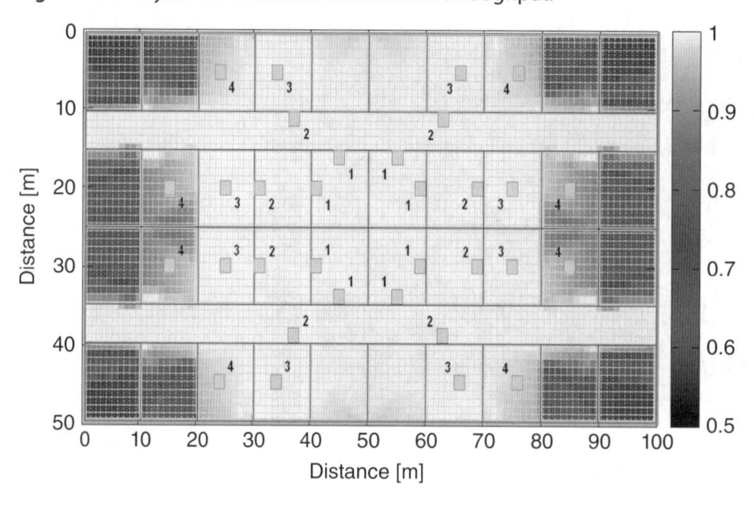

Figure A.51 Layout A2 combined with relative throughput.

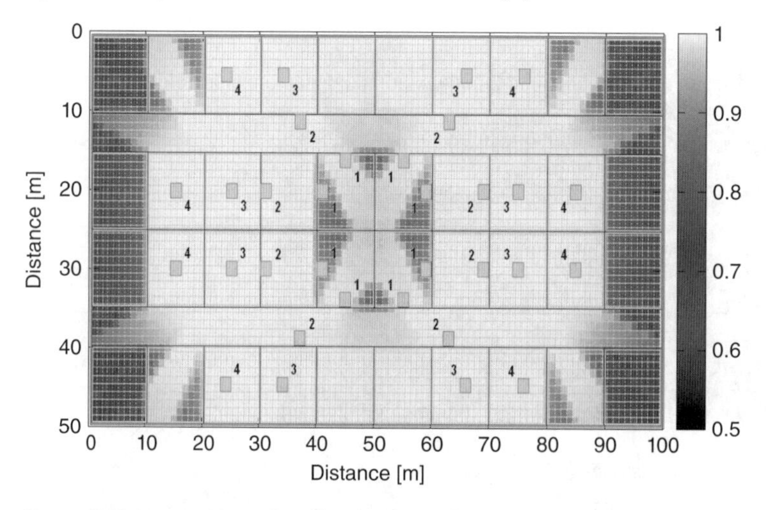

Figure A.52 Layout A3 combined with relative throughput.

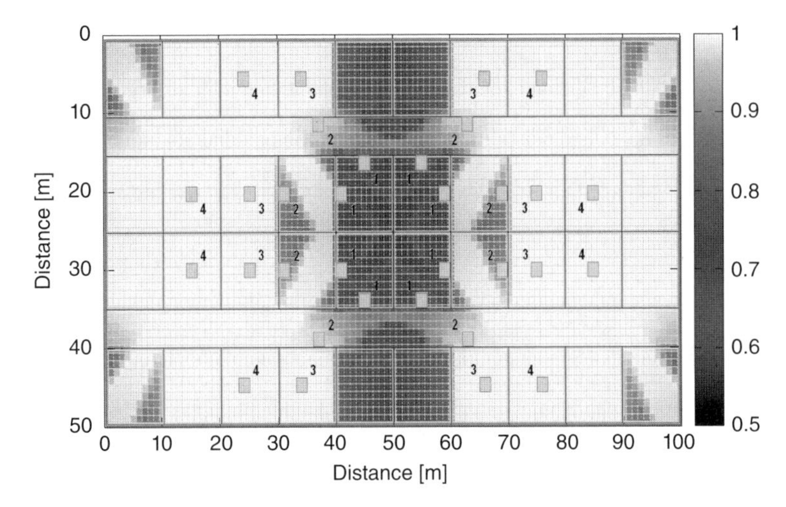

Figure A.53 Layout A4 combined with relative throughput.

Index

Symbols

Autonomic Intelligence Evolved Cooperative Networking, First Edition. Michał Wódczak.